Environmental Economics

Environmental Economics explores the ways in which economic theory and its applications, as practised and taught today, must be modified to explicitly accommodate the goal of sustainability and the vital role played by environmental capital.

Pivoting around the first and second laws of thermodynamics, as well as the principles of ecological resilience, this book is divided into five key parts, which include extensive coverage of environmental microeconomics and macroeconomics. It drills down into issues and challenges including consumer demand; production and supply; market organisation; renewable and non-renewable resources; environmental valuation; macroeconomic stabilisation and international trade and globalisation. Drawing on case studies from forestry, water, soil, air quality and mining, this book will equip readers with skills that enable the analyses of environmental and economic policy issues with a specific focus on the sustainability of the economy. This new edition has been updated throughout and provides further coverage on topics such as energy transition, market organisation and the role of environmental economics in regulatory decision-making including critiques of contemporary policy directives like tradable pollution permits and net zero emissions. Challenges to achieving stabilisation and emission reduction have been expanded to include wars and conflicts such as those in the Middle East and the Russian invasion of Ukraine. This book further reinforces the premise that there are clear limits to growth and that modesty and moderation are superior alternatives.

Rich in pedagogical features, including key concept boxes and review questions at the end of each chapter, this book will be a vital resource for upper-level undergraduate and postgraduate students studying not only environmental economics/ecological economics but also economics in general.

Dodo J. Thampapillai is an honorary professor in the School of Natural Sciences at Macquarie University, where he previously held a personal chair in environmental economics. He is currently a senior fellow in executive education at the Lee Kuan Yew School of Public Policy, National University of Singapore, where he was previously a professor. He also held Visiting Professorships/Fellowships at Simon Fraser University, Swedish University of Agricultural Sciences, Christian Albrechts University of Kiel and the Australian National University. Dodo Thampapillai has been teaching economics and environmental economics for over 50 years leading him towards a serious rethink of how economics is taught and practised.

Matthias Ruth is Pro-Vice-Chancellor for Research at the University of York, York, United Kingdom. He previously served as Vice-President for Research and Innovation at the University of Alberta, Canada and in various other leadership positions in the USA and with academic appointments in numerous fields, including economics, engineering, geography, public policy and urban affairs. His research focuses on dynamic modelling of natural resource use, industrial and infrastructure systems analysis and environmental economics and policy. His theoretical work heavily draws on concepts from engineering, economics and ecology, while his applied research utilises methods of non-linear dynamic modelling as well as adaptive and anticipatory management. Applications of his work cover the full spectrum from local to regional, to national and global environmental challenges, as well as the investment and policy opportunities these challenges present.

Environmental Economics
Concepts, Methods and Policies

Second edition

Dodo J. Thampapillai and Matthias Ruth

LONDON AND NEW YORK

from Routledge

Designed cover image: Getty Images

Second edition published 2025
by Routledge
4 Park Square, Milton Park, Abingdon, Oxon OX14 4RN

and by Routledge
605 Third Avenue, New York, NY 10158

Routledge is an imprint of the Taylor & Francis Group, an informa business

© 2025 Dodo J. Thampapillai and Matthias Ruth

The right of Dodo J. Thampapillai and Matthias Ruth to be identified as authors of this work has been asserted in accordance with sections 77 and 78 of the Copyright, Designs and Patents Act 1988.

All rights reserved. No part of this book may be reprinted or reproduced or utilised in any form or by any electronic, mechanical, or other means, now known or hereafter invented, including photocopying and recording, or in any information storage or retrieval system, without permission in writing from the publishers.

Trademark notice: Product or corporate names may be trademarks or registered trademarks, and are used only for identification and explanation without intent to infringe.

First edition published by Routledge 2019

British Library Cataloguing-in-Publication Data
A catalogue record for this book is available from the British Library

ISBN: 978-1-032-52824-3 (hbk)
ISBN: 978-1-032-52826-7 (pbk)
ISBN: 978-1-003-40857-4 (ebk)

DOI: 10.4324/9781003408574

Typeset in Times New Roman
by codeMantra

Contents

List of Figures	xi
List of Tables	xv
List of Boxes	xvii
Preface	xix
Acknowledgements	xxiii

PART I
Nature-Capital (KN) and the Economy   1

1 KN and Economics   3
  Environmental Economics and Economics 3
  KN and Limits to Growth 6
  KN Issues of Contemporary Times 8
  Climate Change – The Issue of This Century 10
  The Organisation of this Text 12
  References 14

2 The Economic System Revised   16
  The Economic System: Standard Version 16
  The Economic System and Materials Balance 19
  The Economy and the Entropy Law 23
  Further Adaptations of the Economic System 25
  Equilibrium in the Adapted System 27
  Concluding Remarks 28
  Review Questions 29
  References 29

PART II
Microeconomics and KN   31

3   The Market Model and Its Failure   33
    The Functions of the Market 33
    KN Goods and Market Failure 42
    Price Mechanism and Property Rights 43
    Some Further Issues 48
    Review Questions 48
    References 48

4   Public Goods and Externalities   49
    Public Goods 49
    Externalities 52
    Public Goods and Externalities 54
    Residual Externalities 58
    Review Questions 59
    References 60

5   Economics of Renewable and Non-Renewable Resources   61
    Objectives of Resource Management 61
    Conflicts between EG and KN-Q Objectives 66
    Conflicts Due to Intergenerational Concern 72
    Some Resource Management Policies 78
    Review Questions 82
    References 83

6   Economics of Non-Renewable Resources with Renewable Services   85
    A Simple Framework for Pollution Control 87
    Standards 88
    Taxes and Charges 92
    Emissions Trading 94
    Net Zero Emissions 97
    Property Rights 98
    Other Incentives and Disincentives 99
    Review Questions 100
    References 100

7   Consumer Demand and KN   102
    Utility Functions, Indifference Curves and Demand 104
    Adapting the Standard Theory for Effects Pertaining to KN 107
    The Market Demand Curve and KN Effects 112
    Elasticity of Demand and KN 113

Contents vii

Concluding Remarks 116
Review Questions 117
References 118

8 Production, Costs Supply and KN   119
The Production Function and KN 121
Production and Assimilative Capacity 123
Isoquants, Substitution and Input Mixes 128
Analysis of Costs 132
Implications 135
Review Questions 138
References 138

9 Market Organisation and KN   139
Perfect Competition and Sustainability 140
PC Versus PCS and Preference Relations 143
Monopoly 144
Oligopoly 149
Monopolistic Competition 151
Perfect Information and Full Employment of KN 153
Concluding Remarks 155
Review Questions 156
References 156

PART III
Macroeconomics and KN   157

10 Some Important Concepts in Macroeconomics   159
National Product 159
Real Values and Price Level Indexes 160
Final Expenditures 161
Economic Performance and Goals of Macroeconomics 162
Measurement of NP 163
The Relationship between NP and Inflation 165
AD–AS Framework 168
Concluding Remarks 174
Review Questions 175
References 175

11 KN: Investment and Depreciation   176
KN and the Economy 176
Distinguishing KN Investments from Depreciation of KN 178

viii   Contents

    Sustainable Income 180
    A Conceptual Framework for $D_{KN}(t)$ 184
    Some Empirical Evidence 191
    Concluding Remarks 192
    Review Questions 192
    References 192

12 Environmental Macroeconomics: Short-Run Analysis-I   194
    Equilibrium Income without D_{KN} 197
    Equilibrium Income with D_{KN}: Linear Framework 199
    Equilibrium Income with D_{KN}: Non-Linear Framework 203
    Income Determination and Policy Analyses 204
    Concluding Remarks 210
    Review Questions 210
    References 211

13 Environmental Macroeconomics: Short-Run Analysis II   212
    The Standard Framework 212
    Display of the Standard Framework 218
    The Environmental Macroeconomic Framework 220
    Concluding Remarks 228
    Review Questions 228
    References 229

14 Environmental Macroeconomics: Long-Run Analysis   230
    Introduction 230
    The Steady State 231
    The Harrod–Domar (H–D) Growth Model 233
    The Swan-Solow Model 235
    Endogenous Growth Models 240
    An Illustration – The South Korean Economy 243
    Concluding Remarks 248
    Review Questions 248

15 International Trade and Globalisation   254
    Comparative Advantage and Specialisation 256
    The Factor Endowment and Trade Framework 256
    Trade Framework Based on the Market Model 260
    The Market for Foreign Exchange 263
    Global Issues and the Need for a Global Paradigm 265
    Concluding Comments 267
    Review Questions 268
    References 268

Contents ix

PART IV
Valuation 269

16 Valuation of KN: Microeconomic Basis 271
    The Basis for Valuation 271
    Methods Based on Demand: KN as a Consumption Good 273
    Methods Based on Demand: KN as an
        Input in Production 278
    Methods Based on Opportunity Costs (OC) 280
    Concluding Remarks 290
    Review Questions 291
    References 293

17 Valuation of KN: Macroeconomic Basis – I 296
    Use Value of KN in Macroeconomics 296
    Illustration of the Use Values of KN and the
        Scarcity of KN 299
    Concluding Remarks 304
    Review Questions 305
    References 305

18 Valuation of KN: Macroeconomic Basis – II 307
    Stocks and Flows of KN 307
    The Perpetual Inventory Method (PIM) 309
    Adapting the PIM for KN 310
    Empirical Illustration 312
    Concluding Remarks 313
    Review Questions 314
    References 314

PART V
Policy 315

19 Policies for Sustainable Development 317
    Policies Aimed at Reducing Damages 319
    Policies Aimed at Minimising KN Damages towards Zero
        and Restoring KN 331
    Overarching Policies – Social Capital (KS) –
        A Precondition 337
    Concluding Remarks 338
    Review Questions 339
    References 339

20 An Environmental Economics for Sustainability 342
   The Premises for Sustainability 342
   Building KN into Economic Models 343
   The Need for a New Kind of Policy 345
   References 347

Index 349

Figures

2.1	The Economic System – Standard Version	17
2.2	The Economic System Revised for Leakages and Injections	18
2.3	The Economic System Revised for Materials Balance	19
2.4	The Economic System – Revised for Materials Balance and Entropy	26
3.1	Aggregating Individual demands	35
3.2	The Cost Structure of a Firm	38
3.3	Aggregating Individual Supplies	39
3.4	Market Equilibrium	40
3.5	Imperfect Competition – Government Intervention	45
4.1	The Continuum of Goods	50
4.2	Quasi-Public Good Characteristics of a National Park	51
4.3	Illustration of an Externality	54
4.4	Public Goods and Externalities	55
4.5	Open Defecation and Assimilative Capacity	58
5.1	Illustration of Profit Maximisation (Without KN Costs)	67
5.2	Illustration of Profit Maximisation (With KN Costs)	68
5.3	Illustration of Coase Theorem	69
5.4a	Market for a Non-Renewable Resource – No Conflicts between Generations	74
5.4b	Market for a Non-Renewable Resource – Conflicts between Generations	75
5.5	Market for a Renewable Resource over Time	77
6.1	Equilibrium between Marginal Costs of Abatement and Pollution	87
6.2	Taxes and Standards	89
6.3	The Basis for Emissions Trading	95
7.1	Utility Surface for $\{U(Q_A, Q_B)\}$	105
7.2	The Derivation of the Demand Curve	106
7.3	Varying Endowments and Indifference Curves	109
7.4	Two Distinct Demand Curves	111
7.5	Income Effect and Substitution Effect	112
7.6	The Relationship between Price Elasticity of Demand and Revenue	114

xii *Figures*

8.1	The Standard Production Function and the Zones of Production	121
8.2	The Production Function Relating to KN	123
8.3	Family of Production Functions and Increasing Fragility of KN	125
8.4	The Display of Entropy and Diminishing Marginal Returns	126
8.5	Utilisation Vs Augmentation of KN and Time	128
8.6	Isoquants in Standard Theory	129
8.7	Isoquants for KN and KM-L (K)	130
8.8	Cost Structure of a Firm	133
8.9	Average and Marginal Cost in Standard Theory	134
8.10	Average and Marginal Cost That Tends to Infinity	134
9.1	PC vs PCS	142
9.2	Pricing and Welfare Losses Due to Monopoly	146
9.3	PC and PCS vs Monopoly	147
9.4	Natural Monopoly	149
9.5	Oligopoly	151
9.6	Monopolistic Competition	152
10.1	Standard View vs Sustainability View of Steady State	162
10.2	Aggregate Demand	166
10.3	Aggregate Supply	167
10.4	Aggregate Supply and Productive Capacity	168
10.5	Macroeconomic Equilibrium	169
10.6	(a) OPEC Oil Crisis (1973–1974). (b) Entry of North Sea Oil (1987–1992). (c) The Asian Financial Crisis. (d) The Consumption Tax in Australia. (e) The Global Financial Crisis	170
10.7	Illustrating the Inflation Effects of the War in Ukraine	174
11.1	Time Paths of Y and (Y-D_{KN}) for China and India (1992–2015)	182
11.2A	The Depreciation of the Mine without Re-Investment of Savings	183
11.2B	The Depreciation of the Mine with Re-Investment of Savings	183
11.3	KN-Depreciation Function – Linear and Discontinuous (Residual Externalities Absent)	185
11.4	KN-Depreciation Function – Linear and Discontinuous (Residual Externalities Present)	186
11.5	KN-Depreciation Function – Exponential	190
12.1a	Steady-State in Standard Macroeconomics	195
12.1b	Steady-State in Environmental Macroeconomics	195
12.2	Basis for Income (Y) being determined by Expenditure (AD)	196
12.3	Equilibrium Income within an Elementary Linear Keynesian Model	198
12.4	Sustainable Income within an Elementary Linear Keynesian Model	202
12.5	Non-linear System and Sustainable Income within a Keynesian Model	204
12.6	Time Paths of Y* and Y** for United States (1980–1991)	206
12.7	Effects of KN Restoration in Indonesia	207

12.8	Impact of Reforestation (RF) and Sewerage Treatment (ST) on η in Vietnam	207
12.9	Simulated Time-Paths of Y* and Y** for Mongolia	209
13.1	The Australian Economy (2009) – Standard Macroeconomic Framework	219
13.2	The Conceptual Basis for the Estimation of KN	222
13.3	The Australian Economy (2009) – Environmental-Macroeconomic Framework Compared with the Standard Framework	225
14.1	Swan-Solow Model Capital – Accumulation and the Time-Path	236
14.2	Revised Swan-Solow Model Capital – Accumulation and the Time-Path	239
14.3	Factors Influencing Economic Growth	243
14.4	Comparison of Income Paths (Y) and $(Y - D_{KN})$ 1970–2016	245
14.5	KN Utilisation (1970–2016)	246
14.6	The Ratios of KM (k^*/k^{**}) for Steady State Equilibrium in Both Standard and Revised S-S Models (1970–2016)	246
14.7	KN Utilisation in Both Standard and Revised S-S Models (1970–2016)	247
14.8	Rates of Growth – All Three Models (2011–2016)	247
15.1	Factor Endowment and Trade – Standard Framework	257
15.2	Factor Endowment and Trade – Revised Framework	259
15.3a	Market for an Export Good	260
15.3b	Market for an Import Good	261
15.4	Market for Foreign Exchange	264
16.1	An Overview of the Valuation Methods for Environmental Outcomes	272
16.2	The Demand for the KN by TCM	276
16.3	Indifference Map from a Game Theory Model	278
16.4	Illustration of the Dose-Response Method	280
16.5	The Value of KN in Varying Degrees of Irreversibility	284
16.6	An Illustration of the Cost-Saving Method	285
16.7	A Specific Strike Section Through Lower Delta-Plain Facies in the Liverpool Plains	289
17.1	Time Trend of Y_{KN} in Australia (1990–2020)	301
17.2	Time Trends of P_{KN} and P_{KM} in Australia (1990–2020)	301
17.3	(KN/KM) in Australia (1990–2020)	303
17.4	(Y/KN) in Australia (1990–2020)	304
18.1	Price of KN in Australia (1990–2020) – Based on KN_S	313

Tables

6.1	The NEPM Ambient Air Quality Standards	90
11.1	KN Investment Compared with Depreciation of KN	178
13.1	Selected Macroeconomic Indicators and Estimates for Australia (2005–2009)	214
13.2	Environmental Macroeconomic Indicators and Estimates for Australia (2005–2009)	221
14.1	Illustration of Model Estimates for South Korea	244
16A.1	The Ramsey Model	292
16A.2	The Ramsey Model – Game 1	293
16A.3	The Ramsey Model – Game 2	293
17.1	The Valuation of KN in Australia	300
17.2	An Alternative Efficiency Indicator for Utilised KN in Australia	304
18.1	Estimation of KN Stock	312
19.1	The Policy Mix	318

Boxes

1.1	The Concept of Sustainable Income	5
1.2	Some Early Ideas in Neoclassical Economics	7
2.1	First Law Efficiencies	20
2.2	The Longevity of Wastes	21
3.1	Property Rights for KN Goods	46
4.1	Misuse of Public Space and Externalities from Open Defecation	57
5.1	Japan's 3/11 Earthquakes and Mining – A Tenuous Link?	62
5.2	Are Forests Renewable?	79
5.3	Israel Sets the Standard in Water Recycling	81
6.1	Controversy over Standards – Treated Nuclear Wastewater	90
6.2	Emissions Trading in China and the USA	94
6.3	An Extreme and Flawed Adaptation of Trade in Pollution	96
7.1	KN-Friendly Attributes of Consumer Goods	108
7.2	The Elasticity of KN-Friendliness	116
8.1	Thermodynamic Constraints on Production Processes	120
8.2	How Does One Define Substituting between KN and KM–L?	131
8.3	The Murray River – A Case of Fragile KN in Australia	137
9.1	Perfect Information and Truth Telling	154
11.1	Mine Depreciation in Guinea (West Africa)	184
11.2	Some Selected Examples of KN Depletion That Supported Economic Growth in Australia	187
12.1	The Resource Rent Tax	208
13.1	The Global Financial Crisis and the Environmental Crisis	226
15.1	Trade as the Priority and KN as an Afterthought	255
15.2	Seemingly Nonsensical Trade Patterns	262
15.3	Palm Oil and Green House Emissions from South East Asian Peat Soils	266
16.1	The OC of Preserving Old-Growth Forests	281
16.2	The OC Preserving Sensitive Environmental resources in Mining – Liverpool Plains in New South Wales, Australia	289

Preface

The central theme of this book is that nature is capital without which no economy could survive. Hence the theme translates to the following premise:

> The sustainability of environmental capital is an essential pre-requisite for the sustainability of an economy.

This premise is exposited in the text by recourse to the first and second laws of thermodynamics as well as the principles of ecological resilience. The recognition of the premise warrants revisions to concepts and methods in all areas of economics as taught and practised today.

The main aim of this book is to inform readers of these revisions. For instance, this book makes explicit that the theory of production in economics cannot be confined to manufactured capital and labour. They must also include environmental capital – which we abbreviate as KN throughout this text. This inclusion renders significant changes to the exposition of primary concepts in microeconomics, macroeconomics and the economics of trade and exchange. The text illustrates that the inclusion of sustainability in the basic benchmark of microeconomics – the theory of perfect competition – provides a completely different basis for welfare maximisation. The recognition of sustainability also means that the system of national income accounting must encompass a system of environmental accounts. The revised system of accounting paves the way for formulating environmental-macroeconomic frameworks, the applications of which demonstrate that macroeconomic stabilisation must also embrace elements of environmental policy. Further, the recognition of sustainability in theories of trade and exchange affords different bases for the concept of comparative advantage and the determination of exchange rates. This book is intended to equip readers with skills that will enable the analyses of environmental and economic policy issues.

The persistence of environmental and social problems is due, at least in part, to the flawed premises in mainstream economics that have evolved especially during the past 4–5 decades and have been used to support or drive policy decisions. Many economists contending that environmental economics is in essence the application of economics to environmental issues have exacerbated the problem. In this text,

xx *Preface*

we uphold the belief that *environmental economics is a revision and reformulation of contemporary economics to acknowledge the fact that the natural environment is the core of any economy.*

This text is organised in five parts as illustrated below.

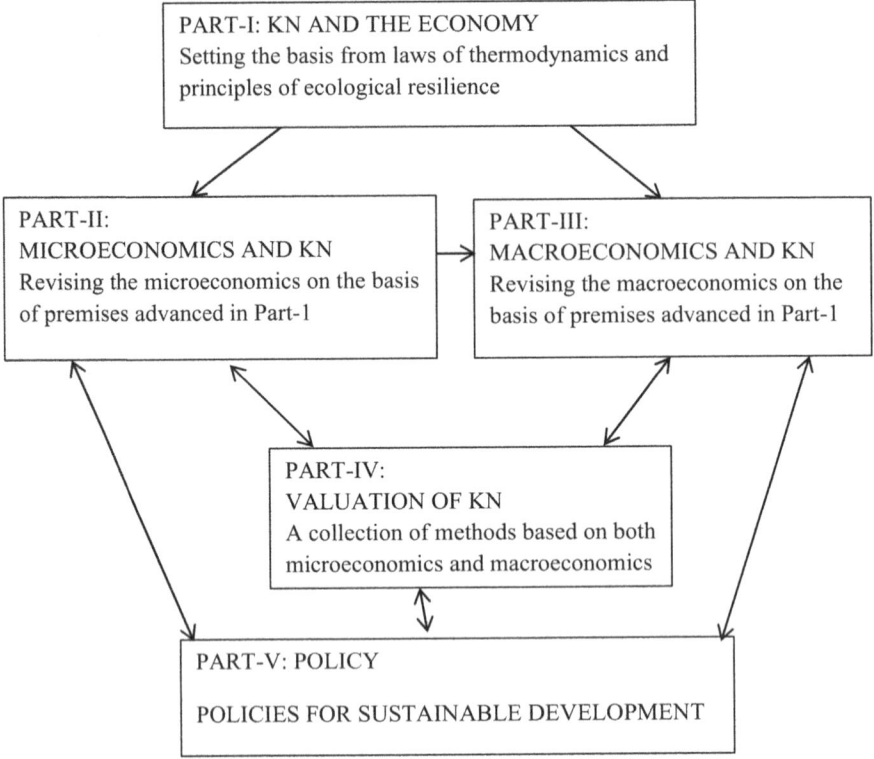

This text is designed for the delivery of environmental economics at an introductory level as well as an advanced level. The possible scheduling of topics for offering the subject at both levels is illustrated in the table below.

A typical semester usually consists of 13 weeks of instruction. In the format suggested in the table below, it is possible to use at least two weeks for discussion forums and empirical analyses such as formulating analytic frameworks. It is possible to create an extra week or two for such purposes by covering a greater amount of content in certain weeks than what is illustrated above. For example, for the introductory unit, it may be possible to cover the content of Chapters 3 and 4 in one week; and for the advanced unit, Chapters 14 and 15 may be covered in a week. Such delivery of material was tested previously at Macquarie University, Swedish University of Agricultural Sciences (Uppsala), Christian Albrechts University of Kiel (Germany), and the National University of Singapore.

Potential Organisation of Chapters for Delivery in One Semester

Week	Introductory Unit	Advanced Unit
1	Chapter 1	Chapters 1–2
2	Chapter 2	Chapters 3–6
3	Chapters 3 and 4	Chapter 7
4	Chapters 5–6	Chapter 8
5	Chapters 7–8	Chapter 9
6	Chapter 9	Chapters 10–11
7	Chapter 10	Chapters 12–13
8	Chapters 11–12	Chapter 14
9	Chapter 12	Chapter 15
10	Chapter 16	Chapters 16
11	Chapter 16	Chapters 17–18
12	Chapter 19	Chapter 19
13	Chapters 19–20	Chapters 19–20

Acknowledgements

We wish to thank our colleagues and students at our respective institutions (past and present), namely:

- Macquarie University, The National University of Singapore, University of Wollongong and University of New England – for Dodo
- The University of York (United Kingdom), University of Alberta (Canada), Northeastern University (United States) and University of Maryland – for Matthias

Dodo:
I begin with a special tribute to my mentors – Professors Thambapillai Jogaratnam, Jack Sinden, Warren Musgrave, Claus-Hennig Hanf and Jack Knetsch for their guidance and nurturing support throughout my career. Sadly, they have all passed away and entered eternal rest. The University of New England paved the way (in the early 1970s) for me to commence a career in environmental economics. I was able to develop this experience further at the University of Wollongong, where natural resource economics was part of the curriculum in economics and commerce programmes. In a previous configuration, Macquarie University had environmental economics as a core module for programmes dealing with environmental management, environmental planning and environmental law. At the National University of Singapore, environmental economics was offered as mandatory module for two cross-disciplinary programmes, namely the master of environmental management and the bachelor of environmental studies. The above-mentioned programmes contributed to the development of the content presented in this text. I remain truly grateful to Peter Phibbs, Paul Morris, Hans-Erik Uhlin, Bö Ohlmer, Lars Drake, Thomas Hahn, and the late Claus-Hennig Hanf for facilitating workshops and seminars to test much of the content presented in this text. Special thanks are due, for reviews of content and support, to Geoffrey Harcourt, Jock Anderson, Ramprasad Sengupta, Krishnan Chandramohan, Azad Bali, Ifthikar Lodhi, Seck Tan, Scott Valentine, Peter Phibbs, Euston Quah, Paul Morris, Hans-Erik Uhlin, Bö Ohlmer, Lars Drake, Peter Daniels, Kaliappa Kalirajan, Namrata Chindarkar and Yvonne Chen. I most gratefully appreciate the leadership and administrative support given by Kishore Mahbubani, Hui Weng Tat, Danny Quah, Lau Siu Kit and Kenneth Paul

Tan (National University of Singapore); and Peter Nelson, Neil Saintilan, Damien Gore and Nathan Hart (Macquarie University). Last but never least, I remain forever grateful to my wife Gowry and our children Vinoli and Dilan for their enduring support and affection.

Matthias:

I thank my many master's and doctoral students from over 20 years of teaching environmental and ecological economics courses and their invaluable feedback, most notably Anthony Amato, Mike Asaro, Andy Blohm, Russ Bowman, Binna Daviðsdóttir, Matthias Deutsch, Rebecca Gasper, Vaishali Kushwaha, James Lindholm, Sahar Mirzaee and Junming Zhu, as well as colleagues and mentors including Clark Bullard, Robert Costanza, Malte Faber and Bob Herendeen, and those who have passed and yet remain to leave their mark on this world – the late Hermann Daly, Bruce Hannon and John Proops. Special thanks go to my wife, Rachel S. Franklin, for her unwavering support of my research and her many probing questions and insights that have helped shape it, and to our children Robert, Oliver, Eleanor, Nicolas and Julia for their patience.

We wish to thank World Scientific and the managing editor of the *Singapore Economic Review* for granting us permission to reproduce the content from the following papers as indicated below.

In Chapter 14: Thampapillai, D. J., and Chen, Y., "Environmental Macroeconomics: A Neglected Theme in Environmental Economics – Leave Alone Economics", *Singapore Economic Review,* October 2018 (https://doi.org/10.1142/S0217590818500327)

In Chapter 18: Thampapillai, D. J., "The Size of an Air Shed – A Macroeconomic Stock Estimate", *Singapore Economic Review,* May 2018 (https://doi.org/10.1142/S0217590818500200);

Finally, our most sincere thanks to the editorial team at Routledge – especially Jyotsna Gurung, Rosie Anderson, Matthew Shobbrook and Annabel Harris for their guidance and encouragement. We wish to make a special mention of Annabel Harris for painstakingly reviewing our proposals for both editions and giving us valuable feedback.

Finally, our most sincere thanks to the editorial team at Routledge – especially Jyotsna Gurung, Rosie Anderson, Sashivadana along with typesetters from codeMantra, Matthew Shobbrook and Annabel Harris for their guidance and encouragement. We wish to make a special mention of Annabel Harris for painstakingly reviewing our proposals for both editions and giving us valuable feedback.

Dodo J. Thampapillai and Matthias Ruth

Part I
Nature-Capital (KN) and the Economy

1 KN and Economics

Environmental Economics and Economics

The main theme in environmental economics is that nature is capital. Sustaining a steady stock of environmental capital (KN) is a necessary condition for economic sustainability. This is because KN acts as both a source and a sink for the economy. It is a source of the basic resources the economy needs and a sink for the waste that the economy generates. When and where the contributions of KN to the economy are disregarded, the ability to produce and enjoy the goods and services generated by the economy is undermined.

The objective of this text is to show how economic theory and its applications, as practised and taught today, can be modified to explicitly accommodate the goal of sustainability and the vital role played by KN. But it would be incorrect to presume that economics has relegated KN to irrelevance. Both classical and neoclassical economists have recognised the pivotal role of KN. However, a sense of environmental complacency pervaded economic policy during the three decades spanning 1950–1980. This was mainly due to a misplaced singular emphasis on economic growth and the spread of evidence that appeared to rationally negate the possibility of environmental limits. However, the continued proliferation of environmental problems among both rich and poor countries has rekindled genuine concerns for a proper balance between economic activity and the environment. The attainment and maintenance of this balance is the primary focus of environmental economics.

Many observers tend to associate the natural environment with classical economics rather than neoclassical economics. In all classical theories (Smith 1776; Malthus 1798; Ricardo 1817; Mill 1848), economic growth was explicitly constrained by environmental limits. However, it was not only these classical economists who emphasised the role of nature in the economic enterprise. Early neoclassical economists too, were aware that humans activities are shaped by natural processes and that economic production itself is the outcome of the interplay of KN with other inputs. For example, consider what Alfred Marshall (1891), who is often considered the founder of modern neoclassical economics, had to say: 'In a sense there are only two agents of production, nature and man. But on the other hand man is himself largely formed by his surroundings, in which nature plays a

4 Nature-Capital (KN) and the Economy

great part (p. 116)' and 'The labour and capital of a country, acting on its natural resources, produce annually a certain net aggregate of commodities...'.

These statements show recognition of *nature* and *natural resources* as the ultimate factors of production. The implication is that, if an economy is not endowed with natural resources, then it cannot produce goods and services. This same conclusion was reached nearly a 100 years after Marshall by the Brundtland Commission (1987, p. 37): 'Environment and development are not separate challenges; they are inexorably linked'.

Yet, much of recent economic theory relegated nature to the sidelines, if not eliminating the same from its analysis. In some instances, the elimination of nature was a simple consequence of representing economic processes in mathematical form. For example, one may express the quantity Q of a finished product as a function f of the amount of labour, L_Q, capital, K_Q and resources R (such as energy and materials) required to produce Q:

$$Q = f\left(L_Q, K_Q, R\right) \tag{1.1}$$

Similarly, the amount of resources that need to be extracted for Q may be written as

$$R = g\left(L_R, K_R\right), \tag{1.2}$$

where L_R and K_R are the labour and capital inputs in the extraction process. Substitution of (1.2) into (1.1) yields

$$Q = f\left(L_Q, K_Q, g\left(L_R, K_R\right)\right), \tag{1.3}$$

which gives the impression that Q can be produced without R. Similar arguments of irrelevance to the economic process can be made with respect to the waste assimilation services provided by nature.

In other cases, the misunderstanding of the fundamentally different role that nature plays in economic production from other inputs has shaped economic reasoning. For some economists, natural endowments are simply not essential; for example, Mankiw (2004, p. 246) states:

> Although natural resources can be important, they are not necessary for an economy to be highly productive in producing goods and services. Japan, for instance, is one of the richest countries in the world, despite having few natural resources. International trade makes Japan's success possible. Japan imports many of the natural resources it needs, such as oil, and exports its manufactured goods to economies rich in natural resources.

This is clearly a mistaken view on at least two grounds. First, natural resources are not simply extractable resources whose acquisition and use are unrelated to the performance of other parts of the natural system. Instead, natural resources, such as oil or natural gas, are a collection of linked endowments within a broader ecosystem. Scientists now believe that it would be unwise to isolate ecosystems to local contexts, given their global connectivity. For one thing, when oil or gas deposits are extracted in one location, after-effects (such as earthquakes) can be felt elsewhere. For another, the combustion of these resources in one location will result in emissions that cause climate change, with effects that are felt globally and for centuries to come. Second, the fact that natural resources can be traded does not mean they are infinite. Japan's riches have enabled it to draw on, and draw down, KN in other countries. To the extent that Japan and a large number of other nations have neglected to preserve KN globally, they have impacted their and everyone else's ability to grow and develop their economies in the future.

Although environmental economics became established as a discipline within economics only in the 1960s, several economists had written extensively from as early as the mid-1800s on environmental issues and problems within the framework of neoclassical economics. Important developments in economics owe their origins to theoretical and empirical work on environmental issues. For example, the theory of externalities, which was introduced by Marshall in 1891 and subsequently formalised in welfare economics by Pigou in 1920, arose primarily from the recognition of events such as pollution, which traditionally fall outside the market. Irving Fisher, writing in 1904 on the various definitions of capital, considered natural endowments such as lakes and rivers as capital assets and used them to illustrate the difference between stocks and flows of resources. The concept of user costs was developed by Hotelling (1949) to account for the requirements of future generations. This was taken up by Keynes (1936) to explain the concept of permanent income, more popularly known in contemporary times as sustainable income (see Box 1.1).

Box 1.1   The Concept of Sustainable Income

Hicks (1946) reviews the meaning of income in Chapter 14 of his *Value and Capital*. He suggests that the practical role of income calculations is to indicate the amount individuals can consume without impoverishing themselves. He, therefore, argues (p. 174) that income should be defined as 'the maximum amount of money which the individual can spend this week and still be able to spend the same amount **in real terms** in each ensuing week'. The calculations must be conducted in real terms, to hold the purchasing power of money constant. This procedure will recognise the influences of variations in prices and interest rates on how much can be spent each week on a sustainable basis.

The issue of resource conservation within neoclassical economics dates back to at least Jevons (1866), who expressed concern over the rapid depletion of Britain's coal reserves. More rigorous analyses of conservation issues during the earlier half of the twentieth century can be found in Gray (1914), Schickele (1935), Bunce (1942) and Scott (1954), to name a few. The valuation of environmental goods also employed neoclassical tools and one of the earliest applications was by Hotelling (1949). Concise reviews of environmental valuation spanning the developments of the nineteenth century and the early to mid-twentieth century can be found in Sinden (1967).

Throughout this volume, we will revisit several of these themes that have shaped economic thinking about the roles of KN in production and consumption processes. We will build on these themes to explore their implications for decision-making about KN and policy formation in support of sustainable practices.

KN and Limits to Growth

Given the central role that nature plays in social and economic welfare, why is environmental economics a relatively peripheral subject area within the economics discipline? There are perhaps two sets of reasons. The first is that the formalisation of a theory of economic growth by economists such as Harrod (1939), Domar (1946), Samuelson (1948), Swan (1956) and Solow (1956) confined the explanation of economic growth to labour (L) and manufactured capital stocks (KM). This simplification – that is, the omission of KN in explaining growth – was perhaps premised on the assumption that KN was infinite or on the increasingly mathematical representation of economic production and consumption processes that allowed for the elimination of natural resources and waste absorption akin to what we have seen with equations (1.1)–(1.3). The second set of reasons stems from the belief that technology can persistently offset the scarcity of KN. When the Malthusian notion of 'Limits to Growth' surfaced in the 1970s from the publications of Forrester (1971) and Meadows et al. (1972), the counter-evidence of the role of technology (Nordhaus 1973, the World Bank 1992, Samuelson and Nordhaus 1990) was strong. For example, consider the following statement issued by the World Bank (1992, p. 37) with reference to the fall in long-run prices of non-ferrous metals (driven mainly by technological improvements in the manufacturing sector but also the expanding contribution that cheap fossil fuels have made to resource extraction, processing and shipment) between 1900 and 1991: 'Declining price trends also indicate that many non-renewables have become more, rather than less, abundant'. Samuelson and Nordhaus (1990, pp. 854–5) were much stronger in their condemnation of Malthusian limits:

> The dour Reverend T.R. Malthus thought that population pressures would drive the economy to a point where workers were at the minimum level of subsistence... What did Malthus forget or at least underestimate? He overlooked the future contribution of investment and technology. He failed to realise how technological innovation could intervene—not to repeal the law of diminishing returns but to more than offset it. He stood at the brink of a new era and failed to anticipate that the succeeding two centuries would show the greatest scientific and economic gains in history—a chastening fact, and one to keep in mind while listening to modern Malthusians sing on their baleful dirge.[1]

The net result has been the development of an economics curriculum that has offered very little space for the study of environmental economics – let alone the recognition it deserves. This applies to several contemporary economics texts as well. For example, prominent books such as those by Frank and Bernanke (2009) and Pindyck and Rubinfield (2007) devote no more than a few pages to a discussion of environmental issues. Mankiw (2004) and Taylor (2007) demonstrate that historically technological change has averted scarcity and that new solutions are continually found. We do not dispute here the vast improvement in efficiency that have been observed over the last century and more, or the expanding scope of technology that now reaches further into all aspects of a firm's or household's activities. But we do note that, in many instances, these technological changes tend to squeeze ever more out of the declining stocks. For example, advanced fishing methods that use radar and sonar to locate fish just make it easier to catch fish. The fact that landings of many fish species have increased (for a while), however, simply means that we have become ever more effective at emptying the oceans. It does not mean that the oceans are producing more fish. On the contrary, stocks have declined precipitously, and just looking at landings misses how unsustainable the 'future contribution of investment and technology' (using Samuelson and Nordhaus' words) actually is. The fish we catch do not constitute sustainable income (in the Hicksian sense) because they include a depletion of the stock needed to generate the income.

The appreciation (within the economics curriculum) of the scientifically driven links between KN and the economy is indeed limited. This can lead to the emergence of ineffective policies, mistaken premises and policy tools. Ineffective policies are clearly illustrated by the World Bank, which in the 1970s and 1980s funded several forest-clearing programmes on the premise that growth needs only L and KM. There has been, of course, a reversal of this ideology within the World Bank, which has subsequently undertaken reforestation programmes accompanied by transmigration programmes for relocating people.

Box 1.2   Some Early Ideas in Neoclassical Economics

A common Cambridge response to new ideas in economics has been: 'But it is all in Marshall', thus relegating to the sidelines further in-depth discussion by labelling the ideas as already known, if not old. For example, in his *Principles* Marshall says: 'Man does not create things. He only rearranges matter'. This statement is indeed the first law of thermodynamics, which we will consider in the next chapter – a law that is albeit acknowledged as relevant does not inform explicitly, for example, how modern economics models production processes, giving the impression that only labour and manufactured capital generate output.

Marshall's description of nature as capital can be interpreted as follows. If one were to disaggregate any commodity into its components until one could disaggregate no more, then the ultimate components would be those that come from nature. At about the same time as Marshall's writing, Irving Fisher laid the foundations of capital theory and attributed three properties to capital. These properties are that capital is durable, that it provides a flow of services, and that it depreciates with use. But Fisher used nature as the basis for this conceptualisation. If we do not interfere with nature it can remain intact; that is, durable. Nature, however, provides a flow of services – the air we breathe, the water we drink and the soil on which we grow food. When we use nature for its services, it depreciates. We call this environmental degradation. And returning to Marshall, the environmental degradation diminishes the ultimate components from which other goods and services are produced.

KN Issues of Contemporary Times

The widespread acceptance of, and faith in, technology to avert natural resource scarcity has introduced not only complacency in the handling of environmental matters but also some mistaken premises for guiding policy decisions. The Environmental Kuznets Curve (EKC) is one such mistaken premise. It posits that economic growth will eventually lead to lower emission loads that demand the sink services of KN. The EKC is an inverted U-type relationship between income levels and emissions of specific pollutants (see, for example, Grossman and Krueger 1995, and Shafik 1994). At the early stages of economic growth, emissions increase but from some point onward further growth is accompanied by declining emissions. The threshold at which emissions begin to decline is supposedly in the vicinity of a per-capita income of US$5,000–6,000 (Thampapillai 2007). The EKC evidence appears to lend support to a mistaken premise: economic growth generates the wealth needed to overcome the harms of economic growth. The policy advice that follows from this premise is simple: concentrate on economic growth and environmental problems will be solved through that. The fallacy of that logic and often overlooked empirical and estimation challenges aside, newer research actually finds that the EKC is not an inverted-U relationship at all but follows more of an N-shaped pattern, with temporarily slowed-down emissions for medium ranges of income and rapid acceleration again as income goes up (Franklin and Ruth 2011).

Related to the EKC observation is the growing belief that emissions trading will reduce pollution levels to acceptable ones, such as those dictated by the Kyoto Protocol. It is certainly desirable, if not essential, to work towards much lower pollution loads than we have at present. However, to contend that emissions trading and the anticipated lower pollution loads will resolve global environmental problems to deliver sustainability of economies is simply foolhardy. In this instance, an analogy is warranted. Suppose that a prisoner in a torture chamber has been receiving 100 lashes per day. The question is: will their wounds heal if the prisoner now receives 50 lashes per day?

Both the EKC observation and the Kyoto Protocol overlook two important biophysical realities. First, the restoration of environmental sinks will rest not on the reduction of *marginal* pollution loads, but on the reduction of *cumulative* pollution loads. Global warming and climate change are manifestations of the fact that cumulative (stock) pollution loads have exceeded threshold levels of the sink capacities of KN. The implication of this is that governance regimes must seek very different types of policies and practices. Second, the accumulation of pollutants triggers biophysical changes at global scales that are at first slow but persistent. For example, carbon dioxide and other notable greenhouse gases have a mean residence time in the atmosphere of over 100 years. As a consequence, even if the Kyoto Protocol or its successor treaties were able to bring emissions down to zero in the next few years (which isn't even their goal), humanity has already committed the globe to several centuries of temperature increases and millennia of sea level rise. None of this means, of course, that one should abandon efforts to revert to past and current behaviours, but it does point to the very limited effectiveness of doing so.

While it is true that rich countries are better able to deal with environmental problems than poor countries, there are some basic flaws in this argument.

- First, even if the rich countries had a substantial amount of wealth, many KN damages are irreversible, and there appear to be a growing number of them.
- Second, while the poverty of poor countries does not help them deal with KN problems, there is a circular cause–effect relationship between poverty and KN damage. Poverty prompts KN degradation. At the same time, KN degradation perpetuates poverty.
- Third, KN problems and damage rarely, if ever, have finite geographic boundaries. KN problems in any country can affect the entire globe.

These observations go beyond the need for individual countries to strive for some type of balance between the environment and the economy. There is a need for a global partnership to help resolve environmental problems.

Climate change and global warming occupy centre stage at present as perhaps one of the most global KN challenges humanity faces. But others also demonstrate the global connectivity inherent in KN damages. The following examples are just a few amongst the myriad of cases across the world:

The permanent and growing brown cloud in Asia
Fragmentation of polar icebergs and shrinking of glaciers
The hole in the ozone layer
Trade in toxic wastes and emergence of toxic soils
Pumping of untreated (or partially treated) sewage into the oceans and red tides
Deforestation
Depletion of fish stocks
Pollution and irreversible damage to inland water resources
Dying and drying up of rivers
Increased intensity and frequency of forest fires

10 Nature-Capital (KN) and the Economy

Increased intensity and frequency of earthquakes
Increased intensity and frequency of heat waves, cold spells, floods and droughts

Given how detrimental climate change already is around the world, and given the fact that current interventions do not seem to be of urgency or extent to significantly change the trajectory on which we find ourselves, the following section probes the issues raised above in more detail.

Climate Change – The Issue of This Century

It is well understood that climate has been changing over geological time, with periods of abnormal temperatures, such as ice ages, occurring over the historic record. It is also well understood that since the Industrial Revolution, burning of fossil energy carriers accelerated to fuel economic growth and development. That burning, in turn, resulted in an accumulation of carbon dioxide (and other greenhouse gases) in the atmosphere. With their accumulation comes a change in atmospheric conditions that trap radiation and thus tend to increase temperatures. Changes in the heat balance of the globe manifest themselves, among other effects, in changes of evaporation of water, which then, like a steam heating system, distributes energy across the globe. And since the evaporated water forms clouds, and rain, changes in temperature also result in changes in precipitation patterns. Similarly, oceans heat up, and currents change, again leading to alterations in the way energy is distributed across the globe. While all of this unfolds gradually, the local effects of these changes can be seen in the form of an increased frequency and severity of extreme weather events. Furthermore, sea levels rise (because of the melting of glaciers, for example, and the expansion of water that occurs at higher temperatures), snow pack melts, which makes previously 'white' reflective surfaces darker, that then absorb more heat. Some species (like agricultural pests) find new areas habitable in which they were not prominent before, while others (like the ones used in agriculture) may suffer from droughts or flooding. Others, like forest and fish populations on which local economies rely, find it difficult if not impossible, to survive where they are, and their ability to move at the rates required by climate change, is limited at best. In short, the entire ensemble of geochemical, climatological and biological conditions to which humans have calibrated their behaviours and choices of food, fuel and materials is in significant upheaval, not to mention the impacts all this has on non-human species that are affected by the use of fossil fuels, irrespective of where they are and where that use takes place.

What's all that got to do with the choice of economic tools to guide decision-making? For one, the significant time lag between emissions (root causes of climate change) and effects makes it difficult for people to understand cause-effect relationships. Similarly, some biological processes unfold quickly (pests spread easily under new climate conditions), while others take very long times (changes in the reflectivity of the surfaces of the globe). Other complications come from the intricate and subtle impacts that changes in one part of an ecosystem have on others. The world of biology, as we know it, is already highly varied and complex.

Changing the KN conditions for the species at any given place, and the ways they interact with each other, and then introducing new ones to a region, and losing others, makes matters even more difficult to understand. The upshot is not only akin to a reshuffling of a deck of cards, but with it a change in the rules, without being told what they are. Yet, economics as we know it, has been designed to guide decision-making in situations where the boundary constraints are known (e.g. one's income) and the choices are known (one's consumption options), where the implications of these choices are well-understood (e.g. the externalities of the choices made) and where all one needs to do is find an optimum, usually within a relatively short period. Now we have a situation where there is high uncertainty about boundary constraints, where the full range of externalities cannot be known, i.e. where not even the uncertainty ranges of the opportunity costs of any action are readily quantifiable, and where the ramifications of actions today extend over centuries and global scales. The presumption of market mechanisms so central to modern economics, and the cognitive and behavioural assumptions that underly it, are seriously put into question here.

Most greenhouse gases (GHGs) have mean residence times in the atmosphere well above a century. So even if we were able to restructure our economies and change our behaviours overnight to result in zero emissions tomorrow, humanity has committed itself to hundreds of years of climate change, given the slowness of biogeochemical cycles. This observation raises two points: First, there is clear, present and growing danger from continuing on the current paths of fossil fuel burning and greenhouse gas accumulation in the atmosphere. Reducing emissions – known as climate mitigation – is an imperative. Without it, the problem will become ever more intractable. Second, while mitigation takes place, economies and the societies of which they are a part, must adapt to new climate conditions. Cities must retreat from coastlines, for example, to provide buffers for sea level rise, or agriculture must use species less prone to drought and heat, and energy systems must decentralise so the economies which depend on them, are less susceptible to power outages triggered by extreme weather events. A whole host of adaptations will be required to help secure the standards of living some of us have enjoyed for a while, and that other wish to achieve.

As the example of an increasingly decentralised power sector illustrates, adaptation and mitigation actions may be mutually reinforcing. For example, an increasingly decentralised power system will benefit from using local – typically renewable – resources, such as biobased waste products from agricultural production. Harnessing these local resources may not be economical if they must be transported long distances, given their low caloric value relative to their weight. But a decentralised system may find use of sources previously underutilised, or overlooked altogether. However, there is considerable technological and institutional lock-in that makes the transition to a decentralised renewable power supply system difficult to accomplish. Power plants and energy distribution networks are designed and built at huge cost and to last decades, with expected returns on investment that must be realised in the interest of investors. Regulatory agencies, and their staff, are deployed to oversee building and operating these systems, and they too have some interest in maintaining them as they are, or at least not changing them fundamentally, because that would entail transition costs (new regulation, new training requirements, etc.) that can be difficult to justify with the current setup. The benefits, therefore, of maintaining the status quo, fall on a select set of actors who can speak with powerful and concerted voices to policymakers. The downsides of maintaining the status quo, in contrast, are socialised, with people across the globe and future generations all affected in various ways. Their interests are fragmented and not presented in unison. In short, the climate problem is not just an environmental problem, it is also an indication of the severe shortcomings of existing governance structures – the interplay of institutions, rights, policies and practices – which are so aligned with the short-term, individual interests expressed on the market place, not the long-term global challenges posed.

While this is not a book about political economy, the nature-based perspective on which it builds challenges more than just the choice of economic frameworks and instruments that we use to guide the 'optimal allocation of scarce resources' to meet human needs and wants. It raises fundamental questions about the appropriateness of existing methods and tools to promote economic decision-making that leads to sustainable outcomes, and it calls on all of us – as producers, sellers and buyers of goods and services, and as global citizens, to take responsibility and action. Changing the instruments we use to analyse problems and point towards solutions is an important component in the developments that are needed. This book attempts to head the call for a new kind of environmental economics.

The Organisation of this Text

The accumulation of evidence on the diverse sources of KN damage has prompted the following thesis:

> While KN damage can emerge from unrestricted economic growth, it can also emerge due to poverty, unemployment and general underdevelopment.

Simply put, there is a very fine balance between the KN and the economy. When this balance is lost, both the KN and the economy deteriorate. The loss of this balance can be caused by rapid, unregulated economic growth as well as by the prevalence of large-scale poverty and underdevelopment. This is what prompted the Brundtland Commission (1987) to decree that environment and development are not separate challenges.

That is, economic sustainability is impossible without the sustainability of KN. This is the theme with which this chapter commenced, and it is this same theme around which the various chapters of this text are organised. We begin the next chapter with a review of how economic organisation is currently represented in the literature: by the circular flow model. This model is the cornerstone of all economic analysis, both in microeconomics and in macroeconomics, and it needs to be revised in order to account for, and properly treat, KN as a factor of production.

In short, the second chapter deals with a hierarchical framework that regards KN as the foundation that supports the economy. The description of this foundational role of KN draws on the first and second laws of thermodynamics.

Chapters 3–9 deal with revisions of microeconomics in the context of the revised economic system. The traditional approach in environmental economics has been to deal with KN issues within the frameworks of markets and property rights (Chapter 3); market failure caused by absent markets (Chapter 4) and imperfect competition (Chapter 5); and intergenerational markets for extractive natural resources (Chapter 6). While retaining this framework, we illustrate the modifications of the basic tools of microeconomics that are required when KN is offered explicit recognition. We describe changes to traditional conceptualisations, such as utility functions and indifference curves (Chapter 7) and production functions and isoquants (Chapter 8), in order to provide a basis on which the microeconomic analysis of KN issues should proceed. For example, KN awareness can prompt consumers to distinguish between KN-friendly goods and KN-unfriendly goods, and this can affect the shape and nature of individual indifference curves. Similarly, the recognition of KN as a factor of production could render the standard production function (as exposited in most texts) to be irrelevant, and a basic conceptual tool associated with production theory – the isoquant – can take on a completely different shape. Further, we also show (Chapter 9) that, when sustainability is recognised as an essential condition within the economic system, how the theory of perfect competition and explanations of various market organisations can be overhauled.

Chapters 10 - 15 concern the revision of macroeconomic frameworks. We start with the ideas that underlie environmental accounting, which is an adaptation of the traditional method of national income accounting (Chapter 10). Building on the tenets of environmental accounting, we develop analytic frameworks for KN in macroeconomics (Chapter 11). Such treatment enables us to revise the approaches to macroeconomic stabilisation in the short run (Chapters 12 and 13) and in the long run (Chapter 14). The global connectivity inherent in environmental problems warrants a re-examination of the premises of trade theory and international linkages (Chapter 15). There are at least two important implications that emerge from the revisions to macroeconomics suggested in Chapters 12–15. Traditional theories of macroeconomics and trade fall back on monetary, fiscal and exchange rate policies as the main instruments of short-run stabilisation. We will show that various aspects of KN policy are also instruments of macroeconomic policy. We will further demonstrate that the notion of a 'steady state' in long-run macroeconomic analysis is not continued growth, but rather the attainment and maintenance of a certain sustainable level of income.

The recognition of KN within the analytic frameworks of both microeconomics and macroeconomics requires the development of methods to measure and value the natural environment. Chapter 16 considers methods of KN valuation at the microeconomic level, whilst Chapters 17 and 18 deal with methods at the macroeconomic level. In Chapter 17, we pose the question: 'how much KN do we use as a flow'? The question raised in Chapter 18 is 'how much KN do we have'?

In Chapter 19 we present a framework for policy formulation that draws on the lessons gleaned from the preceding chapters. An important consideration in this chapter is the need for social harmony and the absence of societal conflicts in order to stabilize KN. The final chapter, Chapter 20, is an attempt to present the challenges in terms of conceptualisation, methodology, data, empirical analyses and policy-formulation that lie ahead.

Note

1 This paragraph has been removed from more recent editions of their text.

References

Brundtland Commission, *Food 2000: Global Policies for Sustainable Agriculture*, Zed Books, London, 1987.
Bunce, A. C., *The Economics of Soil Conservation*, University of Nebraska Press, Lincoln, 1942.
Domar, E. D., 'Capital expansion, rate of growth and employment', *Econometrica*, 14(2): 137–147, 1946.
Fisher, I., 'Precedents for defining capital', *Quarterly Journal of Economics,* 18(3): 386–408, 1904.
Forrester, J. W., *World Dynamics*, Wright-Allen Press, Cambridge, MA, 1971.
Frank, R. H. and Bernanke, B., *Principles of Economics*, McGraw Hill, New York, 2009.
Franklin, R. S. and Ruth, M., Growing up and cleaning up: The environmental Kuznets curve redux, *Journal of Applied Geography*, 32: 29–39, 2011.
Gray, L. C., 'Rent under the assumption of exhaustibility', *The Quarterly Journal of Economics*, 28(3): 466–489, 1914.
Grossman, G. M. and Krueger, A. B., 'Economic growth and the environment', *Quarterly Journal of Economics*, 110(2): 353–377, 1995.
Harrod, R. F., 'An essay in dynamic theory', *Economic Journal*, 40(0): 14–33, 1939.
Hicks, J. R., *Value and Capital: An Inquiry into Some Fundamental Principles of Economic Theory*, Clarendon Press, Oxford, 1946.
Hotelling, H., Letter quoted in *Economics of Public Recreation: An Economic Study of the Monetary Evaluation of Recreation in National Parks Service,* National Parks Service, United States Department of the Interior, 1949.
Jevons, W. S., *An Inquiry Concerning the Progress of the Nation, and the Probable Exhaustion of Our Coal-Mines*,Macmillan and Co., London, 1866.
Keynes, J., *The General Theory of Employment, Interest and Money*, London, Macmillan, 1936.
Malthus, T., *An Essay on the Principle of Population, as it Affects the Future Improvement of Society with Remarks on the Speculations of Mr. Godwin, M. Condorcet, and Other Writers*, J. Johnson, St Paul's Church-Yard, London, 1798.
Mankiw, N. G., *Principles of Macroeconomics* (3rd Ed), Thomson South-Western, Mason Ohio, 2004.
Marshall, A., *Principles of Economics,* Macmillan, London, 1891.
Meadows, D. H., Meadows, D. L., Randers, J. and Behrens, W. W., *The Limits to Growth: A Report of the Club of Rome's Project on the Predicament of Mankind*, Earth Island, Universe Books, New York, 1972.

Mill, J. S., *Principles of Political Economy, With Some of Their Applications to Social Philosophy*, John W. Parker, West Strand, London, 1848.
Nordhaus, W. D., 'World dynamics: Measurement without data', *The Economic Journal*, 83(332): 1156–1183, 1973.
Pindyck, R. S. and Rubinfeld, D. L., *Microeconomics*, Pearson – Prentice Hall, Singapore, 2007.
Ricardo, D., *On the Principles of Political Economy and Taxation*, John Murray, London, 1817.
Samuelson, P. A., *Economics*, McGraw Hill, New York, 1948.
Samuelson, P. A. and Nordhaus, W., *Economics*, McGraw Hill, New York, 1990.
Schickele, R., 'Economic implications of erosion control in the corn belt', *Journal of Farm Economics*, 17(3): 433–448, 1935.
Scott, A., 'Conservation policy and capital theory', *The Canadian Journal of Economics and Political Science*, 20(4): 504–513, 1954.
Shafik, N., 'Economic growth and environmental quality: An econometric analysis', *Oxford Economic Papers*, 46(S1): 757–773, 1994.
Sinden, J. A., (1967), 'The evaluation of extra market benefits: A review', *World Agricultural Economics and Rural Sociology Abstracts*, December 1967.
Smith, A., *An Inquiry into the Nature and Causes of the Wealth of Nations* (5th Ed), Methuen & Co., London, 1776 (Original published in 1904).
Solow, R., 'A contribution to the theory of economic growth', *Quarterly Journal of Economics*, 70(1): 65–94, 1956.
Swan, T. W., 'Economic growth and capital accumulation', *Economic Record*, 32(2): 334–361, 1956.
Taylor, J. B., *Economics*, Houghton Mifflin Company, Boston, MA, 2007.
Thampapillai, D. J., 'The scarcity of environmental capital and economic growth: A comparative study of Australia and the United States', *Singapore Economic Review*, 52(2): 251–263, 2007.
World Bank, *World Development Report*, Oxford University Press, New York, 1992.

2 The Economic System Revised

As indicated in the previous chapter, the goal of economic sustainability cannot be attained without also ensuring the sustainability of nature-capital (KN). Note that we use the abbreviation KN throughout this text to represent nature as capital. The dependence of economic sustainability on KN sustainability is given by the fact that the economy does not function in isolation from the KN. The resources and energy that an economy requires originate from KN, and the wastes produced in the economy are exported and assimilated by KN. If KN sustainability must coexist for economic sustainability, then the overall system must permit the identification of an equilibrium between KN and the economy.

There is yet another precondition for economic sustainability, however, besides that of KN sustainability, namely the sustainability of social capital (KS). Without some level of harmonious relations among the people who plan, organise, produce and consume, for example, economic disruptions are likely to ensue, which in turn prevent or undermine economic sustainability. Such harmonious relations are manifestations of premium KS and require a host of ethical, legal, regulatory and other mechanisms to place checks on the decisions and actions by the members of society. Such checks, in turn, foster behaviours that support society's welfare over the long haul. With its focus on economic and KN related issues, as well as their interrelations, this volume does not explicitly deal with these societal challenges. But we do recognise, and wish to emphasise, the paramount importance of the sustainability of KS alongside that of KN for an economy to sustain itself. We briefly revisit the importance of KS in Chapter 19 and reflect on it as well in our discussion of policy options provided in Chapter 20.

We begin this chapter with a description of the economic system as seen in contemporary economics texts. We then consider an environmental economist's approach to adapting this description. As indicated below, this adaptation is made in terms of the effects that fundamental physical laws and biological principles have on the economy.

The Economic System: Standard Version

The multitude of economics textbooks that have appeared to date, especially the introductory to intermediate-level ones, usually include a circular flow model of

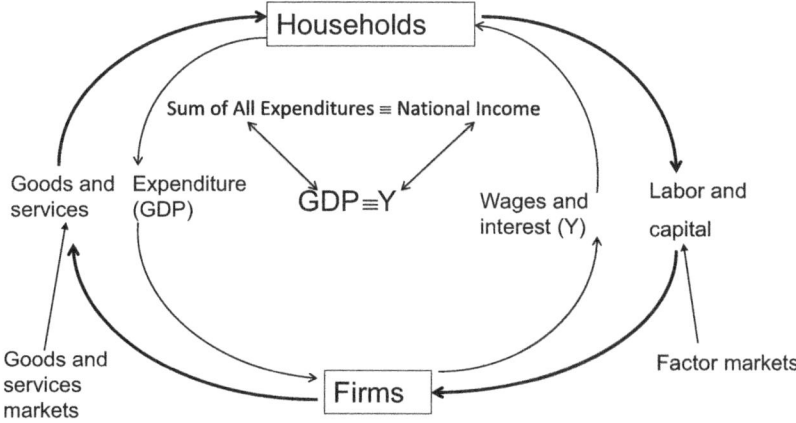

Figure 2.1 The Economic System – Standard Version

the economy. The simplest of these models explains the linkages between two basic entities of any economy, namely households and firms. This is illustrated in Figure 2.1.

Households are units of consumption as well as the owners of labour and capital. The ownership capital by households, as explained below, is based on their savings, which are used by firms to manufacture durable items such as buildings and machinery. Hence, we use the abbreviation KM to represent such manufactured capital so that the distinction from KN is clear. Firms are units of production and supply goods and services. Households demand the goods and services produced by firms and, in turn, firms demand the labour (L) and capital (KM) from households. Hence two types of exchanges occur in this simple system. Firms pay wages and rents in exchange for the households' L and KM, while households expend their income in exchange for the firms' goods and services. For each physical flow – the provision of L and KM for production by firms, for example, or the acquisition of finished products by households – there is a flow of money in the opposite direction, such as wages and rents paid by firms to households, and household expenditures going to firms.

Note that the two types of exchanges do in fact represent two types of markets: markets for goods and services and markets for factors of production, namely L and KM. In each of these markets, prices and quantities of items for exchange are determined by the equilibrium between the demand for, and supply of, the items in question. The study of markets is the subject of Chapters 3–9. Our focus is to illustrate how the recognition of KN and its services will alter the equilibrium in various markets.

At the macroeconomic level, equilibrium is defined by the identity between the sum of all expenditures on final goods and services (more commonly known as Gross Domestic Product – GDP – or Aggregate Demand) and the sum of all wages and rents (more commonly known as National Income – Y – or Aggregate Supply). The microeconomic foundations of this equilibrium stem from the fact

18 *Nature-Capital (KN) and the Economy*

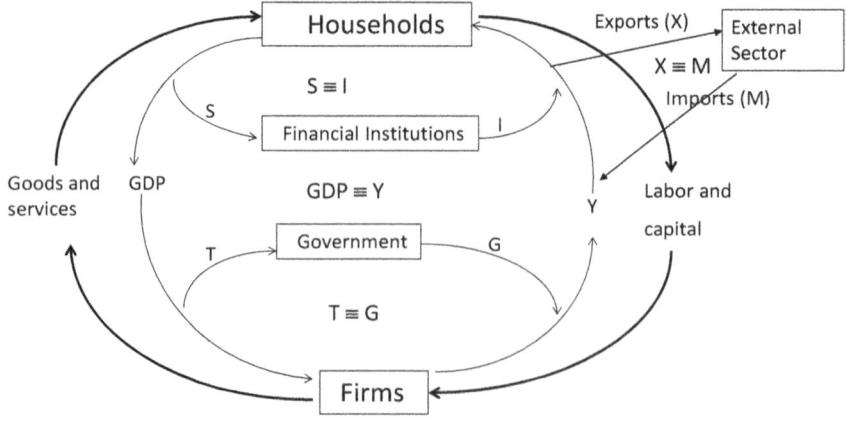

Figure 2.2 The Economic System Revised for Leakages and Injections

that the aggregate values of expenditure and income are derived from transactions in product and factor markets. The standard equilibrium framework in Figure 2.1 can be extended to include equilibria between at least three sets of leakages and injections that are common in most economies, as illustrated in Figure 2.2. The first of these pertains to the activities of governments, which collect taxes (T) from households and firms and then spend these in various parts of the economy. When the leakage (T) from households and firms is exactly offset by government expenditure (G), we have a balanced budget, as typified by the equilibrium (T ≡ G). Macroeconomists often worry about budget imbalances, namely budget deficits (T < G) and budget surpluses (T > G). As we will observe in Chapters 13–15, these deficits or surpluses can influence amongst others the interest rate, which in turn can affect the exchange rate. Hence the ability of governments to manage their budgets can influence balances in the financial and trade sectors. For example, consider the second set of leakages pertaining to the financial sector. Households and firms save some portions of their income with financial institutions, which in turn lend these funds to investors who undertake income-generating activities in the economy. A savings (S) and investment (I) equilibrium (that is, S ≡ I) illustrates that domestic investments are exactly covered by domestic savings. Because savings – investment behaviour is governed by interest rates, one could appreciate the association between the (S ≡ I) equilibrium and the (T ≡ G) equilibrium. The third leakage-injection stems from a country's relationship with the rest of the world through exports (X) and imports (M). A perfectly balanced trade equilibrium (X ≡ M) illustrates that a country's imports are financed by its exports. However, one should note that, in an open economy, deficits in financial and trade markets can be offset by incomes and transfers from foreign locations. Most texts in economics confine the study of equilibria to those mentioned above – that is, equilibria between demand and supply in markets; (G ≡ T) with respect to budgets; (S ≡ I) in financial markets; (X ≡ M) in trade markets; and (GDP ≡ Y) for the economy overall.

The Economic System and Materials Balance

The depiction of Figure 2.2 suggests that the economy functions like a perpetual motion machine – goods and services, and the payments for them, are floating back and forth among the various actors, generating growth in real and financial assets held by households, firms, financial institutions and governments. In order for all this to work, though, materials and energy must be appropriated from KN and waste products are inevitably generated, many of which are released back to KN. In short, all of the flows and stocks associated with the economic process presuppose the presence of adequate levels of KN, and growth of the economy requires that none of the essential resource provision and waste assimilation services of KN run out.

Figure 2.3 illustrates the system in which KN provides the foundational role for the economy. The essential role of KN in any economy is twofold. First, KN acts as a source of raw materials (R) and amenities (A). Note that R is not confined to extractable resources such as minerals, forests and fisheries. It also includes basic essential resources such as air, water and soil. Second, KN acts as a sink (repository) for the wastes (W) that are generated within the economy. In several instances, the source and the sink are the same. For example, the air shed can be described as a 'KN-source' because it provides us with the air we breathe. At the same time, the air shed is also a 'KN-sink' because it acts as a receptacle for the various pollutants that are emitted into the atmosphere.

One of the earliest adaptations of the circular flow model to recognise the source–sink role of KN was in the development of the materials balance approach by Kneese et al. (1970). This approach related the functions of KN to the first law

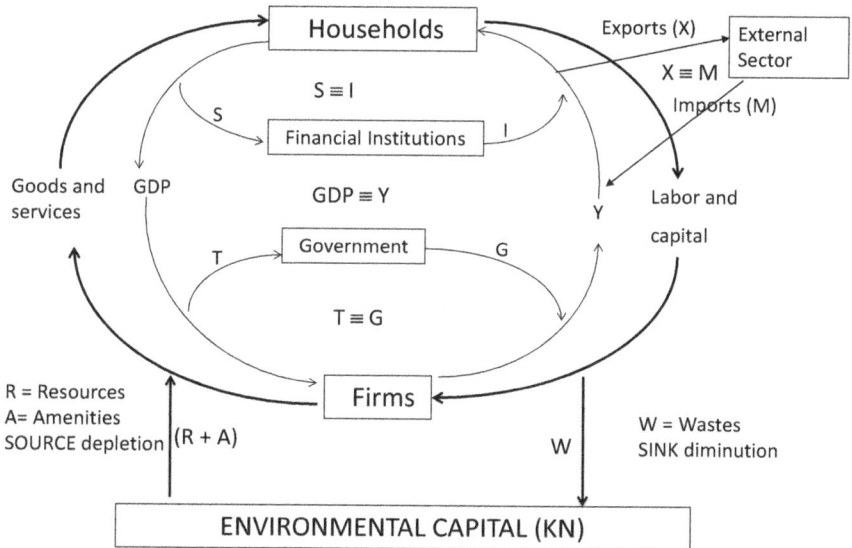

Figure 2.3 The Economic System Revised for Materials Balance

20   Nature-Capital (KN) and the Economy

of thermodynamics – the law of conservation of mass and energy. The first law can be readily expressed for mass as follows:

Matter can neither be created nor destroyed.

The direct implication of this law for any economy is the following equality:

$$\left\{ \begin{array}{l} \text{The sum of matter in R} \\ \text{that enters the economy} \\ \text{from KN-sources} \end{array} \right\} = \left\{ \begin{array}{l} \text{The sum of matter retained} \\ \text{in the economy } + \text{ sum of matter in} \\ \text{W that is returned to KN-sinks} \end{array} \right\} \quad (2.1)$$

At any given period, the sum of resource flows that enter the economy will not equal the sum of residuals because the resource materials used in the economy take time to decay before they are returned to KN-sinks. But the chief implication of this law is that (W) will accumulate in KN-sinks and consequently diminish the source function of KN. This in turn can limit and threaten the sustainability of KN.

Similarly, the first law of thermodynamics can be expressed for energy as

energy can neither be created nor destroyed,

implying a corresponding identity as in Equation (2.1), where energy is received from KN, some of it is stored in the economy (for example, in the chemical bonds of the materials that are produced in the economy), and a considerable amount is returned into KN as waste heat.

Box 2.1   First Law Efficiencies

The first law of thermodynamics is regularly used to compare the efficiencies of different processes. To do so for a power plant, for example, requires knowledge of the heat rate, which is the amount of energy used by an electrical generator or power plant to deliver one kilowatt-hour (kWh) of electricity to the power transmission line that is connected to the power plan, i.e. net of all the energy needed by the power plant itself for its various operations, such as fuel feeding systems, cooling equipment and pollution control devices. In the US, the standard measurement by which heat rates are expressed is in British Thermal Units (BTU). Other countries may use kilo Joules (kJ), for example. One kWh is equivalent to 3,412 BTU.

The first law efficiency of the power plant is its output divided by its heat rate. The following table is for power plants that use steam generators. Note that average efficiencies in an eight-year span only changed marginally, with many of these changes due in large part to the retirement of older equipment.

Note also that these efficiencies are all well below 50 per cent, meaning that more than half the energy going into the system is lost. And since transmission lines are also not 100 per cent efficient, even less arrives at the power outlet into which (inefficient) end-use devices are plugged in.

Energy Source	2007	2014
Petroleum	32.81	33.59
Natural Gas	32.68	32.78
Nuclear	32.52	32.62

Calculated from the US Energy Information Administration, Department of Energy, Average Tested Heat Rates by Prime Mover and Energy Source, 2007–2014 (http://www.eia.gov/electricity/annual/html/epa_08_02.html) – Last accessed 04 October 2018.

The limitation caused by the accumulation of W can be explained as follows. Suppose that an economy begins its operations in a pristine state of KN. The W arising from the initial set of economic activities can be assimilated by KN-sinks without losing any of its characteristics. For example, micro-organisms can decompose biological wastes while plants can absorb carbon gases. This is referred to as the *assimilative ability* of KN-sinks. However, there is only a certain limit up to which the KN-sinks can display this ability. This limit is referred to as the *assimilative capacity*. When the rate at which W is generated exceeds the assimilative capacity, then the KN-sink loses its assimilative ability and is unable to fulfil its usual functions as a waste receptacle. When this happens, the KN-sink, which is polluted with toxic materials, also ceases to be a KN-source for R and A. For example, in cities where exhaust pollution is a problem, clean air is a scarce resource, breathing may become difficult, and the skyline, which is marred by smog, is not a pleasant sight.

Box 2.2   The Longevity of Wastes

For some forms of W, KN can assimilate large quantities. Emissions of carbon dioxide from fossil fuel burning, for example, have been added to the atmosphere at high and growing rates since the Industrial Revolution. Some of that carbon dioxide was readily absorbed by soils and plants, but soon the assimilation capacities of global ecosystems were exceeded, leading to a growth in atmospheric carbon dioxide concentrations and beginning to alter the global climate. For other substances, there are no natural processes to provide waste assimilation at even very low rates of release, such as nuclear materials. Only with time will these break down and lose their adverse impact on the health of ecosystems and the humans living in them. For example, the half-life of uranium 235 – the fissile material used in nuclear reactors – is 704 million years, which means that after 704 million years half of the original radioactivity remains; another 704 million years later, half of that remains, and so on.

But a whole host of other materials are similarly persistent in KN sinks. A wide range of pharmaceuticals, dioxin and dioxin-like compounds, polychlorinated biphenyls, DDT and other pesticides, and plasticisers are found in many everyday products – including plastic bottles, metal food cans, detergents, flame retardants, food, toys, cosmetics and pesticides. Once released, they can affect metabolic processes in organisms and cause developmental abnormalities and sexual or reproductive dysfunction. Because of their low water solubility and short residence time in the atmosphere, these endocrine disruptors tend to bioaccumulate and increase in concentrations from organisms lower in the food chain to those higher up. The effects can be felt for many generations and across many different species.

To summarise, the important conclusion we draw from the first law of thermodynamics is that we need to be conscious of the extent to which KN can act as a sink for various types and quantities of W generated by the economy. If W accumulates and KN loses its assimilative ability, then it ceases to be a KN-source and the economy will be deprived of goods and services on which it relies. Of course, one could argue that increasing scarcity of KN's sink or source function will stimulate the economy activities that find and provide substitutes or that support the restoration of KN. However, all these activities, too, require materials and energy and lead to the generation of wastes.

The outcome of the first law of thermodynamics is strikingly similar to a Malthusian outcome, with the exception that Malthus predicted the limit on the grounds of a growing population. For many economists, the limits imposed by KN are elusive, due to the role of technology. For example, consider what Mankiw (2004, p. 258) has to say about Malthus:

> Where did Malthus go wrong? ... growth in mankind's ingenuity has offset the effects of a larger population. Pesticides, fertilizer, mechanized farm equipment, new crop varieties and other technological advances that Malthus never imagined have allowed each farmer to feed ever greater numbers of people.

The so-called modern Malthusians have in fact nominated the accumulation of W, the filling up of KN-sinks and their resulting potential for losing assimilative capacity, as the major factors in limiting continued growth. The optimism of economists like Mankiw is based on the assumptions that technological change, storage or alternative uses of W, and recycling may either solve the problem of declining

KN sinks or, at a minimum, that one can forever shift the burden from one sink to another. Aside from the fact that KN is finite, there is another physical law that constrains economic activity, namely the second law of thermodynamics: the entropy law. We will consider this next.

The Economy and the Entropy Law

Entropy is a measure of the amount of energy that is *unavailable* for work. In a state of low entropy, there is substantial energy available to do work. Conversely, high entropy means a low ability to do work. Fuelling a car with gasoline, for example, means packing it up with a lot of low-entropy energy. In the process of driving, the low entropy of the gasoline is lost to the atmosphere in the form of waste heat from the engine, friction of the tires on the road and of the body of the car on the air, and dissipation into the atmosphere of carbon and other molecules that make up the fuel. At any part of the journey, it is impossible to drive back by collecting all that has been dissipated along the way, ending up with a full tank of gasoline again. In short, entropy has been irreversibly increased, and undoing that increase would require a disproportionately larger amount of low-entropy resources.

The second law of thermodynamics, also known as the entropy law, states:

All natural processes occur with an increase in entropy.

The economists Georgescu-Roegen (1971) and Herman Daly (1974, 1991, 1992) have been among the main writers who related the importance of the entropy law to the economy. Herman Daly (1992, p. 94) describes the effect of the entropy law on the potential for the economy as follows:

> Economies, like organisms, consist of ordered structures subject to entropic decay, but capable of building up their internal order at the expense of imposing greater disorder on their environment.... At the input end it takes low-entropy matter-energy from finite environment sources, and at the output end it returns high-entropy wastes into environmental sinks. The sources become depleted and the sinks fill up and become polluted. The entropy law is supremely relevant because it says that sinks cannot serve as sources. Absolute scarcity arises from the combination of the first and second laws of thermodynamics, not from either alone.

Sceptics of the entropy law's relevance to the economic system have argued that this law only holds to the extent that the economy 'takes low entropy matter energy from finite KN sources'. Since Earth is not really isolated, so goes the argument, but instead exchanges energy with the rest of the universe, the second law should not constrain economic activity. The high concentration of low entropy energy in fossil fuels, for example, is the very embodiment of the solar energy influx that humanity is harvesting now. As long as the sun shines, we should be fine.

While in principle Earth is indeed an open system, material exchanges with the rest of the universe (cosmic dust and meteors coming in, and the odd satellite and spacecraft going out) are miniscule in comparison to the overall mass of Earth. In short, for all intents and purposes, Earth is a materially closed system. Not only are we a long way from being able to import materials from outer space and export our material waste into space, but building the infrastructure for this to happen is also extremely costly in both material and energy terms. We must live within our material means.

In principle, we could of course harness the influx of solar radiation and use it to replace the use of non-renewable fuels in our economies. But the extent to which we would use that energy for unsustainable practices – continued depletion of materials stocks, waste generation and dumping into the KN sinks and deterioration of other renewable resources such as fish populations, forests and soils for example – will undermine the effectiveness of this strategy.

Technology and new knowledge can offset some of the depletion we observe and counteract the dissipation of materials. This can be achieved by substituting between materials, and by developing new methods of waste treatment and recycling. For example, fibre optics has replaced copper in telecommunications, and there has been a steady decline in the use of metals and energy per unit output in industrial processes. However, we will never be able to create more than there has been to start with (the first law), and as a consequence, the best we may be able to do is close material cycles to minimise waste – a topic we will return to in Chapters 18 and 19.

As Daly and Cobb (1989) argue, new knowledge is never complete. This incompleteness of knowledge calls for caution. Asbestos, when it was first introduced as a substitute for various uses of biomass in insulation and soundproofing, for example, was hailed as a useful and versatile material. Similarly, a host of chlorofluorocarbon-containing refrigerants substituted for the uses of ice in households and industrial processes, allowing for unprecedented efficiency gains in cooling. Years later, when the adverse health effects of asbestos fibres were understood, their usefulness came to an end. Similarly, with the discovery of stratospheric ozone depletion as a result of using refrigerants, international agreements were put in place to curtail their manufacturing and use. The laws of thermodynamics decree that we cannot rely indefinitely on recycling and the treatment of residuals as a source of material inputs to sustain economic growth, and that undoing the damages from using any particular material, energy source or process with another in itself creates waste materials and energy.

But what about increasingly efficient uses of energy? Can't we just use ever more cleverly what we have on Earth and what we receive for free from the sun? Of course, by definition non-renewable resources such as oil and gas are, in essence, finite, since the time frames over which they have been formed extend over millennia. And there is a maximum upper limit to the solar energy that can be received during any given period. That limit is determined by the solar influx itself and by the ability to harness it. The latter, in turn, is a function of the biomass on Earth that can capture sunlight, and the technology that can convert the sun's energy into more useful forms, such as solar collectors that make direct use of solar radiation or turbines that take advantage of the movements of air and water caused by temperature differentials in KN. But to make and deploy these technologies in the first place requires using (finite) material stocks and energy flows. And then there is the release of waste heat generated by the economy. Not all that waste heat can be exported into space, but much of it accumulates in Earth's atmosphere, affecting the global climate, and increasingly diminishing the productivity of ecosystems and diverting human investment towards climate mitigation and adaptation efforts.

With an economy that must do with finite material endowments on Earth and that is constrained by the energy flow onto it and back into space, the potential for absolute scarcity that Herman Daly identifies is real. Only an economy that makes use of materials and energy in ways that leave KN intact can flourish in the long run. Any depletion of fossil fuels must therefore be accompanied by their replacement with renewables at rates that compensate for the depletion of these fossil fuels. And such replacements must make energy available to reuse and recycle the materials that are dissipated in the processes that extract and use fossil fuels and that make and maintain renewable energy systems.

Finding a sustainable balance between drawing on renewable and non-renewable energy sources is very tricky, however. For example, since the 1970s Brazil has adopted ethanol from agriculture as its transport fuel instead of traditional fossil fuels. There are two issues to be noted with this shift. One is that intensive use of agricultural land for ethanol production can render the basic soil infertile. The other is that the combustion of ethanol is not without residuals, although these may be less problematic globally when compared to the combustion of fossil fuels.

The closed nature of the first and second laws both refute the conviction that substitutes for various types of R can be found indefinitely. Further, they reinforce the importance of the need to reuse and recycle materials and to shift towards investments in renewable KN. We will consider this distinction between renewable and non-renewable types of KN in more detail in Chapter 5.

Further Adaptations of the Economic System

The main difference between Figures 2.2 and 2.3 is that the latter has included the role of KN in terms of the materials balance approach. However, this description does not sufficiently capture the effects of the entropy law. We shall now consider a further extension to Figure 2.3, so that we can incorporate both thermodynamic laws and also establish a basis for identifying the equilibrium between KN and the economy. This extension is presented in Figure 2.4.

The basic argument is that KN is an ultimate factor of production. This means that, if we try to disaggregate any good or service until no further disaggregation is possible, we will end up with a set of basic factors that are components of KN: air, water, soil and energy. These basic factors together form numerous resource systems, such as river systems, mangroves, lagoons, coastal estuaries, forests and so forth. The collection of these resource systems – the natural endowments of an

26 *Nature-Capital (KN) and the Economy*

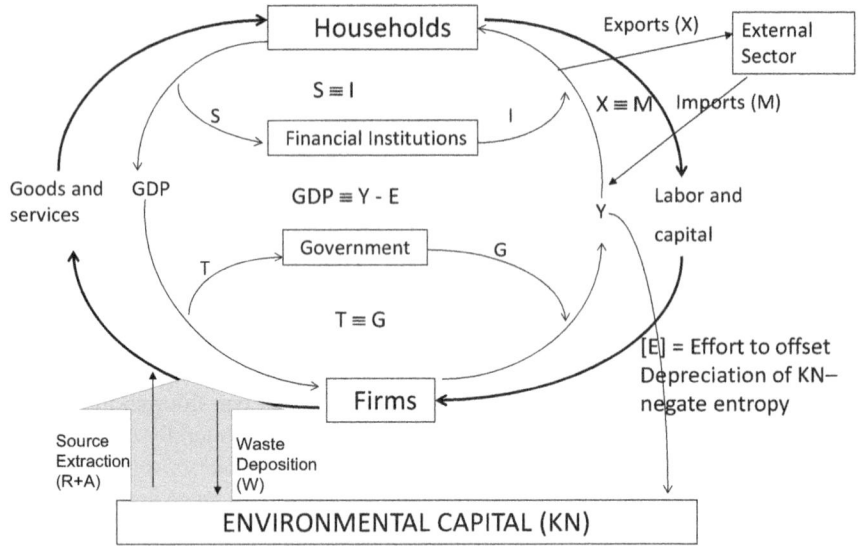

Figure 2.4 The Economic System – Revised for Materials Balance and Entropy

economy – makes up the stock of KN that acts as a source and a sink. In Figure 2.4, the arrow labelled (R + A) from KN to the economy describes the flow of resources and amenities and the reverse arrow labelled W describes the flow of waste and residuals into KN-sinks. These two arrows, (R + A) and (W), embody the first law of thermodynamics. If the time dimension were absent, then the difference between (R + A) and (W) would represent the matter that is retained in the economy within various goods and services. Note that, in contrast to Figure 2.2 for example, no monetary flows accompany the acquisition of R and A, thus leaving KN conceptually outside the purview of the marketplace that governs the allocation of human-made capital. KN is free. Only when its overuse affects the economy would there be a call to manage the resulting 'externalities'. Such management, however, usually comes long after the fact. A host of problems affecting KN illustrate these. We list just three prominent examples:

- The decades-long generation of acid rain and its impact on ecosystems
- Centuries-long accumulation of carbon dioxide in the atmosphere and its resultant impact on the global climate
- Deforestation of the tropics with commensurate impacts on the species diversity that will be felt forever

In the standard economic system (Figure 2.2), economists assume that the total of all income from goods and services is used up in final consumption and investment. We can conveniently assume that the term 'final consumption' also includes government expenditure and net exports. However, if KN represents the ultimate factor of production, then the final output should not be exclusively used up in consumption and investment. A part of it must be ploughed back into KN. In Figure 2.4, the arrow labelled E represents the share of the final output that is redirected to KN. This share of output is for activities that will maintain, and in some instances even expand, the flow of services from KN. The expansion of the services flow can be due to the activities in E that strive to restore previously damaged or lost endowments, and hence can expand the set of resource endowments that make up KN. Others in E are analogous to regular maintenance operations; for example, treating wastes before they are discharged, treating river flows at periodic intervals and so forth. We will consider the components of E in greater detail in later chapters, where we will illustrate the adaptation of standard macroeconomic frameworks. Suffice it to say for the moment that arrow E describes the effort expended by the economy to sustain KN, and this effort falls within the influence of the entropy law of thermodynamics. This is because the presence of entropy results in the amount of effort entailed in E to increase as the economy's utilisation of the KN increases. Further, the effort in E will normally be larger when the flow in (R + A) is dominated by non-renewable resources than when this flow is predominantly made up of renewable resources. As indicated above, technology and knowledge can dampen the rate of increase of this effort.

Equilibrium in the Adapted System

We noted above that the arrow E in Figure 2.4 involves two types of activities:

- Restoring lost (non-functional) KN endowments
- Maintaining existing (functional) KN endowments

The distinction between a functional and non-functional endowment can be illustrated as follows. Suppose that a river cannot be used for any purpose whatsoever because it is affected by toxic algal blooms. This river is a non-functional endowment, and it cannot be counted in the stock of useful endowments owing to its toxicity. When the river is restored (by limiting or removing nutrients that avoid algal blooms), then the restoration is like an investment. The restoration adds to the stock of endowments relative to what the economy started with. But if we start with a river that can be used for various purposes, then that river is a functional endowment that has to be maintained by regular clean-up operations. The distinction between functional and non-functional endowments can be used in defining the equilibrium between KN and the economy.

We will now consider how the meaning of 'equilibrium' in economics would change when the concepts gleaned from the laws of thermodynamics are introduced. So we begin with the system described in Figure 2.2, where equilibrium is defined as:

$$\begin{pmatrix} \text{The sum of all incomes} \\ \text{(National Income)} \end{pmatrix} = \begin{pmatrix} \text{The sum of all final expenditures} \\ \text{(Gross Domestic Product)} \end{pmatrix} \quad (2.2)$$

28 Nature-Capital (KN) and the Economy

As indicated above, this definition does not explain the economy's efforts to maintain KN. So, following Figure 2.4, we can modify the above equilibrium by taking note of the differences between functional and non-functional endowments as follows:

$$\begin{pmatrix} \text{The sum of all incomes} \\ \text{(National Income)} \end{pmatrix} = \begin{pmatrix} \text{The sum of all final expenditures} \\ \text{including restoration of non-functional} \\ \text{endowments less expenditures to offset} \\ \text{the depreciation of natural endowments} \\ \text{(Modified Gross Domestic Product)} \end{pmatrix} \quad (2.3)$$

If an economy did not have any non-functional endowments – that is, if all its endowments were functional – then the equilibrium would become:

$$\begin{pmatrix} \text{The sum of all incomes} \\ \text{(National Income)} \end{pmatrix} = \begin{pmatrix} \text{The sum of all final expenditures} \\ \text{less expenditures to offset the} \\ \text{depreciation of natural endowments} \\ \text{(Modified Gross Domestic Product)} \end{pmatrix} \quad (2.4)$$

These modified definitions (equations (2.3) and (2.4)) describe the equilibrium between the economy and KN. As long as the efforts directed towards KN are successful in maintaining the source-sink role of KN, the economy will continue to function.

Concluding Remarks

So far, we have revised the standard description of the economic system. The framework in Figure 2.4 describes KN as the foundation on which the economy rests and the ceiling to which it can grow. The sustainability of the economy will ultimately depend on how well the foundation and ceiling are preserved and maintained. As can be seen from Figure 2.4, the potential for sustainability rests on two interrelated sources, represented by the arrows labelled (R + A) and (W) on the one hand and by the arrow labelled E on the other. If, in (R + A), the economy shifts from non-renewable to renewable resources, then the quantity and quality of the residuals (arrow W) will become less entropic, and the amount of effort (arrow E) needed to sustain KN will also become less. By the same token, if the economy neglects the application of effort to sustain KN, then there will be an accumulation of W and hence an increase in entropy. As a result, the economy will experience a quantitatively and qualitatively diminished R + A that can be derived from KN.

In Chapters 3–9, we will examine how the associations between the arrows labelled R + A, W and E can be included in the analyses of microeconomic decisions. We will then extend this examination to macroeconomic decisions in Chapters 10–15.

Review Questions

1 Explain why the first law of thermodynamics alone is not sufficient to explain the role of the KN for the economy.
2 Discuss the possibility of developing a concept of equilibrium between the KN and the economy, especially in the context of the second law of thermodynamics.
3 Some commentators such as Fisk (2011) have argued that market prices are capable of capturing the effects of entropy. For example, a rusted kettle has high entropy and is almost worthless relative to a brand-new kettle, which has lower entropy and fetches a higher price (compared to the rusted kettle). Can this observation be generalised to other goods and services? Are more orderly, organised forms of materials always more valued than those that are more dispersed? (Think of a metal, such as gold, as one example, and a milkshake as another.)

References

Daly, H. E., 'The economics of the steady state', *American Economic Review*, 64(2): 15–21, 1974.
Daly, H. E., 'Towards an environmental macroeconomics', *Land Economics*, 67(2): 255–259, 1991.
Daly, H. E., 'Is the entropy law relevant to the economics of natural resource scarcity? — Yes, of course it is! Comment', *Journal of Environmental Economics and Management*, 23(1): 91–95, 1992.
Daly, H. E. and Cobb, J. B., *For the Common Good: Redirecting the Economy towards Community, the Environment, and a Sustainable Future*, Beacon Press, Boston, MA, 1989.
Fisk, D., 'Thermodynamics on main street: When entropy really counts in economics', *Ecological Economics*, 70(11): 1931–1936, 2011
Georgescu-Roegen, N., *The Entropy Law and the Economic Process*, Harvard University Press, Cambridge, MA, 1971.
Kneese, A. V., Ayres, R. U. and D'Arge, R. C., *Economics and the Environment: A Material Balance Approach*, Resources for the Future, Washington DC, 1970.
Mankiw, N. G., *Principles of Macroeconomics*, 3rd edn, Thomson South-Western, Mason, OH, 2004.

Part II
Microeconomics and KN

3 The Market Model and Its Failure

In this and the next three chapters, we will follow the lines of standard microeconomics. A basic premise that underlies microeconomics is that consumers and producers display rational behaviour. The basis for rationality stems from two axioms, namely *self-interest* and *present gain*. That is, individuals would place their own interests first and further prefer to reap the benefits of their efforts sooner rather than later. A further premise in microeconomics pertains to the ability of economic agents to distinguish between 'goods' and 'bads'. *Goods* are those that we generally prefer to have 'more' rather than 'less' of. *Bads*, like pollution, are those that we prefer to have 'less' rather than 'more' of. We return to these premises and axioms in Chapters 7–9 to demonstrate that changes to definitions are warranted when explicit recognition is afforded to the sustainability of KN.

A starting point for our discussion on the market model is Figure 2.3, which illustrates the standard perception of circular flows of goods and services, and associated payments, among households, firms, financial institutions and government. This figure identifies two types of markets, namely those for goods and services, and those of factors of production.

The main aims of this chapter are to:

- Show why some or most of the components of the KN resource flows and amenities (arrows R + A), and waste and residual flows (W) in Figure 2.3 fall outside the scope of the market model
- Explore the avenues that are open to bring these components within the scope of the market model

In other words, we will show why the market fails in the context of decisions involving KN, and then sketch the ways by which this failure could be corrected. This approach will put us in a position to illustrate the types of revisions to standard economics that are warranted when the relevance and implications of the laws of thermodynamics and ecological resilience are accounted for.

The Functions of the Market

The basic entity of exchange is the market. A market emerges from the interaction between consumers and producers. Consumers demand goods and services, and producers supply these goods and services. This interaction between consumers' demand and producers' supply enables the market to perform three functions:

- Helping firms to select the goods to be produced
- Determining the amounts of production for these selected goods
- Identifying the methods of production

These functions are more popularly referred to as: 'What to produce?' 'How much to produce?', and 'How to produce?' To explain these functions more clearly, we need to examine the concepts of demand and supply, as well as the context within which they are often analysed, namely the notion that there is perfect competition among suppliers in the marketplace. We consider these in turn below.

Market Demand

Market demand for a good or service is the aggregate of all individual consumer demands for that good or service. A market usually bears a societal context in terms of a geographic area in which the consumers exist. So, for example, consider the expression 'market demand for apples in Australia'. Hence, this statement refers to an Australian market for apples. Health policy officials will be interested in knowing about this demand because of the effects that apple consumption may have on health. Apple growers might want more details about this market, such as where the apples originate from and about the many varieties of apples. Notwithstanding these nuances, as we show below, the concept of demand provides the basis for estimating the benefits of consumption, and in order to illustrate the basic working principles of the market, we generally abstract from the typical good under consideration or the place where production and consumption occur. Of course, in reality – and specifically with respect to impacts on KN – it does matter a fair deal what is being produced and where, and what happens to the constituent elements of the product after consumption. Similarly, for social impact it matters who produces and consumes, and where those activities take place, as well as who is subjected to the waste products from production and consumption. In short, geographic and social aspects are important, and they in themselves are the subject of related disciplines, such as economic geography and regional science. We will address these issues in this book where necessary, but will largely focus on the underlying mechanisms by which changes in production, consumption and KN interrelate.

The basis for relating benefits to consumption rests on the premise that consumers derive utility (satisfaction) from consuming a good or service. As exposited in most economics texts, the utility from consumption increases – but at a diminishing rate – as more of the good or service is consumed. This utility would eventually decline when consumption exceeds a point of satiation. That is, the marginal utility – the utility derived from consuming an extra unit – will be inversely related to the quantity of consumption. This is referred to as the *law of diminishing marginal utility*. This marginal utility is also manifested in the amount that consumers are willing to pay. An individual consumer's demand for a good or service can be

The Market Model and Its Failure 35

defined in terms of his or her Willingness to Pay (WTP) for consuming an extra unit of the good or service. Given the law of diminishing marginal utility, the relationship between the 'WTP for an extra unit' and the 'quantity of consumption' is generally inverse. Note that WTP for one more unit is in fact the value of one unit. In common parlance, *the value of one unit* is *price*. Hence an individual's demand is usually described in terms of the relationship between price and quantity consumed. That is, for example, when price increases, consumers will tend to consume less, and vice versa. There are of course some exceptions to this general observation, and we consider them in Chapter 7.

In Figure 3.1, we illustrate the derivation of a market demand curve in a purely hypothetical society of only two consumers, A and B. At a price of 50 cents, A consumes two units and B consumes three units. Hence total consumption by society, at a price of 50 cents, is 2 + 3 = 5 units. The market demand curve is derived in this way by adding the individual consumptions for each price. Such method of adding individual demand curves to derive the market demand curve is called 'horizontal aggregation'. As we illustrate in the next chapter, in describing the demand

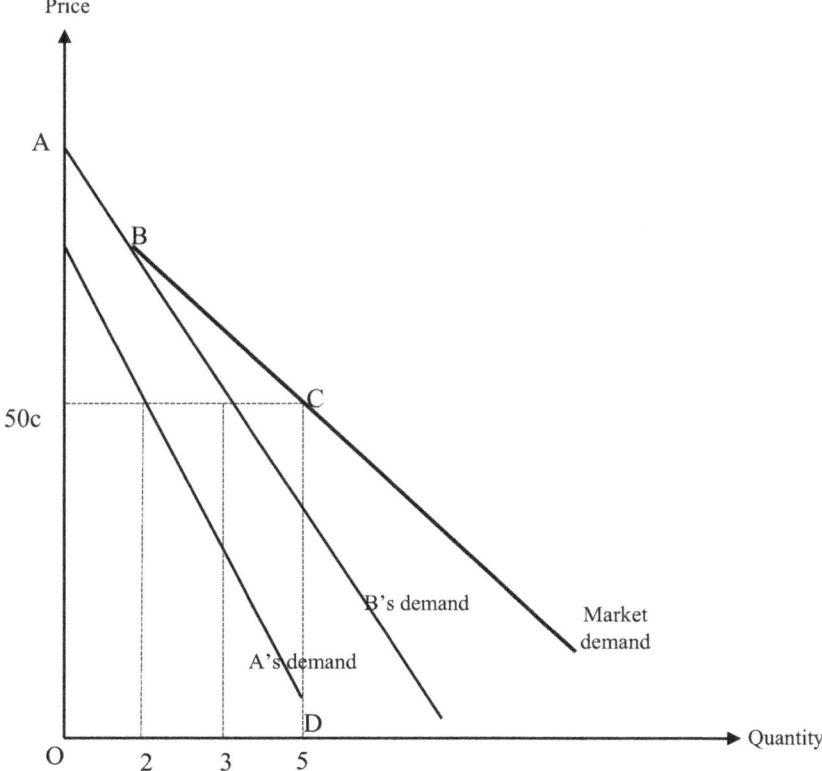

Figure 3.1 Aggregating Individual demands

for KN related goods and services, we would adopt 'vertical aggregation' – that is, total WTP for a given quantity.

As indicated above, for each quantity purchased, an individual's demand shows the amount that he or she is willing to pay for an additional (or marginal) unit. Since market demand is the sum of all individual demand curves in a society, it describes what the society is willing to pay for a marginal unit. So for example in Figure 3.1, when the hypothetical society of only two consumers (that is, A and B) purchases five units, it is willing to pay 50 cents for the fifth unit. What is of importance to us at this point are the following:

1. WTP, being a measure of satisfaction, is also a measure of benefits, and hence the market demand curve forms the basis for defining the benefits to society from consuming or purchasing a specific commodity.
2. The area below the demand curve measures the total value of WTP. Note that this explanation rests on the principles of calculus, where quantity is defined in terms of infinitesimally small units. For example, in Figure 3.1, area OABCD represents the total amount that society is willing to pay for 5 units. This area can be construed as the sum of vertical heights, each of infinitesimally small breadth spanning the distance of five units from the origin. So, we can state that the area OABCD indicates the total benefit to society from consuming five units.

The underlying preference structure of individuals would ultimately determine the shape and position of their demand curve in quantity-price space. A steep demand curve implies that consumers are unlikely to change the quantity they consume by a significant amount when the price of the item changes. Such demand is referred to as inelastic. For example, people who must rely on driving to work, are very likely to display inelastic demand for gasoline. Similarly, people who need specific medications to treat their health condition would display inelastic demand for the treatments. In contrast, a relatively flat demand curve implies that demand is elastic. That is, even a small variation in price could result in a significant change in consumption. Examples of elastic demand can be found with non-essential luxury goods or where close substitutes are readily available.

Several factors such as changes in income, people's attitudes, population size, and prices of alternatives influence both the shape and position of the demand curve in quantity-price space. When the population increases, say due to an increase in net migration, then the demand for several essential items will increase (more individual demand needs to be horizontally aggregated, to use the approach applied above) – that is, the demand curve will shift outwards. Consider the decision by Tompkins County Area Transit (in New York State, USA) to run a bus between the city centre of Ithaca and Cornell University Campus every ten minutes. This decision has most probably shifted the Cornell campus residents' demand curve for gasoline inwards and made it more elastic than before. Prior to 2019, the demand for face masks was relatively low and was confined largely to those in the health profession. But the Covid-19 pandemic, which emerged in 2019, did not only shift the demand for face masks far outwards but also transformed this demand into a near-inelastic one, because wearing a facemask became a mandatory act during the health crisis.

In general, an increase in income can shift the demand curve outwards for most household consumption items. Notable changes in demand due to income fluctuations are observed with items such as restaurant meals and convenience-appliances. Some goods such as high carbohydrate foods, display an inward shift of the demand when incomes increase. Such items are referred to as inferior goods. With improved information and changes in attitudes, it may be possible to transform items that aggravate climate change into inferior goods. We consider these and further aspects of demand in greater detail in Chapter 7.

Market Supply

The concept of supply emanates from activities pertaining to production and producer behaviour. Producers incur costs to supply goods and services that consumers demand. In the short run, the *supply curve is in fact the ascending portion of the marginal cost (MC) function, which lies above the average variable costs (AVC)*. This statement warrants further explanation and hence a small digression into the theory of costs is in order. As explained in standard economics texts such as Hirshleifer et al. (2005), [total cost (TC) = fixed costs (FC) + variable costs (VC)]. Consider now the definition of average cost (AC), namely the cost of producing one unit, that is: $AC = \left[\dfrac{TC}{Q}\right] = \left[\left(\dfrac{FC}{Q}\right) + \left(\dfrac{VC}{Q}\right)\right]$. Note that the first term, $\left(\dfrac{FC}{Q}\right)$ namely the average fixed cost, will continue to decrease as Q increases. However, the second term, $\left(\dfrac{VC}{Q}\right)$, which is in fact the average variable cost (AVC), could initially fall, but would eventually increase because increasing Q entails increased utilisation of factors. As a result, AC and AVC curves are U-shaped and perhaps skewed with the nature of the skew depending on the underlying production process. Whilst AC and AVC pertain to the cost of producing one unit, MC is the cost of producing **one extra** unit. So, when AC and AVC are falling, the MC will also be falling and lie below AC and AVC. Contrarily, when AC and AVC are rising, MC will also rise and would eventually be more than the AVC and AC. These features of AVC, AC and MC are illustrated Figure 3.2.

A producer, who wishes to engage in production, would compare the market price (P) of the product with the MC of production. This is because P represents the revenue from selling one unit of the product, and MC is the cost of producing an extra unit that product. Hence the producer will engage in production only if (P > MC). In Figure 3.2, we present three potential market prices (P_0, P_1, P_2). Should the prevailing price be P_0, then the producer will not attempt to produce this item because revenue (P_0*Q) will always be less than both VC and TC at this price. If the price increased to P_1, then the producer might consider production because the

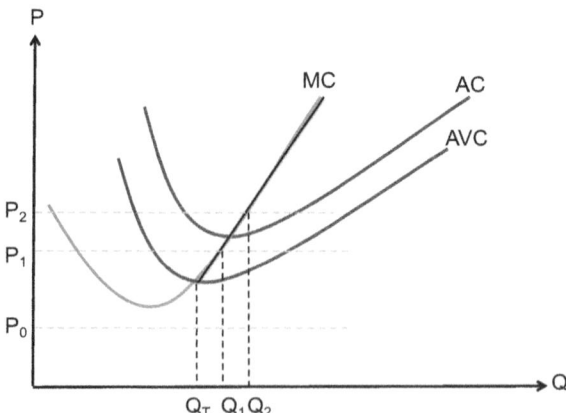

Figure 3.2 The Cost Structure of a Firm

revenue (P_1*Q_1) would cover at least some part of the VC. But if the price P_2, then the producer will engage in production because revenue (P_2*Q_2) will be greater than TC. Hence, producers will engage in production if and when their MC can exceed AC. And they would contemplate production when their MC exceeds AVC. It is for this reason that the supply curve is defined by the MC curve that lies above AVC. When product prices rise, producers are willing to accept higher marginal costs and produce more. Producers will expand production until price equals marginal cost.

It is important to note here, however, that the cost of producing an additional unit is derived on the basis of the prices and quantities of inputs that the supplier acquires on the market for inputs. Not all inputs, however, do have market transactions associated with them. For example, when production involves the use of fossil fuels, the oxygen required for combustion of the fuel typically comes for free. Similarly, and in accordance with the second law of thermodynamics, any production process inevitably results in waste. Many of these wastes, such as waste heat or carbon emissions, typically do not have a price associated with them. Their disposal is free to the supplier, and the marginal cost curves above are unaffected by them. But even if there are prices associated with some inputs, these prices may be affected by non-economic factors and thus lead to marginal cost curves that are skewed. A prominent example here is the case of labour inputs. In societies where children and women have their rights curtailed because of cultural norms and existing power structures, it may be possible to exploit them. The wages paid to children and women may, as a result, be much lower than in the absence of exploitation.

Consider now market supply. This is the aggregate of the individual supplies of each producer. In Figure 3.3, we assume that there are only two producers, C and D, and this figure also shows how the aggregate supply curve can be derived from the individual supply curves. The method of aggregating individual supply curves to form the market supply curve is similar to the type of aggregation that was illustrated above for the demand curve.

The Market Model and Its Failure 39

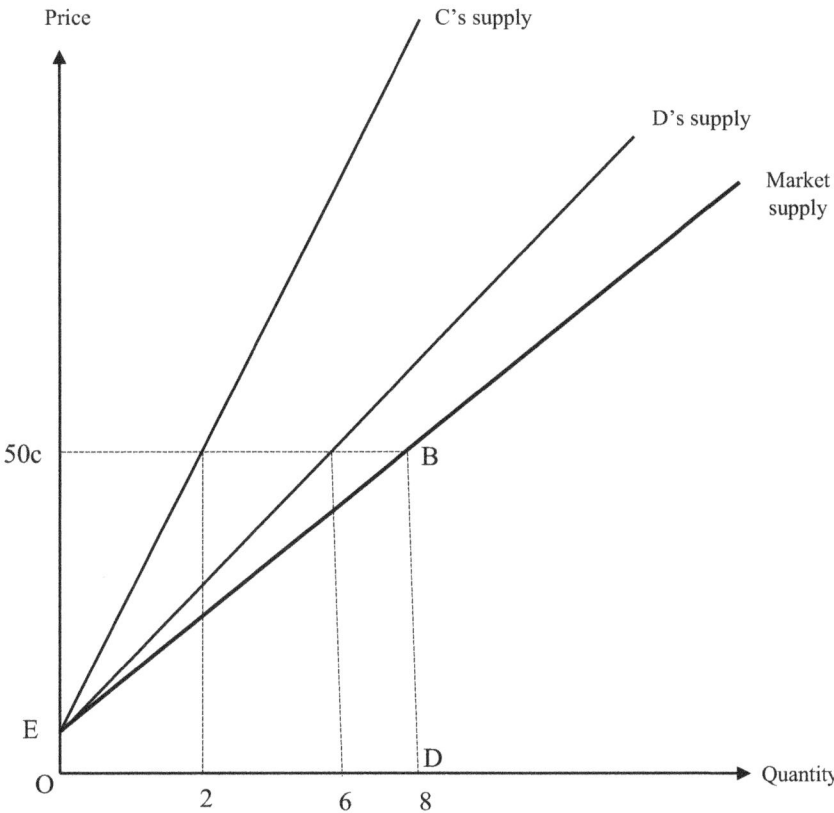

Figure 3.3 Aggregating Individual Supplies

Just as the demand curve is based on the principle of WTP, the supply curve is based on the principle of willingness to accept (WTA). From Figure 3.3 we can infer that producers are willing to accept a cost of 50 cents for producing the eighth unit. Since the supply curve describes the marginal costs, the area below this curve defines the total economic cost of production. Thus, in Figure 3.3, the total costs of producing eight units are defined by the area OEBD.

The position and shape of the supply curve in (Q-P) space is influenced by the underlying cost structure of the producing firms and several other factors including technology and KN-related events – especially climate change. Limited production technology renders the supply curve to be steep – that is, relatively inelastic. Improved technology can alter this – that is, make the supply curve more elastic and even shift it outwards in (Q-P) space. Most of Europe was dominated by food shortages in the immediate post World War II period due to limited supply of agricultural products. But this situation was soon overturned within a decade primarily due to improvements in agricultural production technologies, and surpluses soon became a new problem. As we write this text, climate change has become the overarching problem of the twenty-first century. Floods, fires, extreme weather events and the spread of agricultural pests and diseases have reduced supply, increased cost of production, and shifted the supply curve inwards for many goods and services. We consider these changes in greater detail in Chapter 8 and focus on the role of KN as a factor of production.

The Market

The interaction between demand and supply in a hypothetical economy is shown in Figure 3.4. An equilibrium between demand and supply indicates that Q* units would be produced and consumed. At this equilibrium,

$$\left(\text{Consumers' WTP for an additional unit}\right) = \left(\text{Producers' WTA for an additional unit}\right) = P^* \quad (3.1)$$

But also note that, at this equilibrium, net benefits to society are at a maximum. This will become clear when the concept of net benefits is defined.

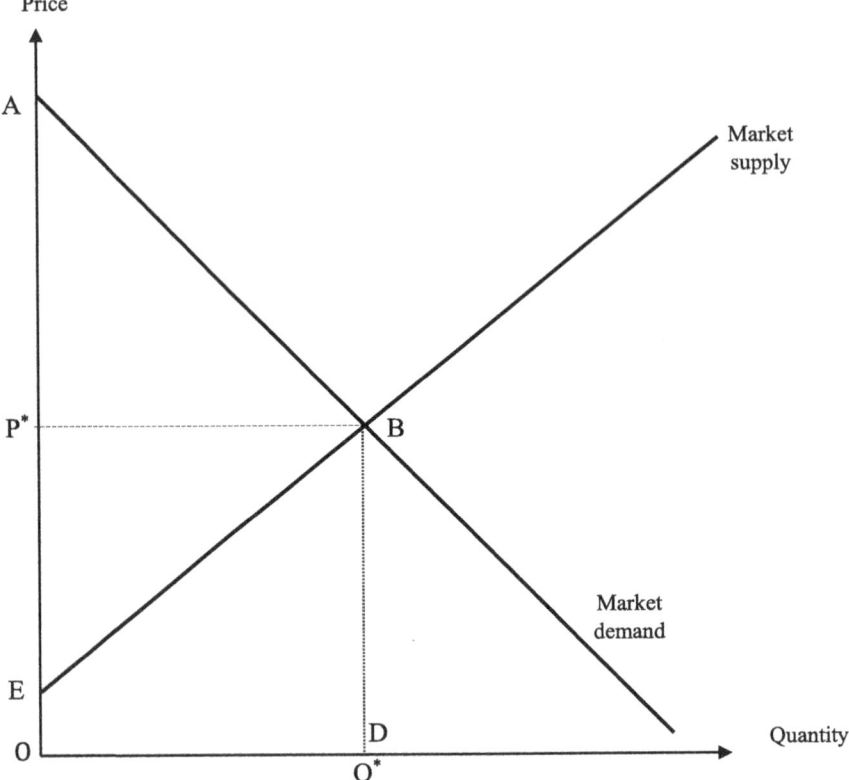

Figure 3.4 Market Equilibrium

The net benefits of producing and consuming Q* units are:

(Total benefits from consuming Q^* units) − (Total cost of producing Q^* units)
= (Area below the demand curve over domain OQ^*) − (Area below the supply curve over domain OQ^*)
= (Area OABD) − (Area OEBD) = Area AEB

(3.2)

If more than Q* units were produced, the difference between the area below the demand curve and the area below the supply curve would be smaller than area AEB. This would also be the case if the number of units produced were less than Q*. That is, for any $(Q, P) \neq (Q^*, P^*)$ in Figure 3.4, the size of net benefits will be smaller than the area AEB.

However, one should note that the attainment of market equilibrium and the maximisation of net benefits are guided by five conditions of perfect competition. These conditions − *anonymity, homogeneity, perfect information, perfect mobility* and *full employment* − are necessary conditions. For example, the *anonymity* condition dictates that every economic agent is a price-taker and will not be able to set the price. That is, influential corporations and the ultra-rich do not exist in perfect competition. The *homogeneity* condition implies that specific commodities cannot be differentiated by their producers (for example, by brand names). *Perfect information* ensures that economic agents can replicate good practices and also make good choices. So, for example if an individual producer has improvised a cost-effective method of production, every producer will know of it and all producers remain on a level playing field. *Perfect mobility* indicates that desirable practices and goods and factors will enter without barriers while undesirable ones will exit. As the term suggests, *full employment* means that every resource is fully utilised and remains an inevitable outcome in the context of the conditions indicated above. This full employment condition serves as the basis for expositing the concept of opportunity costs, namely the reduction in net benefits when a resource is removed from its utilization.

When all of the five conditions work together, the maximisation of net benefits unfolds, and this represents the basis for explaining how a perfect market works. One should note that in reality perfect competition does not exist. The reality of markets is one of imperfect competition. The conditions of perfect competition serve simply as a benchmark to guide policymakers to correct imperfections that prevail in markets. For example, several countries have developed institutional frameworks to regulate and guide competition in markets.

Let us now return to the basic functions of the market mechanism: *what to produce, how much to produce* and *how to produce*. The identification of net benefits is central to the fulfilment of these functions. In a perfectly competitive economy, price is set at the point of equilibrium between demand and supply and, as we just observed above, this is also the point at which net benefit is maximised. The market mechanism would resolve the question of *what to produce* by permitting the production of goods that generate positive net benefits. Clearly, if the total cost of producing any unit of a good is more than consumers' WTP for the good, then the market mechanism would dictate that the good should be produced elsewhere, say overseas. The answer to the question of *how much to produce* is simply that the amount should maximise net benefit, and this happens when demand and supply are in equilibrium. The question of *how to produce* is also resolved by the size of net benefits. Suppose that a given good can be produced by two methods and that one involves higher marginal costs than the other. If demand is independent of the method of production, it is obvious that net benefits will be higher with the method that involves lower marginal costs. Hence the market mechanism signals the choice of techniques with lower marginal costs relative to those with higher marginal costs.

How the equilibrium is actually achieved through the market-based interactions of buyers and sellers, however, remains an open question. The traditional perspective is one in which each economic agent makes an independent decision on what to buy or sell. Then, when interacting with each other, the buyer and seller observe each other's decisions and accordingly make adjustments to their own, as needed. The iterative process of comparing demand and supply for different prices that would result, and adjusting quantities in response to them, will trigger new plans and new prices until they are all in agreement with each other, that is, clear the market. This process may work well in cases of few buyers and sellers and when all the above conditions for perfect competition are met. If there are, however, a large number of buyers and sellers, and especially if they are geographically distributed, or if social, technological or conditions pertaining to KN change rapidly, then there is no guarantee that a (unique) market equilibrium will be found. Rather, while buyers may be able to discern some of the prices at which some of the sellers are willing to sell, and sellers may be able to discern some of the prices at which buyers are willing to buy, neither of them would have information about an equilibrium price (Holcombe 2014; Hirshleifer et al. 2005).

KN Goods and Market Failure

The goods that fall within the scope of the market mechanism are called 'marketable goods' because their market demand and market supply are well defined. All marketable goods have two common characteristics, namely that they can all be:

- Measured in terms of physical quantity
- Valued in monetary terms

Think of the goods that are usually purchased at your home. Bread can be measured in terms of either the number of loaves or the weight of a loaf in kilograms. Clothes can be measured in terms of the number of pieces as well as the sizes they fit. Beverages can be measured in terms of either the number of containers (bottles) or the capacity of such containers in litres. All these items also have specific prices. In a perfectly functioning market, the price of each good will equal the cost of producing the last unit of the good, and, in theory, will result in an equilibrium between demand and supply. While market equilibrium may be a theoretical concept for ordinary goods such as loaves of bread or litres of a beverage, as we mentioned above, equilibrium does even less easily occur for KN goods and services.

Now consider the various types of KN goods. Almost all KN goods – such as clean air, rivers, lakes and forests – cannot be either measured in terms of a physical quantity or valued in monetary terms. However, some KN goods may appear to be measurable. For example, one might say that a lake could be measured by its surface area or its volume of water, and that a forest could be measured by its number of trees. But note that an KN resource cannot be considered as an entity by itself. So a lake is not simply the water it holds, but rather a resource system that includes water and a range of living and non-living items. This inadequacy in terms of measurement implies that market demand and supply curves cannot be properly constructed, and consequently a value in terms of price cannot be adequately ascertained. The goods that present such difficulties are usually labelled as 'unpriced goods' or 'non-marketable goods'. These goods demonstrate the problem of market failure, which is the inability of the market mechanism to fulfil its functions.

Several KN goods are exchanged in the market; for example, iron ore, coal, uranium, timber and the like. However, this does not imply that market failure does not occur with these KN goods. Environmental economists argue that these resources are often undervalued, because their valuation has invariably been based on aspects of the resource in isolation of the larger resource system. Mining firms price their minerals on the basis of the costs of extracting the minerals, which in turn are a function of the inputs these firms buy on the market, and a premium, perhaps, for depleting the mineral assets. These prices do not include, however, the impacts that mining has on the larger ecosystem. The price of a tonne of coal is currently around $120. Does this price adequately reflect the value of trees and wildlife that have been lost or displaced in order to recover the coal? Does it reflect the adverse impacts on rivers and streams as mining wastes leach into them? We shall return to questions of this type in Chapter 4, and illustrate how the functions of the price mechanism are not always fulfilled, even when the market appears to work.

Price Mechanism and Property Rights

All marketable goods will unequivocally qualify for the rules of private property rights. These rules are as follows.

1 The right to ownership can be clearly defined. That is, it is possible to state clearly what is owned, and how much is owned. This also ensures that the good cannot be seized or used by others without the right to ownership being altered. This rule is usually referred to as the *enforceability condition*.
2 People other than the owner can use the good only when the owner transfers the good to others in (voluntary) exchange. Such an exchange also permits the ownership of the good to be transferred. This rule, as implied by its description, is called the *transferability condition*.

44 *Microeconomics and KN*

3   All benefits and costs from using the good are experienced only by the owner of the good. This is called the *exclusivity condition*. Others can experience these benefits and costs only after the transferability condition has been satisfied.

Now think why these rules are satisfied by only marketable goods. For the right of ownership of a good to take effect, the good itself should be clearly defined. That is, it should be possible to specify the physical quantity of the good, and that definition of quantity must encompass all aspects of the good. For example, a carton of milk embodies everything that is contained in the carton of milk. Now compare this with clean air or the flow of water in a river. The fact that clear boundaries cannot be drawn around many KN goods results in the enforceability condition being violated. The transferability condition can be met only as long as there is an incentive for the transfer to take place. This incentive is price. So, with many KN goods, transferability cannot take place (or does take place inefficiently) because the price of the good cannot be properly defined. One can also easily imagine why the exclusivity condition does not work with KN goods. Even if you did own a lake or a river, it could be difficult to prevent others from deriving some benefits from it, at least in terms of enjoying a view. Finally, note that all three conditions are interrelated. If one of the conditions is violated, then it follows that the other conditions will also be violated.

Should the conditions of perfect competition prevail, then a system of private property rights and the price mechanism would function in unison. Such functioning results in the allocation of resources to the production of goods that maximise net benefits to society as a whole. As indicated above, in reality, we do not have perfect competition. Imperfect market arrangements such as monopoly, oligopoly, monopolistic competition and government intervention can hinder the maximisation of net benefits to society. Similarly, imperfect or asymmetric information about the quantities and qualities of goods and services impedes the functioning of the market.

Recall that in Figure 3.4 we considered a society consisting of only two producers and two consumers. Although this model is overly simplistic, it is implied that it would work for the whole of society, and hence area AEB in Figure 3.4 could be regarded as the net benefit to the whole of society from producing Q^* units. Also recall that area AEB represents the maximum net benefit, and this occurs in perfect competition when producers' marginal cost is equal to consumers' marginal WTP.

If, for example, the government intervened and set the price at P_L in Figure 3.5, which is well below P^*, then consumers would demand Q_L but producers would supply only q_L by equating their marginal cost to P_L. Therefore a shortage of $(Q_L - q_L)$ would occur in the market. Shortages often drive prices up due to hoarding by sellers, and the price could settle at P_u when only q_L was made available in the market. In this context, the total WTP for q_L units is the area below the demand curve at q_L, namely area $OADq_L$. The total cost of producing q_L is the appropriate area below the supply curve; that is, area $OECq_L$. The resulting net benefit (Total WTP − Total cost) is area ADCE, and this is clearly smaller than the equilibrium net benefit of area ABE.

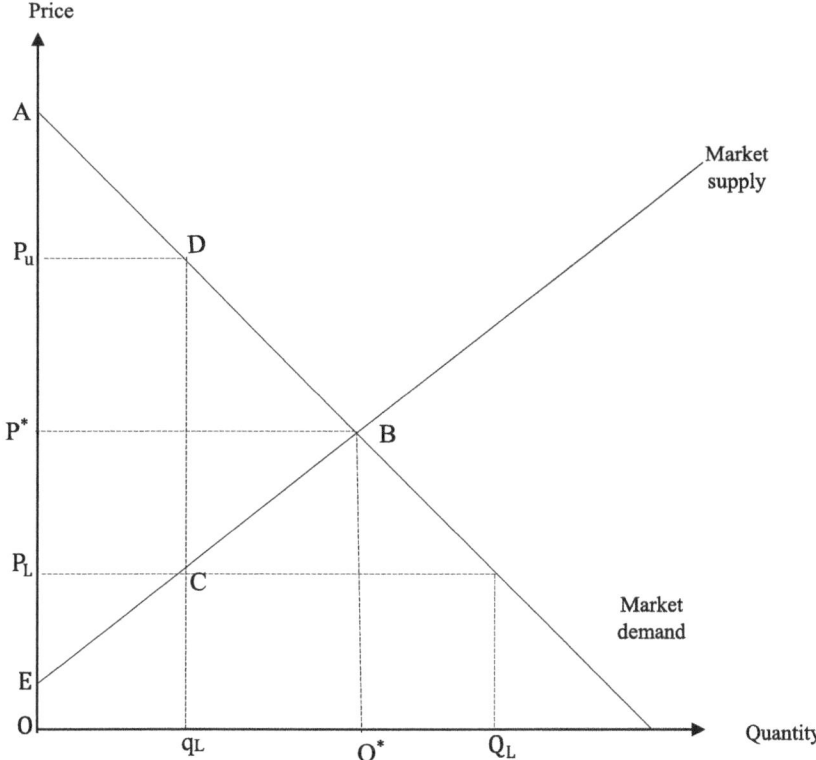

Figure 3.5 Imperfect Competition – Government Intervention

The contexts of monopoly, oligopoly and monopolistic competition can display far greater loss in net benefits than that displayed in Figure 3.5. This is because the main driver of these contexts is the maximisation of profits. And as we illustrate in Chapter 9, the initial determination of prices is governed by the equality between marginal cost and marginal revenue, instead of marginal cost and demand.

The shortcoming caused by imperfect competition is further aggravated by the inability of the price mechanism and the system of private property rights to cope with KN goods; that is, market failure. The aim of national policy is to choose strategies that will enhance overall net social benefits. Therefore, the strategies that governments may consider fall into two categories: those that reduce the extent of imperfections to make the markets more competitive; and those that attempt to remedy the effects of failed markets.

In general, there are two broad approaches to correcting the problem of market failure caused by KN goods.

- The first is to somehow measure and value KN goods so that a market for them can be created.

46 *Microeconomics and KN*

- The second approach is to allocate the private property rights of KN goods to persons either individually or collectively so that a market will emerge from such an allocation.

With respect to the second approach, the argument is simply as follows. When the ownership of a capital asset is vested in an individual, that person will adopt strategies to ensure that the value of the asset is maximised. With almost all items of capital, the maximisation of asset value also coincides with the maximisation of the life of the asset. KN goods can be regarded as items of capital. When their ownership is vested in private individuals, then those owners will adopt strategies to maximise the value of KN; that is, sustain KN. This reasoning is the basis for the property rights (privatisation) approach to restoring market failure. Proponents of this approach argue that the Black Forest in Germany has survived centuries of use primarily because the forest was collectively owned by a group of families with well-defined rules of ownership and utilisation. Blyth and Kirby (1987) illustrate the positive role of private ownership with reference to Australia, where a significant proportion of farmland is under state ownership on leaseholds. They indicate that farmers who own their land outright or those who have no uncertainty regarding the renewal of leases carry out soil conservation works to a much higher standard than those whose ownership and lease renewal are uncertain. For these reasons, some governments have been prompted to adopt commercial-type approaches in the management of KN goods. For examples see Box 3.1.

Box 3.1 Property Rights for KN Goods

Although the privatisation of KN goods may sound attractive in theory, in practice it is difficult to get the private sector to take charge of them. Therefore, in many instances, governments have set up corporations or similar entities to commercially manage these KN goods. A good example is in Laboe, which is a coastal suburb in the port city of Kiel in northern Germany. Located at the mouth of the Kiel fjord that opens into the Baltic Sea, Laboe is a popular tourist destination not only for Germans but also for many other Europeans. Certain beaches have been fenced off and the local government operates these beaches just like a private firm. To gain access to these beaches and their facilities, people pay an entry fee. The commercialisation of the beaches has proved successful in two ways. First, the levy of entry fees has reduced congestion on beaches that would have otherwise been sensitive to excessive use and the income from entry fees has been used to maintain them with a high level of care. Second, this high level of maintenance and the growing popularity of beaches have prompted the port authorities to tighten the controls on the ballast discharges of ships that use the Kiel harbour.

In Australia, Waste Services NSW is a government corporation that has taken over the responsibility for managing several of our KN resources (river systems, oceans and landfills) that serve as receptacles for liquid and solid wastes. The commercialisation of the management of KN resources for waste disposal has facilitated the emergence of recycling and site remediation as ancillary commercial activities.

The commercialisation of KN goods is increasingly being viewed in a global sense as well. In May 1999, Duke University hosted the Fourth Annual Cummings Colloquium on Environmental Law. The theme of the colloquium was 'Global Markets for Global Commons: Will Property Rights Protect the Planet?' Some of the main questions explored in this colloquium were: Can new global property laws and market-based institutions be successfully designed and implemented to constrain overuse of global KN? Can global KN resources be 'propertised' in effective ways, so that global KN conservation and stewardship is 'internalised' into global markets and can be 'purchased' by its beneficiaries? Can property rights in global KN be created under international treaty law, when traditional property rights – even to goods now traded in global markets – have evolved locally? What institutions are needed to create and supervise new global KN property markets, ensuring both efficiency and fairness?

However, in practice the privatisation approach presents several difficulties and is encapsulated in questions such as the following: Who should own the property rights? If they go to the highest bidder, will not the prevailing income inequalities of society be aggravated? Furthermore, if a private owner were to maximise benefits to him or herself from the ownership of a KN resource such as a forest, and if the 'depletion value' of certain components of this resource system – for example, timber – appreciate faster than the preservation of others, then it may be in the "best interest" of that owner to liquidate the KN assets and invest the returns elsewhere in the economy. But of course, the returns on investments elsewhere in the economy will ultimately depend on what happens to forests and other KN systems that contribute to a stable climate, healthy air and water, rich species diversity and other 'boundary conditions' that allow societies and economies to flourish. There is, therefore, a considerable underlying problem when investment decisions uncouple (conceptually and practically) KN and economic performance, and the various models developed in this book address aspects of the analytical and real-world challenges this creates.

Regardless of whether the strategy is to restore the market or change the allocation of property rights, the valuation of KN goods is central to both approaches. For example, the valuation of KN goods permits us to define the demand for, and supply of, KN goods – or at least individual aspects of them, such as timber from a forest. With the allocation of private property rights, the valuation of KN goods permits recognition of asset values. This approach is more amenable to capturing a wider range of attributes of a KN good, such as the beauty of a forest and its biodiversity.

Some Further Issues

As indicated, conditions of perfect competition considered above merely represent a benchmark for illustrating the possibility of maximising net benefits to the community as a whole. Hence, the discussion so far has consisted of correcting the departure from perfect competition, and considering how to include, within markets, those KN goods and services that typically fall outside markets. But note that sustainability of KN is not listed as a condition for the perfect competition benchmark. Rather, the traditional perspective on how markets function, disregards the broader biophysical context in which they operate. In cases when KN is not used sustainably, the long-term performance of the economy is undermined. The recognition of sustainability as an added condition would lead to a significantly different outcome in terms of maximising net benefits. For example, market expansion due to outward shifts in demand and supply can mean the net benefits as illustrated in Figure 3.4, namely the area bound by supply and demand curves, would expand. Such market expansion would be regarded as something good in terms of the distinction between 'goods' and 'bads' that we started this chapter with. As we illustrate in subsequent chapters, market expansion may very well be in a 'bad'.

Review Questions

1 Explain how and when the market and a system of private property rights can work together to maximise net social benefits.
2 Box 3.1 contains examples of beaches and waste receptacles that are commercially managed. Try illustrating the possible shapes of supply and demand for these KN goods.
3 Discuss the following statement: 'The privatisation of KN goods is an important means of achieving sustainability'.

References

Blyth, M. J., and Kirby, M. G., 'Economic aspects of land degradation in Australia', *Australian Journal of Agricultural Economics*, 31(2): 154–174, 1987.

Hirshleifer, J., Glazer, A. and Hirshleifer, D., *Price Theory and Applications: Decisions, Markets, and Information*, Cambridge University Press, New York, 2005.

Holcombe, R. G., *Advanced Introduction to The Austrian School of Economics*, Edward Elgar, Cheltenham, 2014.

4 Public Goods and Externalities

As indicated in the previous chapter, proper valuation of KN goods and services is central to correcting the problems of market failure. In this chapter, we consider the conceptual frameworks for two types of KN goods and services in order to render this task of valuation somewhat easier. Recall Figure 2.3 in Chapter 2 and the two sets of arrows for resources and amenities (R + A) and wastes (W) which link the KN foundation with the economy. Most KN goods that can be categorised as public goods tend to be captured by the arrow R + A. That is, they can be regarded as material, energy or service flows from KN to the economy. Most of the KN-related outcomes that are externalities are generally included in the arrow labelled W. These will be considered later in this chapter.

Public Goods

A *Pure Public Good* is one where a person's consumption of the good is not rivalled by any other person's consumption of the same good. This non-rivalry in consumption means that, if a person consumes a public good and derives benefits from it, there is no reason to suppose that the benefits to other consumers of consuming this public good will be reduced. Pure public goods are indeed rare. The usual textbook examples are a lighthouse and defence. Suppose that several ships are approaching a harbour. The sense of direction and protection that the signals from a lighthouse generate does not diminish for one ship because the other ships' personnel have also seen the same signals. The explanation for the alternative example of defence is similar, in the sense that it extends to the entire society.

From the above two examples, we can note three important properties of public goods. These are as follows.

1 *Non-rivalry in consumption*: The consumption of the good by one person does not diminish the benefits of the good enjoyed by another person.
2 *Non-diminishability in consumption*: One person's consumption of the good does not diminish the total availability of the good.
3 *Zero marginal costs*: This property follows from the first two. If we cannot deny the benefits of the good to anyone and also cannot reduce the quantity of the good, then the cost of providing the good to an extra consumer is zero.

Also, note the distinction between private goods and public goods. Private goods display rivalry in consumption and positive marginal costs for additional users. Indeed, this textbook is a private good. Supposing that there is only one book left on the shelves of the bookstore and you purchase this last copy; then the next consumer-reader is unable to derive satisfaction from this book. That is, with private goods the availability of the good diminishes with consumption. If the publishers wish to make extra copies of this book, they have to bear positive marginal costs.

Many public goods also display the characteristic of non-excludability. This means that it is not possible to prevent someone from consuming the public good, even if that someone is not willing to pay for it. This is clearly the case with a public good such as national defence. Katz and Rosen (1991) argue that excludability and non-excludability are primarily a result of technology and legal arrangements and not of the publicness of a good. For some other authors such as Frank (2010), non-excludability is a characteristic of public goods. That is, it is either very expensive or impossible to prevent people from using a public good. For example, suppose that a defence shield protects a specified but large land area of Australia. Should the residents of Australia perceive imminent danger, then everyone can flock into this shielded area and feel protected. However, a draconian law may be in place to prohibit some inhabitants from entering the protected area. So, despite the publicness of the good, exclusion is possible through a legal arrangement. Similarly, exclusion may take place for mere practical reasons – not everyone may be able to move quickly enough to the area that is protected. However, it is of course true that many goods that display non-rivalry in consumption also display the property of non-excludability.

Figure 4.1 displays a continuum of goods, with pure public goods and pure private goods placed at opposite ends of the continuum. Most KN goods fall in between these two ends and can be labelled 'quasi-public goods' or 'mixed goods'.

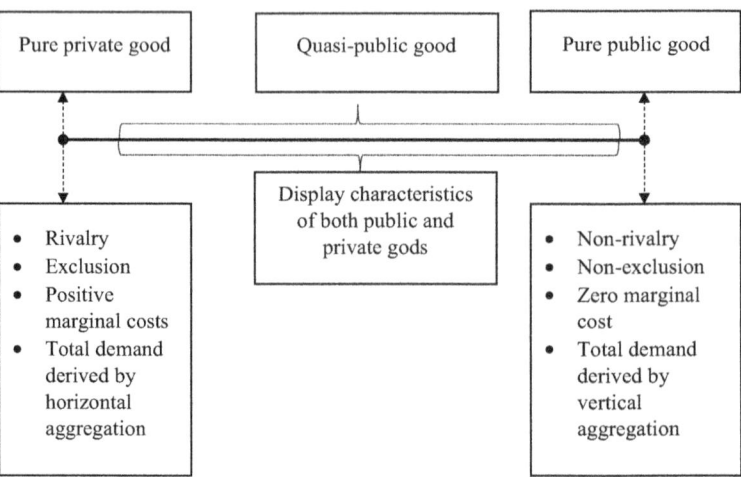

Figure 4.1 The Continuum of Goods

That is, they display public good properties up to a certain point, beyond which they become private goods. Consider a national park. A relatively large number of visitors can enjoy the amenities of a national park without diminishing the enjoyment of each other's use of these amenities. However, this is true only up to a certain point. As illustrated in Figure 4.2, the marginal cost of providing the amenities of a national park is zero up to q_1 visitors. But when the number of visitors exceeds q_1, the marginal cost begins to become positive because of congestion and tends to infinity when the national park gets overly congested. We can offer a similar explanation for the public good property of a range of other KN assets such as lakes, rivers and clean air.

It may now be reasonably clear how the public good property creates market failure. First, public goods present a pricing problem. The market solution is zero price. This is because the supply (marginal cost) curve resides along the horizontal axis, at least for most of the part, as shown in Figure 4.2. With public goods that are gifts of nature, the zero price condition prompts excessive use, resulting in depletion and deterioration. For example, for a long time industrial pollution was readily emitted into the atmosphere and the waterways, because these were perceived to be free receptacles. Up to a certain point, the natural receptacles of air and water are capable of assimilating the pollution. For example, micro-organisms can degrade some pollutants and plants can absorb excess carbon dioxide. However, just as congestion can exclude people from enjoying the amenities of a national park,

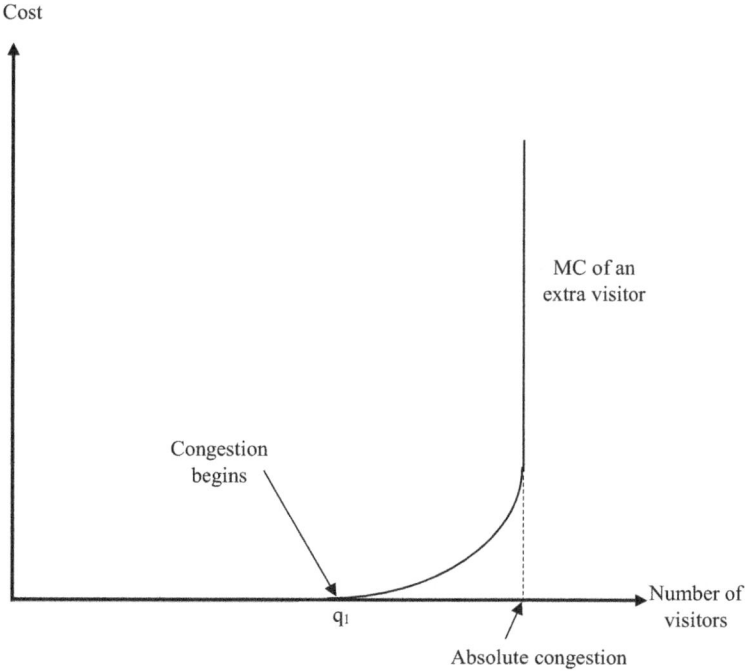

Figure 4.2 Quasi-Public Good Characteristics of a National Park

excessive pollution can exclude, and in many instances certainly has excluded, further use of the KN sinks as a receptacle for pollution.

The second cause of market failure with public goods is also associated with the price of the good being zero. It is also difficult to demonstrate a true demand curve for a public good. This is because people will tend to understate their WTP for a public good. This in turn is because people realise that, once it is provided, they cannot be excluded, so they can have it free. This is often referred to as the 'free rider' problem.

The approaches to correcting the market failure caused by public goods usually involve searching for attributes that can be rendered private. The property rights approach has been common with public goods that command much usage. For example, the management of freeways and expressways has often been auctioned off to private operators to regulate use through the collection of fees (tolls). This approach has been extended, as indicated in the previous chapter, to some KN public goods like forests, beaches and national parks by recourse to entry fees. The general feature of most property rights approaches is to set a reservation price and then auction the rights to manage the public good with some pre-specified rules of management. The highest bidder who acknowledges the pre-specified rules of management and exceeds the reservation price wins the opportunity to manage the public good. However, note that the valuation of the public good is still inevitable because a reservation price has to be set. Other methods of management that apply to certain quasi-public goods are not readily extendable to the case of KN goods and services. For example, knowledge is a public good, but when it is transmitted through copyrighted materials or patented products, some semblance of value can be ascertained.

Externalities

Externalities are side effects of an activity, such as the production or consumption of a good, which are not priced. Consider a factory that produces steel and also emits smoke into the atmosphere. Steel is sold on the market and a price is established for it on the basis of interactions between sellers and buyers. However, markets for the by-product smoke typically do not exist, and as a consequence, smoke is outside the market – external to it. And hence no price gets established for it. Since it causes harm, smoke is considered a negative externality. And because there is no price for smoke, the steel-producing factory, in essence, sells its product at a price that does not reflect the harm it causes and its output will be larger than if the damages were included. Conversely, a bakery producing bread will sell its product to buyers at a price that may be lower than it would be if it were also able to charge people for the pleasant smell that emanates from the bakery to the neighbourhood. The smell created by the bakery is a positive externality. And because there is no price for the positive externality, or an easy way to capitalise on the benefits created for people in the neighbourhood who enjoy the smell but do not buy bread, the supply of bread will be lower than it otherwise would be.

The presence of externalities is ubiquitous. For example, efforts to add runways to airports in London, England, or Frankfurt, Germany, have been a source of controversy for a long time. Additional runway can help airlines and operators of airports save considerable cost, especially when compared to building another airport elsewhere. Apart from reducing air traffic congestion, extra runway capacity, could permit increases in the number of scheduled flights, which in turn could transform the cities and regions of these airports as popular destinations for travellers and people in business, and would generate additional government revenue (due to the collection of more landing taxes and so on). However, adding runways also causes significant disutility to the residents of the suburbs that surround the airport. This disutility is specifically due to the increases in the noise level, which in turn is due to a larger number of aircraft than before. So noise pollution creates interdependence between those who find flights on the additional runway to be convenient (for example, business people) and those who live near the airport. Since this interdependence cannot be easily compensated, noise pollution is an externality.

An externality that is causing much concern in nations around the globe is the salinity of urban water supplies, due to agricultural practices. The classic example is the externality caused by intensive irrigation. For example, prolonged intense irrigation in the Murray–Darling Basin in Australia has resulted in the raising of the water table along the basin, and this in turn has resulted in the increase of salt intake from groundwater sources. The adoption of agricultural drainage practices to alleviate the problem of soil salinity has resulted in the discharge of salts and agricultural chemicals into the river system. This has in turn resulted in the deterioration of the quality of water supplies to the city of Adelaide some one thousand kilometres downstream. In this example, the externality is the contamination of urban water supplies, and this externality creates interdependence between the dwellers of Adelaide and the farmers of the Murray–Darling Basin.

Note that, in the examples given above, the term 'interdependence' was used where the term 'conflict' might have been more appropriate. This is because not all externalities create disutility, as we have mentioned with reference to positive externalities.

Consider now how externalities cause market failure. Figure 4.3 shows hypothetical demand and supply curves for agricultural output from the Murray–Darling Basin. The usual supply (marginal cost) curve does not account for the externality of water pollution. Suppose that we have somehow been able to estimate the monetary value of the harm caused by water pollution. This is shown as the *marginal externality cost* (MEC). So if we add the marginal externality costs to the marginal costs of producing agricultural output, then we have a supply curve that includes the externality. The supply curve that incorporates the externality is labelled *marginal social cost* (MSC) because it includes the costs incurred by all parties, which in this case are the agricultural producers and the urban dwellers. The curve that omits the effect of the externality is the *marginal private cost* (MPC), as it deals with only the costs incurred by one group, namely the agricultural producers.

Now note that the market solution based on the MPC results in excess agricultural output, while that based on the MSC lowers output and raises the price. So the market solution (P_2, Q_2), which is based on the MSC, is a *socially desirable*

54 *Microeconomics and KN*

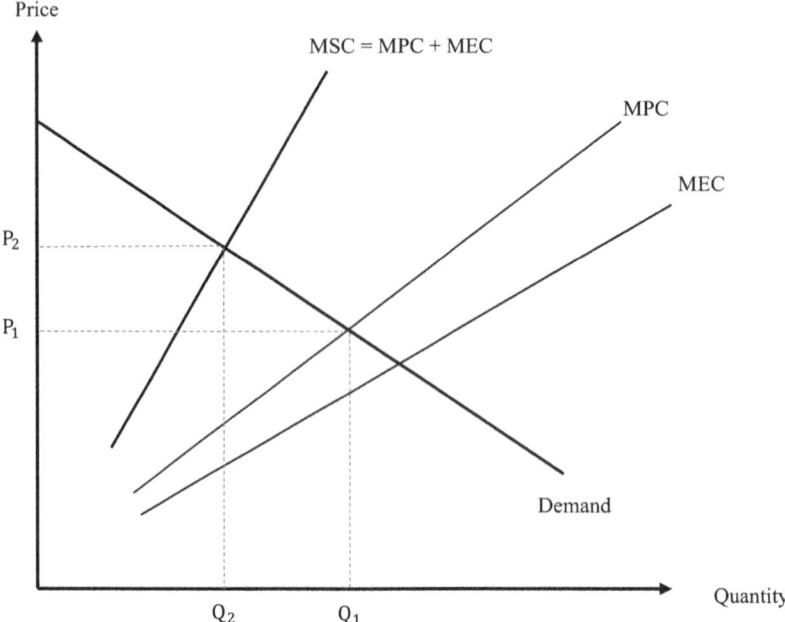

Figure 4.3 Illustration of an Externality

outcome. However, the attainment of socially desirable outcomes is possible only if the externality can be valued in monetary terms and included in the demand and/or supply curves. In this example, we illustrated the inclusion of the effect of the externality in only the supply curve. It is also possible that consumers may lower their WTP for agricultural output due to their awareness of agricultural producers' contribution to water pollution. The inclusion of the effects of the externality into the components of the price mechanism is termed the internalisation of the externality. However, the feasibility as well as the effectiveness of internalisation depends on the types of methods that can be developed to measure and value unpriced effects.

Public Goods and Externalities

To summarise, we commenced the previous chapter with a description of the price mechanism. We illustrated that, under conditions of perfect competition, the price mechanism and a system of private property rights operate in unison, resulting in a market solution that maximises net benefits to society as a whole. However, with KN goods, the price mechanism and the system of private property rights both fail. This is because KN goods cannot be easily measured and valued. This unpriced characteristic of KN goods can also be explained in terms of two sets of properties they possess: the properties of public goods, and the generation of externalities whenever they are utilised.

Public Goods and Externalities 55

Hence, with several KN goods, it is possible to illustrate a relationship between externalities and public goods. Very often the interdependence between persons that represent the externality leads to the degradation of a public good. For example, as consumers we use detergents to wash, clean and brighten our clothes. The phosphorus in these detergents finds its way into in rivers, streams and the ocean, impacting water quality and thus drinking water supplies as well the health of ecosystems, including the scenic beauty of coastal zones and the foods we derive from the ocean. Phosphorus is also found in the residual material that is discharged into waterways from sewage treatment plants. Phosphorus, together with nitrogen, is the major cause of the proliferation of cyanobacteria, which manifest as toxic blue-green sludge. So this blue-green sludge is the externality that causes/arises from interdependence between the members of a community, because the community has used a public good (waterways) to discharge its residues. Note that the externality occurs whenever the public good becomes degraded.

We illustrate this in Figure 4.4. Panel A of Figure 4.4 shows the standard description of a quasi-public good as found in most microeconomics texts. Suppose that, in this instance, the public good is the air shed of a region where certain polluting industries are to be located. These industries would use the air shed by emitting smoke into it; that is, the air shed would be used as a sink for emissions. As the number of users of the air shed – namely the polluting industries – increases, the quantity of emissions also increases. This is shown in panel B. In panel B, the emission quantity of W_{AC} represents the assimilative capacity of the air shed. That is, the air shed is capable of tolerating emissions up to the quantity of W_{AC}

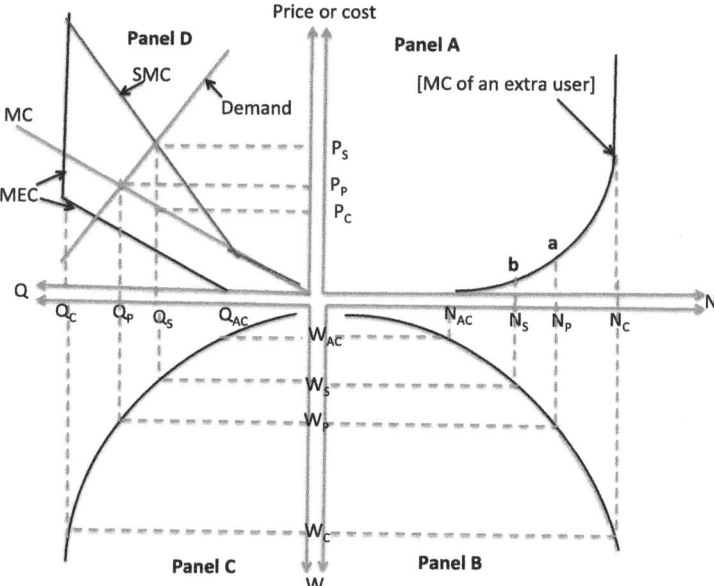

Figure 4.4 Public Goods and Externalities

without losing any of its properties, such as the proper composition of atmospheric gases. Consider panel A. The marginal cost of letting an extra user emit pollution is zero, until the quantity of emissions reaches the assimilative capacity of the upper atmosphere. This corresponds to the number of users being equal to N_{AC}. As the number of users increases, the amount of pollution also increases and at some critical quantity of emission, W_C, which is caused by N_C users, the air-shed becomes fully saturated and ceases to be a sink. At this point (N_C), the marginal cost of an extra user in panel A tends to infinity.

Now consider panels C and D. In panel C, the relationship between the quantity of industrial production and the quantity of pollution is presented, and panel D describes the standard market for the industrial product. Note that, in panel D, the non-recognition of the marginal externality cost (MEC) leads to a market solution of (Q_P, P_P) and is defined by the equilibrium between demand and the industry marginal cost (MC). The social marginal cost (SMC) is the sum of the MC and the MEC, which emerges when the quantity of production exceeds Q_{AC} due to waste production exceeding the level of assimilative capacity. As a result, the SMC deviates from the MC when the quantity of production exceeds Q_{AC}. The external cost and SMC both tend to infinity at production quantity Q_C, which corresponds to a pollution quantity of W_C and user quantity of N_C.

The standard argument for government intervention stems from the fear that, if left unregulated, external costs will be ignored and production will settle at Q_P. This is not too far from Q_C. Further, with increased demand and a mixture of increased production and new entrants, production quantity can in fact reach Q_C, at which level there can be a disaster due to the collapse of KN. Governments would like to reach Q_S by internalising the externality. Note that this involves valuing the externality and being able to quantitatively specify the externality cost function. Should this be the case, governments would impose a tax that is based on the price differential ($P_S - P_C$) or impose a quota restriction on the product market to Q_S. The effect of this on the demand for the public good is readily apparent. When unregulated, the demand for the public good is at point **a** in Panel A. When the government intervenes, this demand is moved to point **b**, which corresponds to quantity Q_S in Panel D. As this conceptual example illustrates, the need for government intervention and regulation stems from the need to protect the air shed and maintain the flow of services it offers. Note that the flow of services is important, not only to the industries that use the air shed, but also to the wider public.

To generalise from the above illustration, an important policy directive to safeguard KN goods is to regulate the use of the public good; that is, to institute controls on the discharges into waste receptacles such as waterways and air sheds. As we will observe subsequently, such controls may be achieved through taxes, regulatory standards, or a combination of both. In some instances, it is also possible to have regulatory controls on the quality of the commodity that is responsible for the externality. This too could be achieved through standards and penalties. It is this type of control that has prompted the emergence of products such as phosphate-free detergents and a range of other consumption goods with KN-friendly attributes. Ideally, one would prefer to have externalities to be fully internalised.

Consider the case in Box 4.1 where the prime minister of India strives for the complete removal of open defecation. Nevertheless, externalities are seldom fully internalised. A residual quantity of the externality does invariably remain – we consider this next.

Box 4.1 Misuse of Public Space and Externalities from Open Defecation

Hygiene is severely lacking in several developing countries due to the absence of proper sanitary facilities. India's prime minister – Mr. Narendra Modi – launched the 'Clean India Campaign' in 2017 to combat the challenges posed by Open Defecation (OD). Millions of people are driven towards OD because they lack proper facilities. As of 2023, the problem has diminished significantly to the extent that PM Modi claimed that India has achieved 100 per cent sanitation. Sceptics of course disagree.

In terms of the concepts covered in this book thus far, we can describe the issue by recourse to the Assimilative Capacity (AC) and Marginal Costs (MC). The costs involve those of cleanup as well as ill health due to exposure to pathogens in the excrement. In this context, AC refers to the ability of the ecosystem of open spaces to deal with human excrement. AC progressively diminishes over time as shown in Figure 4.5. That is, when OD continues, the open space ecosystem will not be able to tolerate even a small amount of excrement. Hence, we present a sequence of Marginal Cost (MC) curves that shift to the left when OD continues and become steeper as they shift; that is $MC(t_0)$, $MC(t_1)$,..., $MC(t_N)$. Note that by time period (t_N), AC is completely lost and $MC(t_N)$ tends to infinity.

The conceptualisation highlights the gravity of the problem. But what is the solution? Prime Minister Modi initiated an inter-faith dialogue urging religious leaders to promote sanitary behaviour in their religious discourses. The World Bank (Neal et al. 2016) draws lessons from behavioural economics and suggests tools for 'nudging' behaviour. For example, building community latrines in places that people visit as part daily routine may make it more readily possible people to avoid OD. Some economists seem content to estimate the marginal benefits (MB) and marginal costs (MC) of constructing latrines and thereby determine the optimal number of latrines to be constructed. The underlying premise is that the MB would be downward sloping whilst the MC would be upward sloping. But will the MB of constructing latrines comply with the standard principles diminishing marginal utility? It is of course possible that the MB of constructing latrines could in fact increase over a larger domain of the number of latrines constructed.

Singapore solved the issue well before it became rich by creating a Housing Development Board to construct dwellings. The motto of the first prime

58 Microeconomics and KN

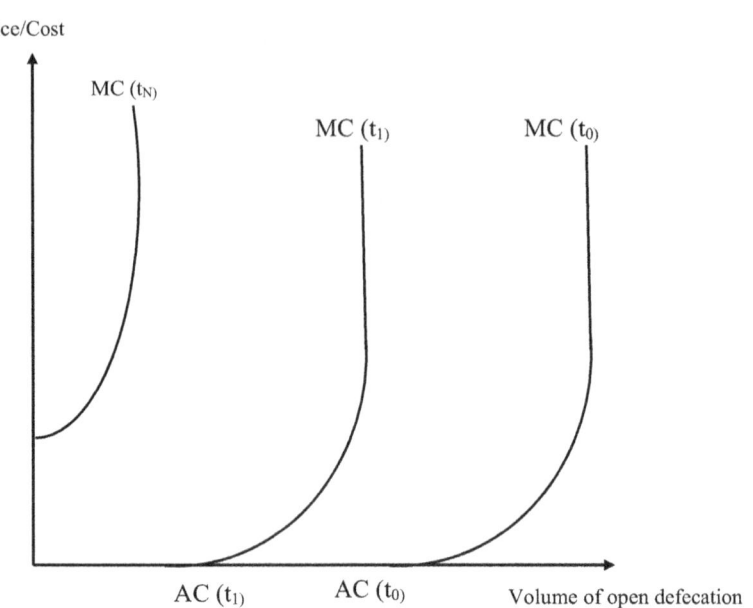

Figure 4.5 Open Defecation and Assimilative Capacity

minister of Singapore – Mr. Lee Kuan Yew – was that every person must have a dwelling and that every dwelling must be sewered. Singapore borrowed and also used a fair share of government revenue for construction of dwellings. Construction activity created employment and income and emergence of ancillary industries. Within a short period of time, Singapore became a hygienic city. One could argue that cleanliness and hygiene have been a contributing factor to Singapore's success. Some might say that Singapore is just a small dot on the landscape. But Deng Xiaoping, the then Senior Vice Premier of China, is claimed to have said after his first and only visit to Singapore in 1978, that he would plant a 'Thousand Singapores' on the landscape of China.

Residual Externalities

To illustrate this concept, consider Panels C and D in Figure 4.4. When the externality is not recognised, that is, the MEC schedule is ignored, then the market clears in Panel D at (Q_p, P_p). This clearance corresponds with the discharge of waste amounting to W_p in Panel C. The recognition of the externality results in the market clearing in Panel D at (Q_s, P_s) and the discharge of waste in Panel C reduces to W_s. Hence a quantity of waste amounting to $(W_s - W_{AC})$ remains in the KN sink and this represents the residual externality. Had the externality been fully internalised, then $(W_s = W_{AC})$ and the quantum of the residual externality would be zero. Such complete internalisation does not occur for at least three reasons:

1. Methods of valuing unpriced effects have several defects – we discuss these in Chapter 16 – and hence the MEC schedule is invariably an inadequate representation of the costs associated with the externality.
2. Scientific knowledge on the exact level of W_{AC} is typically incomplete.
3. Governments are often reluctant to permit markets associated with W to shrink too much for economic reasons.

But the presence of residual externalities is not only real, but these residual externalities accumulate and grow. The evidence of accumulation with reference to air pollution is unequivocal in the scientific literature; for example, see Ramanathan et al. (2001). The accumulation of residuals such as GHGs lead to a string of other externalities, the most prominent of which is climate change involving increased frequency and intensity of extreme events like cyclones, devastating floods, and catastrophic forest fires; for example, see National Science Foundation (2011). The externality of climate change comprises complex perpetuating outcomes such as homelessness and psychological trauma.

As with the climate change example given above, residual externalities can perpetuate a vicious cycle of adverse outcomes. A significant externality that engulfed the entire globe was the Covid-19 health pandemic that most likely emerged from a wet market transaction in the Chinese city of Wuhan. This is a significant residual externality because initial attempts to control the externality – virus infection – failed, using standard public health measures, such as isolation. The infection spread across the world rapidly and remains an issue even today. However, as Thampapillai et al. (2021) argue, the origins of Covid-19 can be attributed to inadequate regulatory controls of markets – especially those pertaining to animal husbandry products. Further, other externalities pertaining to climate change and health did also enhance the spread of virus. For example, externalities such as air pollution and improper sewage disposal lower the human immune response and thereby make people susceptible to the virus.

In the next chapter, we will extend the discussion of public goods and externalities to the context of managing specific types of natural resources. These resources are usually classified as renewable and non-renewable. Although such resources have clearly defined markets, the fact that they are part of a larger body of public goods is often overlooked. Further, as indicated, the externalities stemming from the extraction of these resources invariably spill over geographic and temporal boundaries.

Review Questions

1 Consider Figure 2.4 in Chapter 2. Why do most of the items that could be included in R + A border on being classified as public goods and those that could be included in W be regarded as externalities?

2 How do public goods and externalities cause market failure?
3 Provide an example that illustrates the relationship between a public good and an externality. Use an appropriate supply–demand framework to show how the overuse of the public good becomes an externality.

References

Frank, R. H., *Microeconomics and Behavior,* McGraw-Hill, New York, 2010.

Katz, M. L. and Rosen, H. S., *Microeconomics*, Irwin, Burr Ridge, IL, 1991.

National Science Foundation, *Link Between Air Pollution and Cyclone Intensity in Arabian Sea,* News Release 11–234, 2011.

Ramanathan, V., Crutzen, P. J., Lelieveld, J., Mitra, A. P, Althausen, D., Anderson, J., Andreae, M. O., Cantrell, W., Cass, G.R., Chung, C. E., Clarke, A. D., Coakley, J. A., Collins, W. D., Conant, W. C., Dulac, F., Heintzenberg, J., Heymsfield, A. J., Holben, B., Howell, S., Hudson, J., Jayaraman, A., Kiehl, J. T., Krishnamurti, T. N., Lubin, D., McFarquhar, G., Novakov, T., Ogren, J. A., Podgorny, I. A., Prather, K., Priestley, K., Prospero, J. M., Quinn, P. K., Rajeev, K., Rasch, P., Rupert, S., Sadourny, R., Satheesh, S. K., Shaw, G. E., Sheridan, P., and Valero, F. P. J., 'Indian Ocean experiment: An integrated analysis of the climate forcing and effects of the great Indo-Asian haze', *Journal of Geophysical Research: Atmospheres,* 106(D22): 28371–28398, 2001.

Thampapillai, D. J., Bali, A., Lodhi, I. A., and Thampapillai, Dilan J., *Covid 19 and Its Contagion,* https://scholarbank.nus.edu.sg/handle/10635/201009 (posted 30 September 2021).

5 Economics of Renewable and Non-Renewable Resources

In this chapter, we will continue to use Figure 2.3 as the basis for our illustrations. In that figure, arrows for resources and amenities (R + A) and waste flows (W) were regarded as a composite service flow from the KN to the economy. The aim of this chapter is to illustrate how the market model, and its adaptations to include public goods and externalities, can be applied to the case of specific KN resources. These resources have been traditionally classified as non-renewable and renewable. As the term implies, non-renewable resources are not renewed at timescales of relevance to human decision-making. As a consequence, they are finite and hence can be exhausted. Renewable resources, in contrast, possess the property of regeneration, and thus with careful management can be sustained over time. However, it is important to note that all natural resources – renewable or otherwise – are integral components of complex ecosystems. Fish are part of a food web that connects not only across oceans but also across the ocean, land, and air interface. For example, when sea birds feast on fish and then deposit guano on land where it may provide valuable nutrient input into terrestrial ecosystems. Copper and many other trace elements that are considered non-renewable resources, are found within our bodies and moved through ecosystems via a complex set of biophysical cycles. Such integral linkages among the living and non-living components of the ecosystem, and the underlying non-renewable and renewable resource endowments, are typically disregarded in economic models of natural resource usage. As a consequence, important aspects of resource use are neglected and thus the opportunity cost of using these resources are inappropriately conceived and quantified. The recognition of these linkages would allow a more accurate calculation of the marginal social costs that we considered in the previous chapter.

Objectives of Resource Management

Although the distinction between renewable and non-renewable resources is easily made, two important points should be acknowledged.

1 The dividing line between a resource being renewable or non-renewable is necessarily fine. This is because the over-utilisation of a renewable resource can easily transform it into a non-renewable one. The Blue Whale and certain species of

fish are classic examples of this. Besides, the utilisation of renewable resources can have effects that far exceed those impacting their regenerative capacities. For example, a forest is traditionally regarded as a renewable resource with renewability being retained with rotational harvesting. But, as Wheeling (2019) illustrates by recourse to research from the University of Michigan, changes to canopy density of forests impact the hydrologic cycle leading to reduced soil moisture and prolonged periods of drought, which in turn affects tree growth, and, more broadly, biodiversity in forests.

2   The utilisation of non-renewable resources without adequate safeguards can also transform renewable resources into non-renewable ones. This is due to the range of harmful externalities that are associated with the use of several non-renewable resources. Recall the concept of residual externalities considered in the previous chapter. The reliance of almost all economies across the world on fossil fuel sources for energy generation has resulted in the accumulation of Greenhouse Gases (GHGs) in the upper atmosphere. As a result, the capacity of an air shed to provide clean air has been constrained. Besides, the accumulation of GHGs has been cited as the source of global warming, which in turn contributes to the increase in frequency and severity of extreme weather events such as hurricanes, fires, floods, and droughts; for example, see Emmanuel (2017). Some externalities such as earthquakes stemming from mining for energy resources are also gaining appreciation only recently; see Box 5.1.

Box 5.1   Japan's 3/11 Earthquakes and Mining – A Tenuous Link?

On Friday the 11th of March 2011, Japan experienced one of its worst ever disasters. Note the use of the plural in the opening sentence. It was not just one event. But rather it was a cascading sequence of events and consequences – which are most probably being played out even now.

The sequence of events: An earthquake of magnitude 9 on the Richter-Scale (RS) occurred off the Oshika Peninsula, the east coast of Tōhoku at 14:46 Japanese Standard Time; Lipscy et al. (2013). This earthquake triggered a nearly 14 metre (46+ feet) tsunami that inflicted a wave of damage in its wake. The damage is impossible to properly describe – let alone quantify. The earthquake itself was preceded by a foreshock of around 7.2 on the RS on the 9th of March and then followed by four aftershocks also of magnitude around seven on the RS. A dominant causal factor of damage triggered by the tsunami was the breakdown of the nuclear power reactor at Fukushima and the associated health costs and issues that stemmed from it. In the chaos that ensued various countermeasures were adopted and the failure of these measures resulted in a nuclear meltdown and radiation exposure to those in and around Fukushima. Within 4 hours of the tsunami, the then Prime Minister Naoto Kan declared a nuclear emergency and evacuation orders were issued to residents within a 3-kilometre radius of Fukushima.

As Hiranuma (2014) explains following the Fukushima disaster, Japan did an about-turn on its energy policy. Up until 2011, nearly 25–30 per cent of Japan's baseline power came from nuclear energy. In 2012, the share of nuclear power had fallen to 1.7 per cent. Coal, natural gas and oil got back on the menu, and these are greenhouse gas intensive. So, Japan is now on the active list of greenhouse gas emitters.

Was the 3/11 earthquake really a natural event? Japan sits on a seismically active region. Some records date back to 684 BC. For example, before measurement metrics were refined, death tolls as high as 30,000 were reported in 1498. More recently, earthquake measurements have been increasingly refined. The United States Geological Survey (USGS),[1] reports that between 1950 and 2015, Japan has had 451,668 earthquakes that exceeded 2 on the RS. The number of earthquakes that exceeded three and four on the RS on the same time period are 176,281 and 54,662 respectively. This statistic is alarming because this means that Japan has some 840 earthquakes on average a year in excess of 4 on the RS.

But the earthquake statistics from the USGS unveil another observation. That is, the frequency and intensity of earthquakes have increased over time – especially more recently. For example, in the 65-year period spanning 1950–2015, there have been 590 earthquakes in excess of 6 on the RS. But 292 of these have happened just in the 14 years spanning 2000–2014. Going further with such observations, there were 49 earthquakes in excess of seven on the RS over 65 years (1950–2015); but 32 of these earthquakes happened between 2000 and 2014. The observations in this vein are reinforced if one considers earthquakes in excess of 8 on the RS. There were only six observed between 1950 and 2014. But five of these six happened in the 14 years (2000–2014) including the earthquake that triggered the Fukushima disaster.

What have the increasing frequency and intensity of earthquakes got to do with mining as the caption of this box indicates? This observation may be due to events outside Japan. Seismic scientists (Van Eijs et al. 2006; Yerkes and Castle 1976) have now shown that mining activities induce earthquake events – not necessarily in the near vicinity of the mining – but rather in locations that are vulnerable. As reported by Pérez-Peña (2015), the United States Geological Survey has now established earthquake contours based on mining activities.

It is an undisputed fact that the intensity of mining has expanded on a massive scale in China, Mongolia, and the Russian Federation all the way up to Northern Siberia. If Japan has a role in this induction of seismic activity, it would be through the demand for raw materials that stem from mining and through direct investments in such mining. For example, Japanese investments are at work in the Tavan-Tolgoi deposits in Mongolia. One should also not overlook the extent and intensity of mining over the West Coast of the United States. And this too could influence on the earthquake activity in Japan.

The above observations have direct consequences not only for policies that seek to foster economic growth (EG) and KN quality (KN-Q) but also for issues of intergenerational concerns (IGC).

The EG objective is deemed necessary for economies to function, and resource utilisation is an inevitable component of this objective. It is concerned with achieving increases in real national income and employment and removing poverty. However, a singular pursuit of the EG objective, alongside an over-reliance on non-renewable sources, can potentially exhaust our entire stock of natural resources. This has been the case for most of modern economic history, particularly since the industrial revolution of the 18th century and most notably since the economic expansion observed globally since the Second World War. Operating in a world of seeming plenty, where technological progress opened new opportunities for resource use, firms and households have undervalued KN resources and consequently exploited them beyond sustainable levels. Technological progress that affects resource use often manifests itself in two ways – new resources previously not accessible or usable by the economy are brought into the mix of others being exploited; and increases in the efficiency of exploitation itself allows for ever more resource to be squeezed out of the system. The latter is particularly evident, for example, in the case of fisheries. Increases in the size of fishing boats and nets, combined with highly sophisticated sonar and radar, as well as advanced processing capabilities on the ships themselves, leave few fish populations untouched from ever more effective industrialised approaches to fishing. One of the more bizarre outcomes is an often (temporary!) increase in catch, which is used as a sign that fishing efforts can be increased and regulation of the fishery should be limited.

Other examples abound. For instance, the unconstrained dumping of untreated effluent into oceans and waterways, in order to cut cost of waste disposal, has seriously affected the renewability of our water resources and the fish and other species we extract from them. Investments in stringent emission control technologies were often perceived as too costly, largely because the cost of not controlling pollution was disregarded in decision-making, and where that cost was recognised, it was typically spread across a diffuse set of individuals, including future generations, who had no voice in decision-making. The list of result of such mindsets and actions is long, ranging from acid rain, to eutrophication, to accumulation of endocrine disruptors, to climate change, to species extinction and beyond. Since these adverse impacts of resource use surface at some point as binding constraints on economic activity, it is imperative that the list of management objectives be expanded to include KN-Q and IGC.

The importance of the KN-Q objective becomes evident from our foregoing discussion. The scope of this objective is indeed broad. In many situations we need to restore and enhance the quality of KN resources. For example, we need to clean up polluted waterways, and we need to enforce strict emission controls so that cities can be free of smog. In other situations, we may have to preserve KN resources. For example, the unrestricted harvesting of forestry and fishery resources can threaten their renewability, and so their preservation becomes important. Further, as indicated, we cannot deal with the KN-Q objective in terms of individual resources, but rather as a system of resources. Although we deal with renewable and non-renewable resources under separate headings in this chapter, we need to be aware that resources constitute complex interactive systems. For example, a forest is not merely a collection of trees that make up a renewable resource stock; it is also a complex ecological system of trees, various species of fauna, soil and other endowments. Hence, the scope of the KN-Q objective is to restore, enhance and preserve the overall quality of natural resource and ecological systems.

As the term implies, the IGC objective arises from possible conflicts between generations in terms of their resource requirements. This conflict has been perceived especially in the context of fossil-fuel based energy resources. The twentieth century was dominated by the concern that most of our energy resources are non-renewable and will eventually be exhausted. For example, the British economist Lecomber (1979) emphasised that at the then (1970s) and projected rates of resource utilisation, even a million-fold expansion of total non-renewable energy, could extend resource life by not more than 200 years. Therefore, Lecomber concluded that before long, either radically different sources of energy or a slow-down of growth would be inevitable. However, the inevitability of needing radically different energy sources has surfaced for reasons that outstrip the finiteness of fossil-fuel based resources. That is, climate change driven mainly by GHGs that stem from the use of such resources. This necessity is evident in the succession of conferences sponsored by the United Nations Framework Convention on Climate Change; for example, see Climate Partner (2022).

The relevance of the IGC objective is of course not confined to energy resources. Deforestation is another global problem. According to Global Forest Watch (2023), between 1990 and 2020, the world experienced a net loss of 101 million hectares of forest cover. If one were to also include the extent of disturbed forests, then the loss over this 30-year period amounts to nearly 400 million hectares. Nadkarni (1987) estimated that some five million hectares of forest were cleared in India between 1950 and 1970. Taking a more recent time span, namely 2000–2020, the net loss in India (inclusive of disturbed forests) remains around 5.3 million hectares. In Indonesia and Malaysia, the rate of rainforest clearing was previously estimated to be 7.5 square kilometres per day (Anderson and Thampapillai 1990), and though somewhat abated, as per data from Global Forest Watch (2023), rate of deforestation continues at a significant rate. As Hecht (1985) pointed out, deforestation has occurred at a massive rate in Brazil – nearly a million hectares a year in the eastern Amazon basin. This trend has continued unabated. For example, Moran (2011) refers to the legislation introduced in Brazil's Senate in December 2011 paving the way for legally clearing 220,000 square kilometres (22 million hectares) of rainforest. In the Central African country of Malawi, the acute shortage of timber was already evident in the 1980s (French 1986).

There are numerous other examples. The need for international conferences and agreements on whaling has been prompted by the need to prevent the extinction of whales. Issues with soil erosion, across the world, stem from the concern that

66   *Microeconomics and KN*

it may not be possible to sustain agricultural output unless fertility and quality of soils are conserved. For example, there are serious concerns among agriculturists that traditionally productive areas, such as the Darling Downs in Australia, the Prairies in Canada and the Steppes in Central Asia, need the adoption of urgent conservation measures if they are to remain viable farming areas in the future. In short, the IGC objective strives to ensure that the availability and productivity of resources are sustained over generations.

There is a substantial amount of complementarity between the objectives of KN-Q and IGC. That is, preservation of KN resources also ensures that these resources are made available to future generations. However, these two objectives together are often in conflict with the EG objective. In the next section, we first consider frameworks that deal with the relationships between the objectives of EG and KN-Q. After that, we will introduce some frameworks that also include the IGC objective.

Conflicts between EG and KN-Q Objectives

We shall study two types of frameworks that enable the analysis of conflicts between EG and KN-Q. The first of these is related to the internalisation of externalities that we considered in the previous chapter – but at the firm level. We shall next consider a related framework that enables the analysis of conflicts between groups of people, which results from imposition of an KN-Q objective in the domain of the firm's operation. In order to facilitate exposition, we shall study these frameworks in the context of an example.

The Framework Based on the Market Model

Suppose that a factory is located on a riverbank upstream. This factory produces a consumer good that we shall refer to as M. The manufacture of M also results in the production of a toxic by-product, which the factory dumps into the river. Now suppose that the factory owners have been told to install a water treatment plant upstream to clean the water that flows downstream, because there is a recreational area further down the river. Also assume that the price of M is $20 per unit.

Consider first Figure 5.1, which describes the costs and revenue that are associated with the production of M without any reference to costs associated with KN. Given that the price of M has already been determined, the total revenue (TR) curve is a straight line, and its equation is:

$$TR = 20q \qquad (5.1)$$

where q represents the quantity of M produced. Let us suppose that the total cost of producing M (without KN costs), TC_p, is defined by the equation:

$$TC_p = 2q^2 \qquad (5.2)$$

Economics of Renewable and Non-Renewable Resources 67

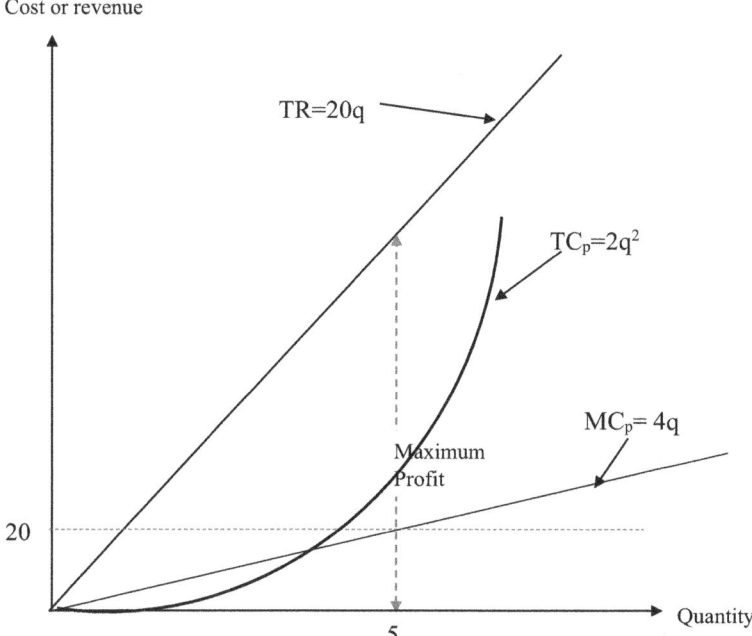

Figure 5.1 Illustration of Profit Maximisation (Without KN Costs)

The marginal cost of producing M, MC_p, can now be defined as the first derivative of the equation for TC_p. That is:

$$MC_p = dTC_p/dq = 4q \qquad (5.3)$$

Note that in Figure 5.1 profit is maximised at the point where the distance between the TR and TC_p curves is at a maximum; that is, when the production of M is equal to five units. Also note that the point of maximum profit is the point at which the horizontal line representing the price of M intersects the marginal cost curve.

Now suppose that the objectives surrounding the production of M are expanded to include KN-Q. That is, the factory owners are told to treat the water that gets contaminated. Suppose that the total cost of water treatment (TC_w) is defined in terms of the quantity of M produced as:

$$TC_w = q^2 \qquad (5.4)$$

We chose here, for simplicity, to have TC_w be the square of the quantity q in order to reflect a common phenomenon of treatment costs rising with increases in the amount of pollution that needs to be treated. If we in turn assume that treatment amount to be proportional to output of the desired product M, then equation (5.4) sufficiently captures treatment cost profiles for our illustrative purposes.

68 *Microeconomics and KN*

The marginal cost of water treatment, (MC_w) is:

$$MC_w = dTC_w/dq = 2q \qquad (5.5)$$

Given that the management objectives now also include KN-Q, we need to define the total cost of producing M as:

$$TC = TC_p + TC_w = 2q^2 + q^2 = 3q^2 \qquad (5.6)$$

The marginal cost of producing M in this context is:

$$MC = dTC/dq = 6q \qquad (5.7)$$

We can clearly see in Figure 5.2 that maximum profit has fallen relative to the situation when treatment costs were ignored, and that the owners of the factory have to cut back production. That is, by incorporating the KN-Q objective by way of the treatment costs, the production of M falls to 3.3 units and presumably the amount of pollution falls to a tolerable level.

However, analysis of this type presents a problem, in that the owners of the factory will not usually include KN-Q in their list of objectives unless they are compelled to do so. Further, making the owners of the factory bear the treatment costs also implies that those who own the factory do not have any ownership

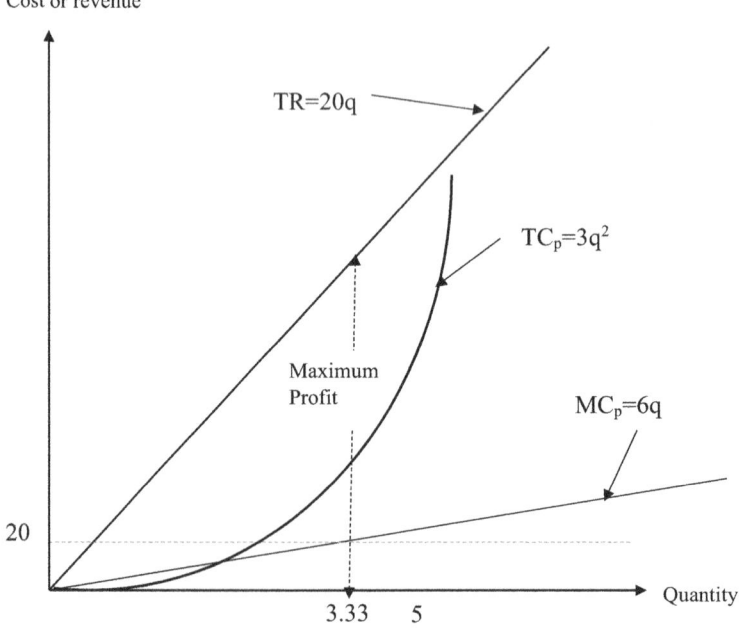

Figure 5.2 Illustration of Profit Maximisation (With KN Costs)

Economics of Renewable and Non-Renewable Resources   69

rights over the river. But then, if the downstream users also have no ownership rights over the river, the owners of the factory can demand that the downstream users must contribute to the treatment costs. The factory may do so by raising the price of M. This is possible if the factory is the only producer of M, or if there is a limited number of producers. Given geographic constraints, that is typically the case.

What we can infer from the foregoing is that the recognition of the KN-Q objective can invariably result in conflicts between different groups. In this example, the conflict would be between the owners of the factory and those who use the river downstream.

The next framework that we consider attempts to address the conflicts between groups.

The Framework for Conflicts Between Groups

We shall continue to use the example of the factory upstream and the recreational area downstream. The source of the conflict is primarily the question: 'Who should pay for the treatment of water that is contaminated?' Economist Ronald Coase (1960) was the first to illustrate a framework for the analysis of this type of conflict.

Consider Figure 5.3, which shows two curves: the marginal cost of controlling water pollution (MCP) and the marginal benefits from pollution control (MBP). The MCP curve, unlike in the analysis above, is now being defined in terms of the

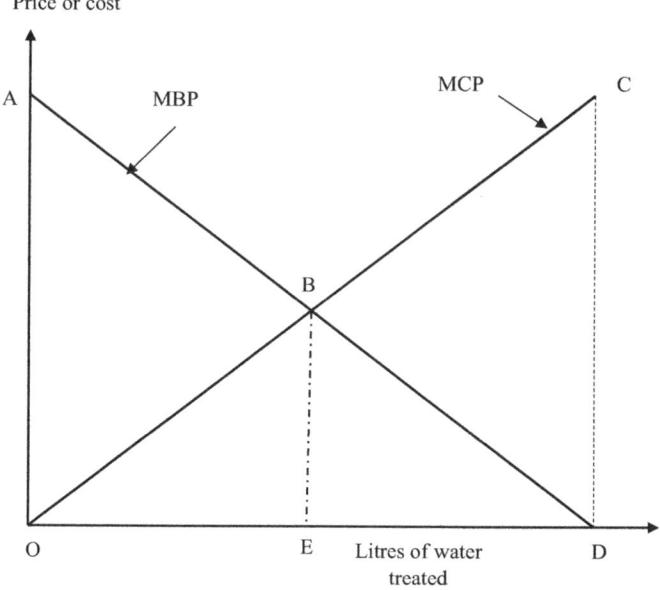

Figure 5.3 Illustration of Coase Theorem

quantity of water that is treated and demonstrates the costs that have to be incurred by the factory to control the pollution.

The MBP curve describes the marginal benefits that accrue to those who participate in recreation and can be derived by applying some of the methods that will be described in more detail in Part IV of this text. One of these methods is the travel cost method. It estimates the costs that people are willing to incur to enjoy an KN amenity. These could include the cost for fuel and parking, the income that is foregone when visiting the resource (i.e. the opportunity cost of leisure), and so on. The total travel costs serve as a proxy for the WTP for recreation. The number of visits similarly serves as a proxy for the quantity of recreation. In our hypothetical case, if we assume a 'one-to-one' relationship between the number of visits and the quantity of water treatment, then we can assume the MBP curve to be the demand curve for recreation measured in terms of the quantity of water that is treated. Even if such an assumption were unreasonable, it is possible to ascertain a relationship between the number of visits and the amount of water to be treated. The underlying premise is that, when the quantity of water treated is small, the opportunity for recreation will be limited, and therefore the WTP for recreation will be high. In contrast, when a relatively large amount of water has been treated, the facilities for recreation will not be so scarce, and therefore the WTP for recreation will be relatively low. So, this premise is consistent with the law of diminishing marginal utility.

Consider first a situation where the river is fully owned by the factory. In this case, the factory will refuse to implement pollution control unless it is paid for effecting such measures. Those who engage in downstream recreation will have to pay the factory the amount of money that is required to achieve the desired level of water treatment. If the downstream users want OD litres of water treatment, then they will have to pay the factory an amount that is equal to the area below the MCP curve up to OD litres, namely area OCD. Remember that the area below a marginal cost curve measures the total cost, and the area below the marginal benefit curve measures total benefit. Hence, OD units of pollution control will give the downstream users a total benefit that is equal to the area below the MBP curve; namely area OAD. If the downstream users are prepared to put up with a level of pollution control that is less than treating OD litres of water, then they incur two types of costs:

- Costs in terms of recreational benefits that are sacrificed
- Costs of pollution control that have to be paid to the factory

For example, if the downstream users were prepared to live with ED litres of contaminated water (that is, if only OE litres of water are treated), then their costs are:

- Area BED in terms of recreation benefits that they have to sacrifice
- Area OBE in terms of treatment costs

Given that the downstream users incur these two types of costs, it also becomes evident that the maximum cost savings to these users occurs when they settle for pollution control that is equal to treating OE litres of water. This maximum cost saving is represented by area BCD, and is defined as follows:

$$\begin{pmatrix} \text{The cost of total} \\ \text{pollution control} \\ \text{(OD Litres)} \end{pmatrix} \text{minus} \begin{pmatrix} \text{The cost of} \\ \text{pollution control} \\ \text{for up to OE litres} \end{pmatrix} \quad (5.8)$$

= Area OCD − (Area OBE + Area BED)
= Area BCD

Should the downstream users choose to pay for any other level of pollution control, their cost savings will be smaller. An alternative solution is that the downstream users settle for paying for OE litres of treatment, as it yields them maximum net benefit. The net benefits to those engaged in recreation would be defined as:

$$\begin{pmatrix} \text{The area below} \\ \text{the MBP curve} \end{pmatrix} \text{minus} \begin{pmatrix} \text{The area below} \\ \text{the MCP curve} \end{pmatrix} \quad (5.9)$$

Area OABE − Area OBE = Area OAB

Note that area OABE represents the total benefits that downstream users get from OE litres of treatment, and area OBE is the total cost of treating OE litres, which they have to pay the factory because the factory owns the river. Their payments for any amount of treatment that is either larger or smaller than OE litres will result in a net benefit smaller than area OAB.

Now consider a situation where the river is fully owned by the downstream users. In this case, if the factory wanted to pollute the river, it would have to pay the downstream users an amount of money that represents the loss of recreation benefits. This is of course possible only if the downstream users are willing to accept such payments. Should this be the case, the largest compensation that the factory can pay is defined by the area OAD, and this will happen if the factory decides to produce units of M that correspond to the contamination of OD litres of water. However, the factory can achieve some savings on the size of this compensation cost if it decides to incur some pollution control costs. For example, when pollution control amounts to treating OE litres of water, the costs incurred by the factory are:

- Area OBE for treating OE litres of water
- Area BED as compensation to the downstream users for the recreation benefits they have to forgo.

The factory will experience maximum cost savings if it treats OE litres of water and compensates the downstream users for the contamination of ED litres of water. This maximum cost saving can be defined as:

$$\begin{pmatrix} \text{Compensation for} \\ \text{contaminating} \\ \text{OD litres} \end{pmatrix} \text{minus} \begin{pmatrix} \text{Cost of treating} \\ \text{OE litres} \end{pmatrix} \text{plus} \begin{pmatrix} \text{Compensation for} \\ \text{contaminating} \\ \text{ED litres} \end{pmatrix} \quad (5.10)$$

= Area OAD − (Area OBE + Area BED) = Area OAB

The factory will experience a cost saving that is smaller than area OAB, should the level of pollution control be anything other than OE litres.

What we have seen from the above discussion is that the treatment of OE litres of water is the optimal level of pollution control, regardless of who has the ownership rights to the river. But the claim here is that, if property rights were bestowed to one of the parties, then an optimal solution would ensue. However, in practice, there are several difficulties in achieving the optimal level of pollution control. To begin with, it is difficult to bestow the ownership of an endowment such as a river on a group of individuals or a firm. This difficulty is compounded when there are several stakeholders along the banks of the river. Further, even if the ownership rights were vested with one group among those involved in a conflict, it is not certain that an optimal solution may be reached by negotiations involving compensation. This is because of the time lags that are usually observed with compensation, the complexities that surround KN issues of water quality and associated ecological attributes, as well as cause–effect relationships between water quality and harm to humans. Significant issue is the continuous nature of production and hence pollution, and as we indicated in Chapter 4 the potential possibility of the pollutants to accumulate. Quantifying any of these complexities is ripe with uncertainties and open to interpretation and conflict. However, regardless of these difficulties, the analysis of the framework provides some useful basis for policy. If we are able to construct the MCP and MBP curves, then we can use area OBE (plus some safety margins) as the basis for the tax on pollution, or we can use OE litres of treatment as a standard for legislation. The derivation of these taxes and standards would of course become difficult if there were many polluters and many affected parties.

Conflicts Due to Intergenerational Concern

As indicated, the intergenerational concern (IGC) objective pervades issues that address the question of resource management over time. The objective is pertinent in the context of managing non-renewable as well as renewable resources. For example, the South Pacific Forum has repeatedly expressed an opposition to drift-net fishing because such fishing results in the overexploitation of fishery resources, resulting in the possible decline of supply in the future. For similar reasons, many conservationists oppose the methods of clear felling in forest management. Therefore, the application of the IGC objective to the management of renewable resources is often directed at achieving *sustainable production*.

With non-renewable resources, as indicated earlier, the concern of exhausting our supplies has been overtaken by the need for transitioning to cleaner sources with reduced GHG emissions. Nevertheless, the issue of exhaustibility does remain. For example, consider the transitioning from fossil fuels to renewables that are often more intermittent and thus require batteries (and other means) for energy storage. The production of batteries requires the mining of lithium, which is a finite resource. Similarly, the generation of solar energy requires the utilisation of panels, which depends on the mining of another finite resource, namely sand rich in silicon dioxide. Hence the future generations may have to face varying degrees of challenges to fulfil their resource requirements.

The economists' approach to the management of resources over time, regardless of whether they are renewable or non-renewable, involves the application of what is called *optimal control theory*. We shall illustrate how an important concept that is employed in this theory can be used in the adaptation of the basic market model that we are familiar with. The main concept that warrants familiarity, for our present purposes, is that of the *marginal user cost,* namely the net benefits which the future generation needs to forego. We closely follow the frameworks developed by McInerney (1976) for both renewable and non-renewable resources. So far, in the market models that we have considered, we have not dealt with the issue of time. We have in fact assumed that the timing of a decision and its ramifications over time are not of relevance. In introducing time into the frameworks, McInerney made the following assumptions:

1. There are only two time periods, namely the present (T_0) and the future (T_1).
2. All pertinent market information for both time periods is known. That is, demand (D_0 and D_1) and supply as manifested in the marginal extraction/harvest costs (MEC_1 and MEC_2). For example, if D_0 is known, then it is possible to estimate D_1 in terms of factors influencing demand such as population and income growth.
3. The total cost (TC) function for resource extraction is linear and upward sloping. The linear assumption for TC implies that MEC is constant. Both the MEC and TC functions tend towards infinity at the point of resource exhaustion or extinction.
4. The information for the future (T_1) is presented as present values. That is, D_1 and MEC_1 are presented as discounted values using appropriate discount rate.[2]

Within this context, resource extraction will need to be brought in synch with two counter-acting factors — the impatience to use the resource (i.e. the impatience to consume the products generated with the aid of the resource), which would drive up production in the present, and the higher cost (and price) that will need to be faced by producers (and consumers). The latter unfolds because depletion makes the extraction of the next unit of the resource ever more expensive. As a result, extraction gets shifted to the future (when presumably technology and substitutes reduce the need for the resource, and by when investments from the proceeds from resources sales of the present can be ploughed back into the economy for efficiency gains, for example)

A Framework for Non-Renewable Resources

Suppose that societies in T_0 as well as T_1 must manage with a finite resource stock of size OS. We present two types of market configurations in each of Figures 5.4a and 5.4b. The market information for both time periods is presented in the same figure. To explain the market in T_0, resource extraction is measured from Left to Right and the relevant demand and supply curves are respectively by D_0 and MEC_0.

74 Microeconomics and KN

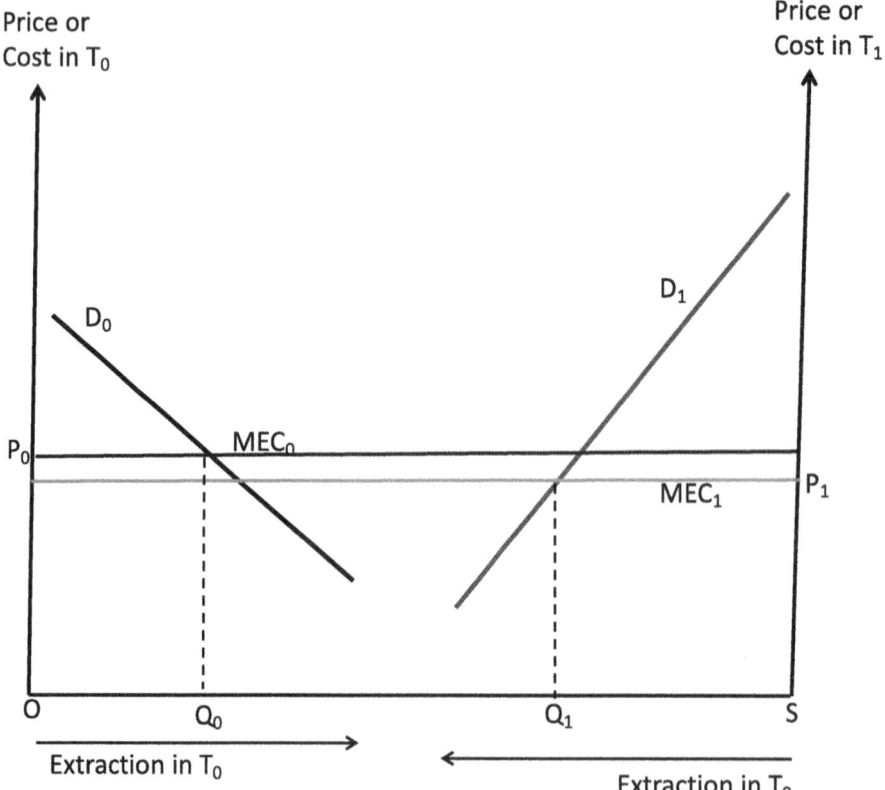

Figure 5.4a Market for a Non-Renewable Resource – No Conflicts between Generations

For the market in T_1, resource extraction begins from Right and proceeds to the Left and the relevant future demand and supply curves are respectively by D_1 and MEC_1. McInerney (1976) assumed that MEC values remain unchanged between time periods. That is, the linear TC function for extraction remains steady with a constant positive gradient. Nevertheless, MEC_1 is shown to lie below MEC_0 because of discounting to obtain present values.

Consider Figure 5.4a. Here the resource requirements and prices resulting from market equilibrium for each period is defined as follows:

$$T_0: (D_0 = MEC_0) \rightarrow (OQ_0, P_0)$$
$$T_1: (D_1 = MEC_1) \rightarrow (SQ_1, P_1)$$

The important observation with reference to Figure 5.4a is that there is no conflict between the societies of T_0 and T_1. This is because the total resource requirement across the two time periods is less than the size of the resource stock; that is:

$$\{(OQ_0 + SQ_1) < OS\}$$

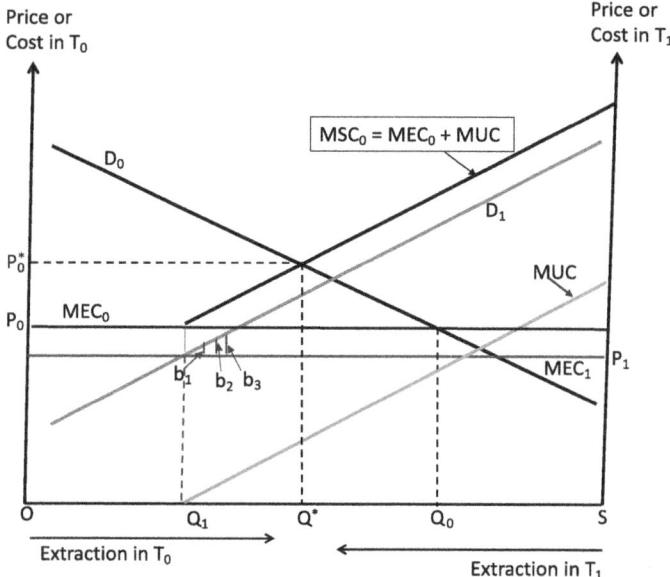

Figure 5.4b Market for a Non-Renewable Resource – Conflicts between Generations

In contrast, there is a clear conflict between the societies of T_0 and T_1 in the market context displayed in Figure 5.4b. Resource requirement and price stemming from market equilibrium for T_0 is (OQ_0, P_0); and T_1 is (SQ_1, P_1). But in this instance, the total resource requirement across the two time periods is more than the size of the resource stock that is available, namely $\{(OQ_0 + SQ_1) > OS\}$.

From Figure 5.4b, we observe that when resource extraction during T_0 exceeds OQ_1, this extraction will deplete the resource requirement of T_1, namely (SQ_1). That is, T_0 extraction in excess of OQ_1 imposes a cost (loss) on the future generation in T_1. This cost, referred to as the User Cost, is the net benefits that the future generation needs to forego because of resource extraction in the present.

> The marginal user cost (MUC) defines the net benefits that the future generation must sacrifice for every unit that is extracted during the present period.

The MUC curve is in fact derived from the demand and supply curves for T_1. For purely illustrative purposes, shown in Figure 5.4b, three vertical lines of height b_1, b_2 and b_3 between the demand and supply curves for T_1. The following explanation is not precise in a mathematical sense, in that we need to explain the MUC in terms of areas rather than heights. However, for illustrative purposes we shall deviate from precision and define the MUC in a discrete sense as $\{MUC = D_1 - MEC_1\}$. When the future requirement, SQ_1, is left intact, the MUC is zero. The vertical distance b_1 defines the loss of net benefits in T_1, when the first unit of T_0 extraction exceeds OQ_1. Similarly, we can envisage the heights b_2 and b_3 to represent the loss of net benefits in T_1, when the second and third units of T_0 extraction exceed OQ_1.

76 Microeconomics and KN

So, the MUC curve is derived in this way. In a strict mathematical sense, the area below the MUC curve in Figure 5.4b is equal to the area that is bound between the demand and supply curves of T_1; namely D_1 and MEC_1. Also note that the MUC curve tends towards infinity – when the quantity of T_0 extraction approaches OS. That is, should the size of T_0 extraction be OS, then the resource stock will be completely depleted, and since a non-renewable resource cannot be reproduced, the cost imposed on the T_1 generation approaches infinity.

Figure 5.4b also illustrates the basis for resolving the conflict between the societies of T_0 and T_1. If the objective of society in T_0 were to simply maximise present income and nothing else, then it would extract OQ_0 units of the resource and leave the society of T_1 to contend with $(OS - OQ_0 = Q_0S)$ units. Alternatively, if the IGC objective were also included, then the cost imposed on the future society would also become pertinent. Hence, the quantity of extraction in T_0 would be based on the equilibrium between present demand (D_0) and the marginal social cost (MSC) that includes the MUC in addition to MEC_0. In this context the MSC is defined as:

$$MSC = MEC_0 + MUC \tag{5.11}$$

Recall that the supply curve describes the marginal cost of extraction. So, until OQ_1 units of extraction in T_0, the MSC curve is synonymous with the MEC_0. When this extraction exceeds OQ_1, the MSC curve begins to slope more steeply upwards, because the MUC gets added to the MEC for every extra unit that is extracted in excess of OQ_1. Hence, when the IGC objective is recognised, the optimal extraction in T_0 will be based on the equilibrium between the MSC and D_0, resulting in the extraction of OQ^* units. That is, the society in T_1 will have Q^*S units of the resource. There will also be a price increase from P_0 to P_0^*. Note the trade-off in extraction quantities between the societies of T_0 and T_1. The recognition of the MUC and MSC, prompts the society in T_0 to sacrifice $(OQ_0 - OQ^*)$ units and hence the society in T_1 sacrifices $(SQ_1 - SQ^*)$ units. Further, given the assumptions made above, it is possible to demonstrate that at the optimal level of resource extraction, the marginal net benefits in T_0 and T_1 are equal to one another. That is, $\{(D_0 - MEC_0) = (D_1 - MEC_1)\}$.

A Framework for Renewable Resources

We present a slightly adapted version of McInerney's (1976) framework for a renewable resource in Figure 5.5. This is similar to that presented above for a non-renewable resource but includes the capability of resource regeneration. The MUC is an important component here as well and emerges due to the loss of future net benefits because of present consumption. The MUC and MEC both tend towards infinity as the possibility of *extinction* becomes real.

As illustrated in Figure 5.5, the framework is displayed in four quadrants, all of which share a common origin, namely O. Quadrants 1 and 4 display respectively the markets for T_0 and T_1. Quadrant-2 illustrates the regeneration capability at the rate of (ρ) between T_0 and T_1. Although the actual rate of regeneration could be nonlinear, McInerney assumed a linear rate for illustrative convenience, and

Economics of Renewable and Non-Renewable Resources 77

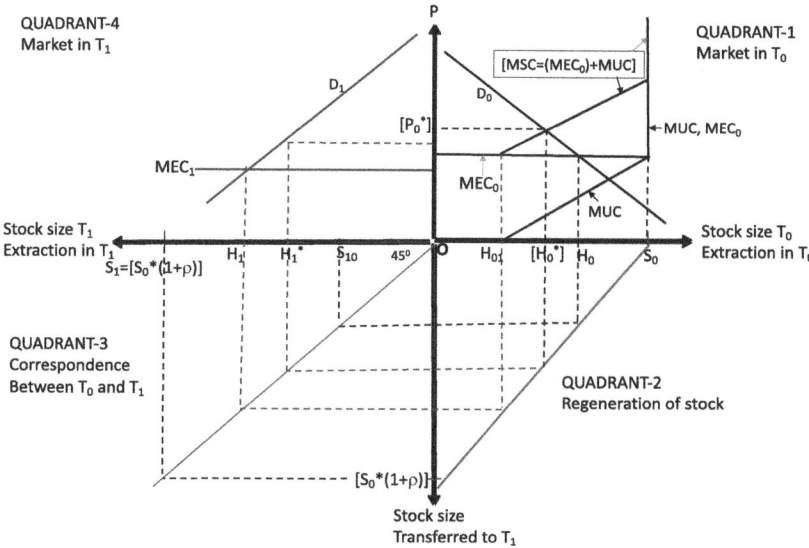

Figure 5.5 Market for a Renewable Resource over Time

we have retained this assumption. Quadrant-3 enables the identity of quantities estimated through regeneration from T_0 to be transferred to the quantity axis of T_1 market in Quadrant-4.

Consider first the market for the resource in T_0 in Quadrant-1. The available resource stock is defined as OS_0. In the absence of any harvest in T_0, the initial stock of S_0 grows, as shown in Quadrant-2, to $[S_0*(1 + \rho)]$. This enlarged stock can then be identified, by recourse to the identity relationship in Quadrant-3, as the amount of resource available for the market in T_1 (Quadrant-4); that is, $S_1 = [S_0*(1 + \rho)]$. The conflict between generations can be explained as follows:

1 Suppose that harvest in T_0 is performed without any consideration whatsoever for the market requirements of T_1. Then the market in T_0 will clear at $(MEC_0 = D_0)$ resulting in the harvest of H_0 units of the resource. This harvest leaves behind $(S_0 - H_0)$ units of the resource in T_0 to regenerate into S_{10} units in T_1.
2 However, the market information for T_1 suggests that the requirement on the basis of $(MEC_1 = D_1)$ will be H_1, which is of course greater than S_{10}. This market requirement of T_1, dictates that a quantity of $(S_0 - H_{01})$ must remain in T_0, so that it can regenerate into H_1 for T_1.

Hence, there is a conflict between the two generations. The conflict is resolved by recognising the MUC in T_0. Note that the MUC becomes positive when harvest in T_0 exceeds H_{01} and tends towards infinity as harvest approaches the stock size. MEC_0 would also approach infinity as the harvest size drives resource towards extinction. We can now introduce the MSC the definition which is the same as that given in Figure 5.11, namely, $(MEC_0 + MUC)$. As shown in Figure 5.5, the equilibrium between D_0 and MSC determines the optimal harvest size of $[H_0^*]$ at price $[P_0^*]$.

Hence, the recognition of the IGC objective enables a reduction in the danger of extinction by prompting a reduction in harvest size in both time periods. T_0 generation gives up $(H_0 - H_0^*)$ units, whilst the sacrifice to be made by T_1 generation is $(H_1 - H_1^*)$.

Some Resource Management Policies

So far, we have considered some simple frameworks that enable an analysis of extraction and harvest decisions of natural resources over time. Although more complex frameworks can incorporate the influence of time more explicitly, for our purposes the treatment given here is sufficient. It is also useful to note that the MSC would become larger if we also included a marginal cost that impacts KN or a marginal pollution control cost. This would be the case when the list of objectives involved EG, KN-Q and IGC. That is, the inclusion of the KN-Q objective in addition to the IGC objective would prompt further conservation of the resources. Also, the conflicts between the objectives are invariably influenced by technological change. Consequently, technology has an important policy role in resource management.

The frameworks that were considered in the previous section addressed the issue of resource depletion and extinction. In these frameworks we recognised that the costs of extinction or depletion are more generally borne by those of the future generation than those of the present. However, our perception of the problems concerning depletion and exhaustion can be altered if we recognise technological change. Some argue that the problem of resource scarcity, due to rapid extraction or harvest, is not exactly a problem because technology can reverse the scarcity. For example, consider Figure 5.5. Recall that the conflict between generations emerges when present harvest exceeds OH_0. Now suppose that technological change results in a cheap and safe substitute for the resource, and consequently the demand for the resource shifts significantly inwards such that quantity demanded is less than OH_0. Hence technology has the potential to remove the conflicts between generations.

Renewable Resources

Technology can also delay the onset of potential extinction of renewable resources. In Figure 5.5, we saw that the potential for extinction emerges when harvest size approaches OS_0. Suppose now that, due to advances in genetic engineering, the point of potential extinction shifts to the right. The development of aquaculture has been one of the measures that have been widely adopted to offset the potential threat of extinction of renewable fishery resources. However, aquaculture is not without difficulties. For example, as Bell (1992) illustrates, accidental leakages from aquaculture sites can significantly contaminate land and water resource systems. Furthermore, considerable quantities of marine life are used to provide food to fish in aquaculture operations, thus shifting the depletion pressures from one species to others. And wild fish populations, especially in ocean settings, can suffer considerably when exposed to the pharmaceuticals used to manage the dense, farmed fish populations, or when any of those fish escape into the wild.

Economics of Renewable and Non-Renewable Resources 79

The literature on fishery management measures is extensive (Hundloe 1997; Dankel et al. 2008; Singh-Renton and McIvor 2015) and can be extended to the context of other renewable resources as well. These measures include *input controls, property rights and taxation* including *output controls*. In Figure 5.5, (OH_0^*) is the permissible size of the harvest of a renewable resource. It is the determination of this quantity (OH_0^*) that represents the greatest difficulty. If (OH_0^*) has been estimated, then the measures taken in an attempt to ensure that the harvest size does not exceed this quantity are as follows:

- Input controls deal with restricting the number of persons (and other resources) engaged in harvesting so that the size of the harvest does not exceed (OH_0^*). Usually this is implemented through the issue of a limited number of licences, mesh size limitations, seasonal restrictions and the like.
- Property rights deal with the rights of ownership or management being endowed to persons either individually or cooperatively, with the expectation that careful rules of management will be developed and exercised. Of course, the difficulties that were raised in Chapter 3 will remain.
- Tax is the extra levy on resource use to restrict its harvest to (OH_0^*). If we regard the MUC as the extra levy that has to be imposed on the harvesting of a resource, then the tax on a renewable resource will be equal to the price differential ($P_0^* - MEC_0$) in Figure 5.5.
- Output controls are strict quotas that are issued to the harvesters to ensure that the quantity removed does not exceed (OH_0^*). Sometimes output controls are implemented in conjunction with input controls.

Each of these measures contains different degrees of operational difficulties, such as unequal distributional effects between groups of people, the ability to monitor and enforce the measures and the ability to adjust the measures if and when conditions pertaining to KN, technology or knowledge change. Some of these changes have been extreme and cast doubt about the renewability of resources that have been traditionally labelled renewable – especially forests as explained in Box 5.2, but also many fish populations, as mentioned above.

Box 5.2 Are Forests Renewable?

Several forest products are often labelled renewable because forests possess regenerative capacity. A forest product that has recently received much publicity is wooden pellets used as a source of energy through combustion. Canada exports a significantly large quantities of wooden pellets to the United Kingdom, Japan and South Korea. Voegele (2023) reports that such exports in 2022 were approximately 3.3 million metric tons – an increase of 5 per cent compared to the previous year. The manufacturers of these pellets have labelled them as not only renewable but also Carbon-Neutral. This claim rests on the belief that the CO_2 released from the combustion of the pellets could be potentially reabsorbed by growing trees planted as replacement for those harvested.

However, the benefits of renewability and carbon neutrality, which manufacturers try to associate with wooden pellets, become diluted in the context of additional information assembled as follows.

- The production of more than 3 million metric tons of wooden pellets, implies that forest areas of at least 30,000 hectares are cleared each year in Canada – notwithstanding reforestation programmes Canada.
- A study from Simon Fraser University (Benner and Lertzman 2022) indicates that the forest products industry has targeted mainly old growth forests in their logging activity.
- Aron et al. (2019) demonstrate that the reduction in canopy density stemming from forest logging can lead to hydrological disturbances involving prolonged periods of drought. Such disturbances are likely to higher with loss of canopy in old growth forests than newer plantation forests.
- Richardson et al. (2022) display the potential association between wild forest fires and droughts.
- Trenberth (2005) explains a disturbed hydrologic cycle is invariably one where prolonged periods of drought are followed by intense precipitation, floods and storms.
- Canada (like most other countries) has witnessed catastrophic fires and floods. The fires of the summer of 2023 extended from the East to West of Canada and involved a loss of 18 million hectares of forest.

When the above information are reviewed together the renewability and carbon neutrality of utilising forest resources become nebulous concepts.

Non-Renewable Resources

The policy measures for non-renewable resources will also be similar to those for renewable resources. However, some important and distinguishing issues need to be taken into account. With several non-renewable energy resources such as coal and other fossil fuels, the concept of user costs extends beyond the mere depletion effect of the resource. This is because the externalities, which are associated with the extraction and use of non-renewable resources, are capable of reducing the net benefits to future generations and therefore are components of user costs. For example, the continued use of fossil fuels for power generation and transport has resulted in the build-up of GHGs and the climate disruptions. Hence the policy directive with non-renewable energy resources is now increasingly turning to measures that include the following:

- Promotion of cleaner energy technologies, for example, by reducing or capturing unwanted by-products of fuel combustion and by improving efficiencies

- Development and deployment of alternative energy technologies such as wind power and biomass

Some argue that alternative energy technologies remain too expensive, compared to the traditional energy resources such as coal, and that therefore these traditional sources will continue to be used. But if all externalities of these traditional sources are properly internalised, then the price differentials may not be significant or even be reversed. For example, cities running their buses on clean fuels pay premiums when they purchase these vehicles but can save a disproportionally larger amount of money over time not just in terms of fuel costs but in terms of declining public health expenditures because cleaner buses means cleaner urban air quality and thus fewer incidents of asthma and other respiratory illnesses.

It should also be noted that important policy measures, with reference to non-renewable resources, pertain to modesty in the consumption behaviour of societies. In fact, this consideration extends to renewable resources as well. For example, we could achieve significant resource conservation and considerably reduce user costs if we substituted public transport for private transport. How does this represent modesty in consumption behaviour? While trains and buses run half empty, roads are choked with cars, each of which rarely contains more than one person. However, the shift to public transport may require lower prices for public transport and steeper prices for private transport. A good example of a successful public transport system can be found in the island economy of Singapore, which operates a network of buses and Mass Rapid Transit (MRT) rail. The maximum price on this transport network for a single trip does not exceed two Singapore dollars. Some market analysts may argue that such price ceilings may prompt market inefficiencies. However, such ceilings are established to address the externalities associated with private transport, and thus are a form of social policy measures that promote efficiency.

To conclude, the formulation of policies for natural resource management needs to consider the three objectives EG, KN-Q and IGC concurrently. The measures inevitably have to involve a mixture of proper pricing, regulation, innovation and attitudinal changes. We will examine some of these policy measures in more detail in the next chapter. See the narrative, though dated, in Box 5.3, which illustrates this mix.

Box 5.3 Israel Sets the Standard in Water Recycling

This example is an excerpt taken from the Australian Broadcasting Corporation's *Landline* Program, dated 3 June 2001.

In the book of Isaiah, the prophet foretold that the desert shall rejoice and blossom as the rose. In the past 50 or so years, the modern state of Israel has gone a long way to fulfilling that prophesy. But there's been neither deluges of biblical proportions nor more subtle climatic shifts to moisten the scorched earth.

What's changed the face of the desert is an appreciation of the power of such a vital and scarce resource and determination to make every drop count. And that means using things more prodigal nations would simply flush away.

Israel has only limited sources of fresh water for a population, which is expected to increase by 40 per cent over the next two decades. The Sea of Galilee and the Jordan River are already stressed after years of drought, and the aquifers that supply the bulk of the drinking water are in an equally delicate state. More people, more pollution and less water available to feed not only the growing masses but also an essential export industry is a worrying equation. But in Israel the solution may be not only disarmingly simple but more importantly, sustainable – use the waste to feed the people.

The process of weaning irrigators off fresh water has been one of the remarkable successes of Israeli agriculture over the past few decades. Since 1984 the use of fresh water on farms has halved while the value of production continues to climb. There are a number of factors driving the change: the rapid uptake of technologies such as drip irrigation, the industrialisation of farming where orange groves are being pushed aside for greenhouses, and a firm government policy to encourage the reuse of sewage effluent.

It is fair to say Israel and Australia look at water and effluent reuse from entirely different policy directions. Israel puts a far higher value on its raw resource fresh water, while effluent creates additional usable water for economic growth.

Despite being the so-called driest continent, Australia's fresh water is still relatively cheap and effluent is a problem to be tidied up rather than a productive resource. Perhaps the greatest difference is that in Israel the output of a sewage plant is seen as just part of the overall water resource, a resource that's both productive and safe. In Australia, recycled water is largely an entirely different product from the fresh resource, and that may prove to be the biggest waste issue of all.

Reporter: Steve Letts (source http://www.abc.net.au/landline/stories/s303636.htm)

As of 2024, with the escalation of war and conflict in the Middle East, Israel and her neighbours need to innovate on pathways to peace and harmony for greater social well being. Every rocket fired in the region – for that matter anywhere – causes significant ecological disaster and human misery.

Review Questions

1. Consider the case of a factory discharging effluent into a river system upstream. The effluent adversely affects downstream users of the river. An environmental economist has estimated that the marginal benefits (MB) of pollution reduction are given by:

    $MB = 500 - 0.5Q$

    where Q represents units of water treated prior to reaching the users downstream.

An environmental engineer has estimated the total cost (TC) of water treatment as:

$TC = 0.25Q^2$

Use the above information to illustrate the Coase theorem, and evaluate its relevance in resolving conflicts arising from KN related externalities.
2 Do you think that the marginal user cost is a relevant concept for the management of a country's coal and oil resources?
3 Would the relevance of the marginal user cost be diminished with forestry and fishery resources because they are renewable?
4 Explain how the marginal social cost can be derived when IGC is a management objective for: (a) a non-renewable resource, and (b) a renewable resource.
Explain, with examples of renewable as well as non-renewable resources, how technology can delay the emergence of the marginal user cost.

Notes

1 http://ds.iris.edu/ieb/index.html?
2 The basis for discounting stems from the axiom of present gain indicated at the start of Chapter 3. That is, individuals prefer the present over the future. We argue later in Chapter 9, that from a sustainability perspective, the future is at least as important as the present.

References

Anderson, J. R. and Thampapillai, J., *Soil Conservation in Developing Countries: Project and Policy Intervention*, Paper Number 8, Policy and Research Series, Agriculture and Rural Development Department, The World Bank, Washington DC, 1990.
Aron, P. G., Poulsen, C. J., Fiorella, R. P., and Matheny, A. M., 'Stable water isotopes reveal effects of intermediate disturbance and canopy structure on forest water cycling', *Journal of Geophysical Research: Biogeosciences*, 124 (https://doi.org/10.1029/2019JG005118)
Bell, F. C., 'Prospects for aquaculture in NSW', *Report 2 – Developments in Aquaculture in NSW and their Potential Impacts*, Total Environment Centre, Sydney, 1992.
Benner, J., and Lertzman, K., 'Policy interventions and competing management paradigms shape the long-term distribution of forest harvesting across the landscape', *Proceedings of the National Academy of Sciences*, October 2022 (https://doi.org/10.1073/pnas.2208360119)
Climate Partner, *From Pledges to Binding Commitments – A Review of Key Milestones in Previous COP Conferences*. https://www.climatepartner.com/en/climate-action-insights/complete-review-of-key-milestones-from-previous-cop-conferences, October 2022.
Dankel, D. J., Skagen, D. W. and Ulltang, Ø., 'Fisheries management in practice: Review of 13 commercially important fish stocks', *Reviews in Fish Biology and Fisheries*, 18(2): 201–233, 2008.
Emmanuel, K., 'Will global warming make hurricane forecasting more difficult?', *Bulletin of the American Meteorological Society*, 98(3): 495–501, 2017.
French, D., 'Confronting an unsolvable problem: Deforestation in Malawi', *World Development*, 14(4): 531–540, 1986.
Global Forest Watch, 2023 (https://www.globalforestwatch.org/)

Hecht, S., 'Environment, development and politics: Capital accumulation and the livestock sector in eastern Amazonia', *World Development*, 13(6): 663–684, 1985.

Hiranuma, H., Japan's Energy Policy in a Post-3/11 World - juggling safety, sustainability and economics, *The Tokyo Foundation*, 15 October 2014.

Hundloe, T., 'Achieving environmental objectives by the use of economic instruments: Fisheries', *Environmental Economics Round Table Proceedings*, Environment Australia, Canberra, 1997.

Lecomber, R., *The Economics of Natural Resources*, Macmillan, London, 1979.

Lipscy, P., Kushida, K. and Incerti, T., 'The Fukushima Disaster and Japan's nuclear plant vulnerability in comparative perspective', *Environmental Science and Technology*, 47(May), 6082–6088, 2013.

McInerney, J., 'The simple analytics of natural resource economics', *Journal of Agricultural Economics*, 27(1): 31–52, 1976.

Moran, S., 'Brazilian bill weakens Amazon Protection', *Nature–News*, 7 December 2011 (https://www.nature.com/news/brazilian-bill-weakens-amazon-protection-1.9584)

Nadkarni, M.V., 'Agricultural development and ecology – An economist's view', *Indian Journal of Agricultural Economics*, 42(3): 360–375, 1987.

Pérez-Peña, R., 'U.S. maps pinpoint earthquakes linked to quest for oil and gas', New York Times, 23 April 2015.

Richardson, D., Black, A. S., Irving, D., Matear, R. J., Monselesan, D. P., Risbey, J. S., D Squire, D. T. and Tozer, C. R., 'Global increase in wildfire potential from compound fire weather and drought', *Nature: Climate and Atmospheric Science*, 5: 23, 2022. https://doi.org/10.1038/s41612-022-00248-4

Singh-Renton, S. and McIvor, I., 'Review of current fisheries management performance and conservation measures in the WECAFC area', *FAO Fisheries and Aquaculture Technical Paper 587*, Food and Agriculture Organization, Rome, 2015.

Trenberth, K. E., 'The impact of climate change and variability on heavy precipitation, floods, and droughts', in M. G. Anderson (ed), *Encyclopaedia of Hydrological Sciences*, John Wiley & Sons, Ltd. UK, 2005.

Van Eijs, R. M. H. E., Mulders, F. M. M., Nepveu, M., Kenter, C. J., and Scheffers, B. C., 'Correlation between hydrocarbon reservoir properties and induced seismicity in the Netherlands,' *Engineering Geology*, 84(3–4): 99–111, 2006.

Voegele, E., 'Report: Canadian wood pellet exports up 5% in 2022', *BIOMASS MAGAZINE*, 17 January 2023. (https://biomassmagazine.com/articles/19674/report-canadian-wood-pellet-exports-up-5-in-2022#:~:text=Canadian%20pellet%20producers%20exported%20an,million%20metric%20tons%20in%202020)

Wheeling, K., 'How forest structure influences the water cycle', *Eos, 100*, https://doi.org/10.1029/2019EO134709. Published on 15 October 2019.

Yerkes, R. F., and Castle, R. O., 'Seismicity and faulting attributable to fluid extraction,' *Engineering Geology*, 10(2–4): 151–167, 1976.

6 Economics of Non-Renewable Resources with Renewable Services

As we observed in the previous chapter, the externalities arising from the extraction and utilisation of natural resources are important components of user costs. These externalities emerge from various economic activities that discharge pollutants into a range of KN sinks such as air sheds, oceans and a wide variety of land and subterranean resource systems. A small subset of examples are as follows.

- **Mining** involves the construction of tailing dams to treat the rock ore with chemicals so that the sought-after mineral/metal can be isolated and removed. The residue that remains within the dam is a mixture of toxic materials that have to be further treated and carefully discharged in small quantities through pumping or subaerial sprays. The usual KN sinks are the atmosphere, streams, rivers and oceans and subterranean spaces.
- **Textile and Fashion** industry generates significant waste through the increased frequency of fashion cycles and seasons. Off-cuts of fabric end up in landfills and so does much of the 'out of fashion' clothing. Increased frequency of fashion cycles does also lead to the discharge of dyes and colours – usually into streams, rivers, and lakes in the vicinity of manufacturing sites.
- **Food and Beverage** industries generate, amongst others, a significant volume of plastic waste in the form of containers and packaging. Although municipalities attempt to separate plastic material for recycling, a significant amount of plastic has found its way into landfills and oceans to the extent that marine organisms have been found to contain micro-plastics, and some extreme cases have ingested or become entangled in big junks of plastic products, often causing deformities and suffering. Households and restaurants, in mainly high-income countries, also deposit large quantities of food-waste into landfills.
- **Transport** industry is usually associated with the emission of GHGs into the atmosphere because all forms of transport on air, land and water predominantly use fossil fuel-based sources of energy. Besides the discharge of GHGs, there are numerous vehicle graveyards, where all forms of transport carriers are left to decay on land until they are collected for recycling. Although oceanic carriers have dedicated harbours for shipbreaking and recycling, a significant portion of unused material, engine oils and the like are deposited into the ocean bed.

DOI: 10.4324/9781003408574-8

- **Electricity and Power** generation is at present dominated by fossil-fuel sources such as coal and natural gas. Whilst GHGs from fossil fuel use receive much attention, electricity generation also results in other harmful pollutants such as cadmium, mercury, manganese and lead. Some of these pollutants find their way into soils, rivers and oceans and hence enter the food cycle for humans and other species.
- **Agriculture** plays the important role of feeding nations. Yet, the use of fertilisers and pesticides contaminate waterways and soil with chemicals that disturb the ecosystem balance.

The above examples are, as indicated, a small subset of a very large number of polluters and pollutants. As per our narrative in Chapter 5, of the various pollutants, GHGs have received more attention than others because of their acknowledged role as important drivers of climate change. For example, see Climate Partner (2022) for a synopsis of the Conference of Parties (COP) organised by the United Nations (UN) since 1995. However, note that our narrative in the previous chapter depicts that besides GHGs, a mixture of complex factors involving mismanagement of forests and other land and ocean systems also contribute to climate change.

Almost all KN sinks are, strictly speaking, quasi-public goods – a concept we dealt with in Chapter 4. They are also non-renewable assets. Yet they provide renewable service flows. For example, endowments such as air sheds and lakes are both quasi-public goods and non-renewable assets. Given the dynamics and linkages of the earth's biogeochemical and physical processes, these resource systems provide renewable service flows such as the air we breathe and the water we drink. That is, as we identified in Chapter 2, KN as a *sink* for waste is also a *source* of materials and amenities. The contamination of KN assets by various pollutants can compromise the renewability of the service flows and limit their capability to act as a source. So in this chapter, we examine analytic frameworks that will directly target pollution-related issues. Also note that in several instances, because of the public good properties of KN assets, pollution issues fall outside the control of individual consumers and firms. Hence the measures that deal with the issue of pollution invariably involve some form of government intervention.

We begin with a simple framework in environmental economics that provides a basis to explain how governments tend to intervene. This intervention takes the form of standards, taxes and penalties, the setting-up of mechanisms for pollution trading, and other incentives and disincentives for the use of KN goods and services. When these forms of intervention are not feasible, sometimes governments sell the rights to manage KN goods and services through a system of auctions and bidding. Further, in anticipation of increased government regulation on KN-related matters and perhaps also stewardship, some firms have taken it upon themselves to formulate business strategies that comply with international standards, such as ISO 14000, and best practice management.

All measures described in this chapter rest on the premise that if pollution levels are reduced to some minimum threshold level, then KN sinks can recoup the capability of providing renewable service flows. This is a questionable premise given the legacy levels of pollution stemming from the accumulation of pollutants over time and we return to this issue in later chapters. A new development, since around 2015, apart from 'reducing emissions', has been Net Zero Emissions (NZE). That is, the withdrawal of emissions from KN sinks equivalent to the amount discharged into those sinks. Although the literature on NZE appears to focus on GHGs and the air shed, it is possible that the concept could be extended to a wide range of pollutants and KN sinks. However, as we explain below, the concept has some inherent difficulties.

A Simple Framework for Pollution Control

Figure 6.1 illustrates a simple framework that is widely used in the economics of pollution management. Along the horizontal axis, we measure the quantity of pollution that is discharged in the context of some economic activity. We could also describe the horizontal axis (from left to right) as the quantity of discharge that is permitted. The vertical axis measures the costs associated with pollution. The curve labelled MC_A explains the marginal abatement costs; that is, the additional costs that firms have to incur (for example, in terms of using pollution control equipment and labour) for reducing emissions. The MC_A is high when the level of pollution permitted is low, and vice-versa. The curve labelled MC_P measures the marginal cost of pollution. This includes factors such as adverse health effects and the loss of ecosystem functions and other amenities. The MC_P increases as the level of pollution permitted increases.

Some claim that, when both the MC_A and MC_P are *measured correctly*, it is efficient to operate where $MC_A = MC_P$. This is because at this point the total marginal

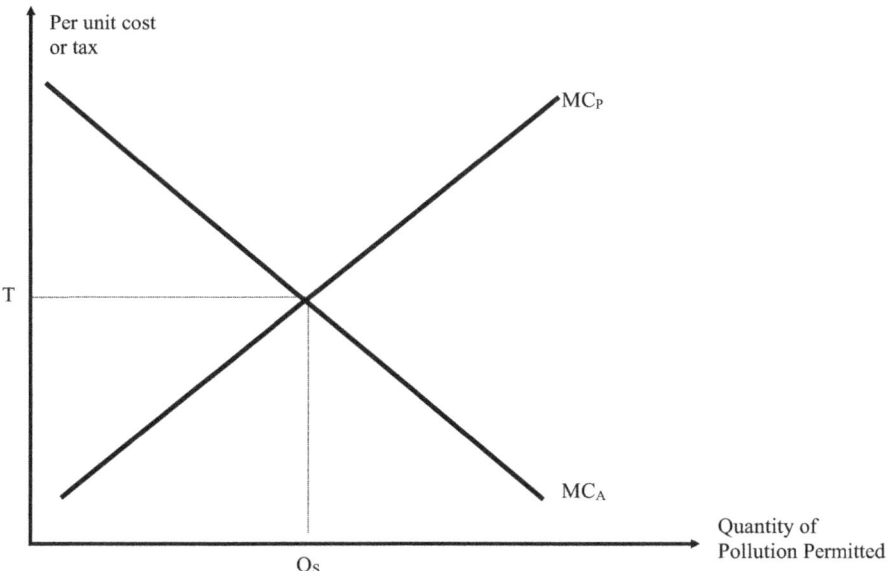

Figure 6.1 Equilibrium between Marginal Costs of Abatement and Pollution

cost of pollution that is ($MC_A + MC_P$) is at a minimum. Note that we used the phrase 'measured correctly'. This applies more to the MC_P than MC_A, because the MC_A is dependent on available technologies and the costs of these are usually known. In contrast, there is considerable uncertainty surrounding the quantification of MC_P. Not all components of MC_P are known, in part because many of these components, such as ill health and ecological damage, are difficult to quantify. Hence it is more likely that one would end up permitting the emission of more pollution than that could be tolerated. That is, if we improve on our information concerning the effects of pollution and our ability to translate these effects into costs, then we will be shifting the MC_P curve to the left, and we will permit a smaller quantity of pollution to be emitted than would be the case otherwise. At the same time, if the technologies of abatement continue to improve, the MC_A curve will shift to the left and we will be able to achieve higher levels of cost-effective pollution control.

Hence, the so-called efficient solution – where $MC_P = MC_A$ – need not necessarily be a socially optimal solution. With improved information and technology, we would be approaching this optimum. The equality of MC_P and MC_A merely represents a compromise between the polluters, usually the firms that bear the MC_A, and society at large, which bears the MC_P. This compromise is attained by minimising total cost. Despite these limitations, the framework in Figure 6.1 provides a basis for government intervention, which we will consider below.

Standards

If measurements have been properly made to apply the framework presented in Figure 6.1, then the government could either dictate Q_S as the standard or impose a tax of T to achieve the limit of Q_S. Alternatively, consider the context when only the MC_A curve is known, and scientific evidence dictates Q_S as the predetermined limit on pollution that the KN sink can tolerate. Then, as illustrated in Figure 6.2, a tax of T per unit of pollution could be levied for controlling pollution at the level Q_S.

A standard is usually a legal limit on pollution that is set by the government. Any violation of this limit can entail a range of punitive measures, including fines. When there are several polluting firms, the limit of Q_S has to be distributed among firms, and ideally, an enforcement agency has to monitor emissions to ensure that limits are not violated. In practice though, the regulatory agency sets the limits, and the polluting firms are expected to self-monitor and self-report. The agency intervenes when complaints and concerns are raised, and in some instances inspections and audits take place. The practice of self-monitoring and self-reporting raises several difficulties, and polluters are invariably tardy to acknowledge and correct their breach of limits set by standards. For example, consider the difficulties faced by the communities in the vicinity of gold and copper mining operations at the Cadia Valley of New South Wales (NSW) in Australia. Following the collapse of a tailings dam that supported the mining activity in March 2018, communities in this region have had to endure metal-dust pollution. And as Wright (2023) reports, it took several independent laboratory tests of water, blood, and hair samples to convince the regulatory body, namely the NSW Environment Protection Authority (EPA),

Economics of Non-Renewable Resources with Renewable Services 89

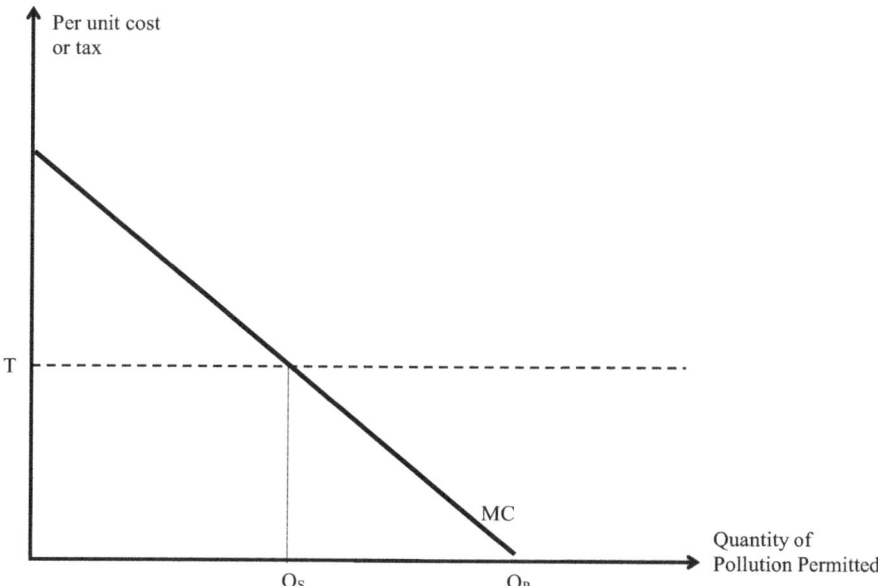

Figure 6.2 Taxes and Standards

that metal contamination in the atmosphere and water had manifold exceeded the acceptable limits. This is indeed not an isolated incident. The literature on environmental science contains a long list of breaches and consequent suffering imposed on individuals and communities. Some notable tragedies include the Exxon-Valdez oil spill in North America, a toxic gas leak in Bhopal in India, and the Chernobyl nuclear reactor accident in the Ukraine; for example, see Semenova (2020).

KN quality standards are now mandated through a range of regulatory bodies in some countries. These are mainly developed countries – especially those belonging to the OECD group. These countries have set up a range of institutions to monitor and enforce compliance with standards. For example, Australia had established a National Pollution Inventory (NPI), which is a database of pollutant emissions and discharges in Australia. This database has been used in the setting of a range of KN quality standards as a National Environment Protection Measure (NEPM). Examples for ambient air quality standards are given in Table 6.1 Similarly, water quality standards limit the salt content of river systems to 60 electrical conductivity (EC) units. Other examples of important pollutants in water quality standards are collectively termed biochemical oxygen-demanding (BOD) materials. These include phosphorus, nitrogen, heavy metals, and faecal coliforms. However, as illustrated by the Cadia mine incident above, greater coordination between polluters, regulators and the community is required.

Note that the basis for these standards rests primarily on human tolerance levels, and as indicated above, on the assumption of KN sinks being able to assimilate minimum threshold levels of pollutants. A looming controversy as of June 2023, pertains

Table 6.1 The NEPM Ambient Air Quality Standards

Pollutant	Averaging Period	Maximum Allowable Concentration	Maximum Days Per Year the Standard Can Be Exceeded
Carbon monoxide	8 hours	9.0 ppm	1
Nitrogen dioxide	1 hour	0.12 ppm	1
	1 year	0.03 ppm	None
Photochemical oxidants	1 hour	0.10 ppm	1
	4 hours	0.08 ppm	1
Sulphur dioxide	1 hour	0.20 ppm	1
	1 day	0.08 ppm	1
	1 year	0.02 ppm	None
Lead	1 year	0.50 µg/m^3	None
Particles less than 10 micrometres in diameter	1 day	50 µg/m^3	5

Source: Environment Australia, State of Knowledge Report: Air Toxics and Indoor Air Quality in Australia, Canberra, 2001.

to the release into the ocean, of very small quantities of treated radioactive sea water that was isolated in tanks for over 12 years since the 2011 Fukushima disaster. As explained by Nogrady (2023), the scientific community is divided on the safety of this release – see Box 6.1. This is because some of the contaminants do not decay easily and can pose a threat to human and marine life over long periods into the future.

Box 6.1   Controversy over Standards – Treated Nuclear Wastewater

Recall our narrative in Chapter 5 (Box 5.1) of catastrophic events that followed the earthquake and tsunami affecting Japan – and the nuclear power plant at Fukushima. As Nogrady (2023) writes, the earthquake had also triggered the explosion of overheated nuclear reactors, which had to be cooled by spraying more than a billion litres of sea water. The contaminated seawater was treated, collected, and stored in tanks. As of 24 June 2023, Tokyo Electric Power Corporation (TEPCO), which was responsible for the treatment and storage of the contaminated seawater began the release of small quantities of the treated water into the Pacific Ocean. This release, in small quantities, is expected to continue over the next 30 years.

Nogrady's (2023) narrative outlines the premises on which TEPCO plans this release as well the concerns raised by those who question the wisdom of the decision. TEPCO's decision rests on at least two premises:

- The contaminated water has been subject to several advanced treatment and filtration processes over the past 12 years resulting in the removal of most radioactive material.

> - The volume of release over a 30-year period will fall well below the regulatory limits set by Japan and that any residual radioactive material would not pose a threat.
>
> Those who oppose the release indicate the potential presence of radioactive elements that could pose a threat to human health despite the 12-year lapse since the accident at Fukushima. Nogrady (2023) identifies the residual elements as: carbon-14, iodine-131, caesium-137, strontium-90, cobalt-60, and hydrogen-3, also known as tritium. Of these carbon-14 has a half life of 5,000 years and tritium can damage DNA.
> Nogrady's (2023) also indicates that the discharge of tritium into the ocean is common practice of most nuclear power plants.

However, the mandate of many regulating institutions, have been unstable and affected by politics and calamitous events. For example, progress towards stringent targets was affected in the European Union by the global financial crisis of 2008–2009. In Australia and the United States, standards and commitments have changed with changes in politics and power.

Nevertheless, some firms seem to adopt self-regulation. One such measure pertains to compliance with a set of standards issued by the International Standards Office (ISO). Governments have also increasingly begun to recognise firms that obtain accreditation from the ISO for adopting prescribed, enhanced KN management practices. The argument is that most taxes, standards and permits achieve only a limited amount of KN improvement because they target single pollutants or single markets. Also, levying taxes diverts financial resources from firms that they could have otherwise been used to improve their KN performance. Pollution control approaches that come to play as pollution is generated are often labelled 'end-of-pipe' or 'product approaches'. In contrast, the ISO 14000 series of 'management systems approaches' provide a more proactive method of dealing with firms' performance with reference to KN. ISO 14001, for example, specifies the requirements for developing a KN management system that can be certified by an external party. This includes designing and implementing the business system. When necessary, the ISO 14001 certification also requires the adoption of the following:

i ISO 1421–24 (KN Labelling): This defines the criteria for labelling products and defines which products could be labelled, and hence is useful for business managers who are considering product differentiation.
ii ISO 1441–44 (Life-Cycle Assessment): This provides the basis for firms to evaluate the outcomes of their decisions within the framework of the 'cradle to grave' concept; that is, from the use of raw materials to the disposal of the product at the end of its life.

92 *Microeconomics and KN*

iii ISO 1410–12 (KN Auditing): This provides the basis for auditing the KN management system and the criteria for identifying the persons who could qualify as auditors who are able assess the status of KN.
iv ISO 1431 (KN Performance Evaluation): This provides guidelines for a business to set up a system to evaluate its own KN performance through a range of measures and indicators.

ISO 14001 certification is given to a firm that develops a KN management system to proactively control factors that would generate damage KN if left uncontrolled. The certification rests on the premise that the firm is able to identify causes of undesirable KN related events and to control them as a precursor, rather than taking action after the event occurs. So, in the context of gold mining, the construction of a tailings dam for cyanide treatment in a high rainfall or snowmelt area would need to satisfy stringent engineering design standards. The management plan must show how the firm proposes to avoid the collapse of the dam and also detail the immediate contingency measures it has in place should a collapse occur. In order to obtain the certification, a gold mining firm would have to show how it carried out ISO 1441–1444 and at the same time specify how it will put into practice ISO 1410–1412 and ISO 1431.

Taxes and Charges

Some economists argue that a tax on pollution could achieve a greater level of pollution abatement than a standard would. To illustrate this argument, we need to first consider how a pollution tax is supposed to work.

In Figure 6.2, we consider an individual firm and its marginal abatement cost (labelled MC_A). Suppose that this firm's production practices involve maximising profits at an emission level of Q_P units of pollution. We have assumed that at Q_P units of pollution, MC_A is zero, implying that profit maximisation is achieved when no pollution is controlled.[1] If the government now levies a tax of T per unit of pollution emitted, then the firm is expected to respond as follows:

1 The firm will make a tax payment of $(T*Q_S)$ for the first Q_S units without abatement. Because $(T \geq MC_A)$ over the domain of pollution levels $(0 \leftrightarrow Q_S)$, it is cheaper to pay the tax than to abate pollution up to Q_S.
2 It will clean up the remaining $(Q_P - Q_S)$ units of pollution at its own abatement costs because it is cheaper to abate this quantity rather than to pay taxes for them; that is, for $(Q > Q_S)$, $(MC_A < T)$.

This response of paying taxes for the pollution quantity Q_S, and then abating the remaining quantity $(Q_P - Q_S)$, appears cheaper than paying a tax of $(T*Q_P)$ for the total quantity (Q_P) of pollution. We can also see how a greater level of pollution gets abated with a tax than the context where a standard is not accompanied by a tax. Suppose that the standard was specified for the firm as Q_S. Then the firm can pollute up to Q_S, and adopt abatement only when its pollution exceeds Q_S. The firm will abate $(Q_p - Q_s)$ units on its own, whilst the government requires its tax collection to abate further units of pollution.

Some of the KN taxes that presently operate in many countries are:

- Effluent charges
- Emission taxes
- Load-based licensing schemes covering air and water pollution
- Product taxes on ozone-depleting substances

Effluent charges, emission taxes and load-based licences can be easily related to the framework in Figure 6.2. That is, the polluter's liability is determined directly in terms of the quantity of pollution and the predetermined tax rate. For emission taxes to work, it should be possible to measure the emission levels such as discharges into air sheds, water and land resources. Usually, the emission taxes are applied in the context of point and stationary sources of pollution, due to the ease of monitoring. With non-stationary sources of pollution such as moving vehicles, it is easier to apply a product tax so that liability is transferred to the price of the product generating the pollution. For example, gasoline carries a product tax because it is combusted by vehicles and generates soot and other air pollutants that affect local air quality and thus public and ecosystem health, as well as emissions of GHGs that affect global climate. In the case of product taxes, those commodities that cause more harm to KN would command a higher price. For example, until around 2002, in most countries the price of leaded petrol was higher than that of unleaded petrol due to product taxes. Since 2002, leaded petrol has been completely phased out.

The potential to achieve greater efficiency through taxes on pollution depends very much on how these tax earnings are used in the wider community and also on how firms respond to the taxes. For example, Hamilton, Hundloe and Quiggin (1997) argue that KN based taxes, if used sensibly, can benefit society in a wide range of ways, including an increase in employment and the promotion of cleaner production methods. In a modelling exercise Hamilton, Hundloe and Quiggin (1997) illustrate that, if KN based taxes are returned to the community and labour and payroll taxes are removed, then a net gain in employment that ranges between 100,000 to 150,000 can be achieved. In the context of being faced with taxes and penalties, many firms have responded by improving their production methods. For example, steel makers in Australia began reusing wastewater for cooling their slag heaps following the introduction of penalties for water pollution (James 1997).

Taxes, like standards, suffer from operational difficulties. Both require well-advanced methods of monitoring, measurement and enforcement. Further, these require the existence of well-established institutions. Almost all countries now have a national agency dedicated to KN issues, but considerable variation exists among them in terms of their mandates and capabilities. As indicated in the example in Box 6.2, China has embarked on an aggressive programme to monitor and maintain inventories of emissions and other pollutants so that emissions trading can progress effectively.

> **Box 6.2 Emissions Trading in China and the USA**
>
> Emissions trading schemes (ETSs) are being rapidly developed in China. As of November 2014, there are seven ETS pilots in operation, respectively in Beijing, Tianjin, Shanghai, Chongqing, Guangdong Province, Hubei Province and Shenzhen. Building on the experiences of the pilot systems, the Department of Climate Change, under the National Development and Reform Commission, has stated publicly that a national emissions trading scheme will be established in 2016.
>
> Wang (2015)
>
> China has become the world leader in renewable energy deployment due to strong government support, such as the use of subsidies to renewable energy production. However, experts are concerned that existing sources of revenue to support renewable energy, primarily the renewable energy surcharge on electricity consumption, are not sufficient to cover the cost of implementing national renewable energy targets. In this context, the International Institute for Sustainable Development (IISD), in collaboration with the China National Renewable Energy Center (CNREC), has embarked on a project to review the possibility of earmarking the revenue from emission trading to subsidise renewable energy in China.
>
> Source: https://www.iisd.org/sites/default/files/publications/emissions-trading-in-china-renewable-energy.pdf
>
> In the USA, where scepticism continues to prevail in national-level policy circles about climate change and what to do about it, seven states had come together in 2005 to establish a 'Regional Greenhouse Gas Initiative' (RGGI). RGGI is a cap-and-trade scheme for carbon emissions from power generation among generators in the north-eastern part of the USA (for details see Ruth et al. 2008). Receipts from sales by the government of emission allowances are used to stimulate a transition towards renewable energy sources, as well as energy efficiency improvements not only in the electricity sector but across a wide range of energy-intensive activities of industry and households as well. Rules by which emissions allowances are allocated to generators and how revenues are distributed, for example, are specific to each of the states participating in RGGI, with markets serving as key institutions to reconcile decisions made within states, across states, and across time.

Emissions Trading

To illustrate the principles underlying emissions trading, suppose that an industry consists of only two firms. Further, assume that the government's target is to achieve an overall standard of Q_S. If this standard is equally divided between the two firms, then the pollution limit on each business is $(Q_S/2)$ and the government

Economics of Non-Renewable Resources with Renewable Services   95

sells to each firm pollution permits to the value of $\{T_0*(Q_S/2)\}$, where T_0 is the per unit value of a pollution permit. This is illustrated in Figure 6.3. As in the previous section, we will also suppose that both firms maximise their profits if they are each allowed to pollute up to Q_P units of pollution, and of course Q_S < {total maximum potential pollution load = $(2Q_P)$}.

In Figure 6.3, MC_{A1} shows the abatement costs for Firm 1, which uses more expensive, older technology than Firm 2, whose abatement costs are represented by MC_{A2}. Figure 6.3 also specifies that each firm faces a per unit pollution permit price of T_0. Following the line of reasoning presented under 'Taxes and Charges' above, Firm 2 would prefer to discharge Q_{P2} units of pollution with its pollution permits and control $(Q_P - Q_{P2})$ units of pollution using its own technology. By the same token, Firm 1 would prefer to control $(Q_P - Q_{P1})$ units of pollution using its own technology and discharge Q_{P1} units of pollution at a per unit cost of T_0. Given that each firm has pollution permits for controlling $(Q_S/2)$ units of pollution, we find that Firm 2 has a surplus of permits $((Q_S/2) - Q_{P2})$, while Firm 1 has a shortage of permits $(Q_{P1} - (Q_S/2))$. As long as $[(Q_{P1} + Q_{P2}) < Q_S]$, Firm 1 can decrease its total costs by buying permits from Firm 2. This in essence is the principle underlying

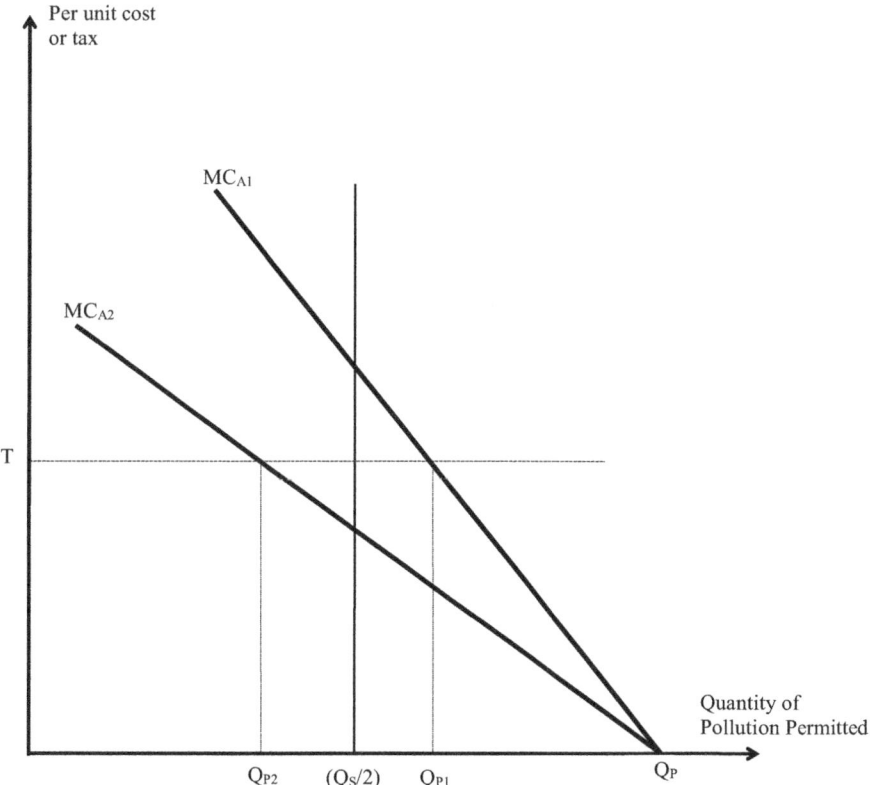

Figure 6.3 The Basis for Emissions Trading

emissions trading. Less efficient firms need not necessarily leave the industry when facing stringent KN standards. They can continue their existence as long as there are some more efficient firms in the industry, and this extended time period enables them to make the necessary adjustments.

Tradable permits have been in operation in the United States (US) for nearly three decades and have been implemented by the US Environment Protection Agency under two programmes: *bubbles* and *offsets*. A bubble refers to a specific geographic region; trading within the bubble is aimed at reducing the pollution levels of that specific region. The offset programmes apply to new firms, generating new emissions; these are permitted to operate only as long as they are able to reduce emissions from existing sources. Large firms have also used both the bubble and the offset frameworks for internal trading. That is, they trade pollution permits among various branches of the same firm.

The significant advantage of emissions trading has been the savings in abatement costs. Hahn and Hester (1989) estimate the cost savings for hydrocarbon emissions in one corporation alone to be around US$ 60 million. However, despite these stated advantages, there are difficulties in this system that pertain to monitoring, detection, and enforcement. It is also important to appreciate that the standard Q_S, which indicates a given tolerance level, will not remain stationary, but most likely shrink over time. This is because emissions trading, like the measures considered above, allows the continuation of discharges that could accumulate in KN sinks. Such accumulation constrains the provision of service flows from KN assets. Real difficulties emerge when the concepts of trading in pollution are stretched to extreme limits as shown in Box 6.3.

Box 6.3   An Extreme and Flawed Adaptation of Trade in Pollution

An extreme and flawed adaptation of the theory of pollution trade can be found in the text (quoted in Enwegbara 2001) of a leaked memorandum from Lawrence Summers as chief economist of the World Bank:

> 'I think the economic logic behind dumping a load of toxic waste in the lowest-wage country is impeccable and we should face up to that.' …. 'I've always thought that under-populated countries in Africa are vastly under polluted.'

The trade in toxic wastes had begun well before the infamous memo from Summers. As detailed in Okaru (1993), in 1988 alone West African countries had imported 24 Million Tons of toxic waste. Further, Okaru (1993) provides a comprehensive list of violations and lack of care in the trade of hazardous toxic wastes. In the context of lax legislations and the lack of stewardship, the notion of residual externalities is indeed meaningless.

In some ways, the offset arrangements represent a precursor to Net Zero Emissions that we consider next.

Net Zero Emissions

To define Net Zero Emissions, one could draw an analogy, in the first instance, from the concept of Net Zero Electricity. If a building draws X Kilo Watts of electricity from the grid, and is then, capable of putting back X Kilo Watts into the same grid – then the building's electricity utilisation is Net Zero. Several builders and architects have embraced this possibility by installing solar panels and/or mini wind turbines on their buildings. Along the same lines, Net Zero Emissions is a state where the emission of X Kilo Tons of a pollutant into a KN sink is countered/negated by the reabsorption of X Kilo Tons from that sink. But the question is – how and where would those Kilo Tons of contaminants be reabsorbed into? Herein lies the problem. The Reabsorption Sink and the Emission Receptor-Sink are not necessarily identical entities – unlike in the electricity analogy. Further, because the UN's current focus is on GHGs, the Emission Receptor-Sink that is mostly discussed is the atmosphere/airshed.

Schumer and Lebling (2022) provide a set of Net Zero Emission options ranging from hardcore engineering to nature-based solutions. The engineering methods include installation of devices for direct air capture. Nature-based solutions for GHGs rest on the premise that the pollutants can be reabsorbed by a range of sinks such as: dense forest stands, phytoplankton in the oceans, mangroves, and peatlands. Not only are these entities different, but they are also diffused (non-point) entities. Therefore, concrete examples with calibrated measurements are difficult to formulate. The language often borders on vague generalities.

Nevertheless, prominent among nature-based solutions are oceans because they carry phytoplankton, which are natural carbon receptors. Unfortunately, significant phytoplankton loss occurs due the dumping of pollution into the oceans. Notable pollutants include sewage (partially treated or otherwise) and plastic. Countries like Belgium, United Kingdom and Singapore have made remarkable progress in tertiary sewage treatment and reclaiming potable water. If this example is replicated across the world – especially coastal cities – then phytoplankton recovery could be large enough to deter climate change. Similar observations apply to rivers, lakes, and streams. The pollution of these bodies water, also disturb the hydrology. A disturbed hydrologic system is a driver of climate change.

As Levin et al. (2023) explain, the Net Zero Emission policy is much more than the mere balancing of emissions and reabsorptions. It also includes reducing emissions to a specific target such that the balancing process would not permit global warming to exceed 1.5°C. The target for emission reduction is around 50 per cent of current (2023) levels. Although several countries have ratified the Net Zero Emission policy, they have set the timeline for emission reduction far into the future – mostly the year 2050. China and India have nominated 2060.

One clear strategy for emission reduction is to move away from fossil-fuels for electricity and power generation as well as transport. In this context, Electric

Vehicles *(EVs)* are naturally an attractive option on face value. This is because EVs do not directly emit GHGs. But the current EV technology depends on power generation and batteries, and the latter in turn require lithium that must be mined. Mining does inevitably involve clearing of vegetation and forests and hence reduces the size of the Reabsorption Sink. However, the batteries have the potential to be recycled – but not without generating toxic wastes and deepening dependency on non-renewable resources.

Direct Air Capture (DAC) technology deserves a comment. There are two aspects to consider. The first is that if there is a technology to capture the emitted volume back from the atmosphere – then there must be a technology to prevent the emission going up and instead reusing it elsewhere. In fact, there is, and this is the content of the circular economy and closed-loop production systems. We consider this in our policy chapter. The second is that there is already an accumulated volume of GHGs in the atmosphere/airshed. Recall our discussion in Chapter 4 of residual externalities that accumulate. This is referred to as legacy emissions. DAC strives to remove/reduce these. The premise is that ambient air contains CO_2 in a diffused form. This air is then passed through the DAC system, which removes the CO_2 through a set absorption devices. CO_2-free air is then released into the atmosphere, and the captured CO_2 is either stored underground or diverted to other uses such as fuels and building materials; (Ozkan et al. 2022). Besides the cost-effectiveness of the method, ecological effects of underground storage need to be understood clearly.

Property Rights

Almost all KN goods and services lack the essential features required for a system of private property rights to function effectively. Recall from Chapter 3 that these are *enforceability*, *exclusivity* and *transferability*. Enforceability means that rights of ownership can be enforced by law, while exclusivity refers to the exclusiveness of the ownership rights once they have been enforced. Transferability refers to the ability to transfer the ownership rights at a price. These three conditions of property rights are satisfied only when the goods concerned can be exchanged through the market system. That is, a system of private property rights and the market mechanism work together in unison when goods are exchanged in the market.

The property rights approach to solving pollution problems is based on the premise that, if ownership rights to KN can be offered (at a price) to a private entity, then a market for the utilisation of KN can emerge. The goal of sustaining this market will in turn necessitate the proper maintenance of KN. As indicated earlier in Chapter 3, this approach has worked in certain situations such as beaches, lakes, forests and even river systems and watersheds. The approach has basically involved auctioning the right to manage the item of KN (say a beach or a forest) and offering this right to the highest bidder who bids above a reserve price. On the downside, there are other difficulties such as the emergence of income inequalities that are triggered by the financial transfer from one industry or firm to another, and considerable transaction, monitoring and enforcement costs. Such costs may consume large portions of government revenues that are generated from the sale of auctioning the rights to manage KN.

Other Incentives and Disincentives

Governments have influenced firms and individuals in several other ways to improve their performance with reference to managing KN. These include income tax credits and subsidies for pollution control, the provision of recycling systems and other favourable KN management strategies. Some of the incentive schemes that are currently operational are:

- Rate concessions by local governments for sustainable land management
- Subsidies and grants for tree planting and vegetation protection
- KN performance bonds for firms
- Beverage container deposit refund schemes

While the role of subsidies, concessions and refunds is self-explanatory, the functioning of performance bonds warrants a brief explanation. The basic principle is similar to that of a rental bond. Firms need to lodge a bond with the appropriate administrative authority, prior to commencing operations, as a guarantee for adopting specific KN protection measures. The bond is forfeited on a proportional basis in relation to the extent of protection measures that have been adopted. The bond is returned in full if the protection measures demonstrate full compliance. In some countries, performance bonds have been used in the context of mining operations. For example, in the United States, the Office of Surface Mining and Reclamation issues such bonds, and mining firms have to demonstrate satisfactory mine rehabilitation measures to recoup their bonds. Besides, in some countries, the finance sector has begun to include KN-related factors in loan and insurance applications. For example, firms that demonstrate safeguards against accidents involving KN can gain lower insurance premiums. Homebuyers could benefit from cheaper home loans when opting for KN-friendly house designs. A newly emerging field is that of Environmental Finance. Daily and Ellison (2012) in their book titled *The New Economy of Nature* outline various ways by which markets can unfold for ecosystem services. These include the sale of 'green bonds' and industrialists paying for nature conservation in order to garner waste treatment services.

In some instances, governments have been compelled to examine their own actions and demonstrate cross-compliance, especially when some of their own actions prompt KN degradation, while punitive measures for such degradation are imposed in other contexts. For example, a farmer may receive a fertiliser subsidy, while at the same time facing a penalty for emitting discharging contaminants that include chemicals that constitute the fertiliser into an adjoining river system.

In the material covered thus far from Chapter 3, we have followed the line of reasoning offered in most environmental economics texts. We now need to examine how the laws of thermodynamics and the principles of ecological resilience introduced in Chapter 2 would lead to a revision of at least some of the concepts and policies presented so far. For example, the content of this chapter dealt primarily with reducing emission loads through various means. This does not mean that KN sinks will not get filled up over time. We turn our attention to such considerations in the next three chapters.

Review Questions

1 Some authors have argued that the enforcement of standards and strict regulation are good for industry, in that industry becomes more efficient and competitive. Do you agree with this statement?
2 Discuss with examples the advantages and disadvantages of a system of marketable pollution permits.
3 Discuss the feasibility and desirability of imposing higher taxes on private transport and using the tax earnings to subsidise public transport.
4 Discuss the statement that Net Zero Emissions cannot be a viable policy option without significant reductions in the current emission regime.

Note

1 This is a false implication because pollution will adversely affect the polluting firm's performance as well.

References

Climate Partner, *From Pledges to Binding Commitments – A Review of Key Milestones in Previous COP Conferences,* October 2022 (https://www.climatepartner.com/en/climate-action-insights/complete-review-of-key-milestones-from-previous-cop-conferences)
Daily, G. and Ellison, K., *The New Economy of Nature*, Island Press, Washington, DC, 2012.
Enwegbara, B., 'Toxic colonialism: Lawrence Summers and let Africans eat pollution', *The Tech*, 121, 2001. (http://tech.mit.edu/V121/N16/col16guest.16c.html)
Hahn, R. W. and Hester, G. L., 'Marketable permits: Lessons for theory and practice', *Ecology Law Quarterly*, 16(2), 361–406, 1989.
Hamilton, C., Hundloe, T. and Quiggin, J., Ecological tax reform in australia: using taxes and public spending to protect the environment without hurting the economy, *Discussion Paper 10*, The Australia Institute, Canberra, 1997.
James, D., Environmental incentives: Australian experience with economic instruments for environmental management, *Environmental Economics Research Paper No. 5*, Environment Australia, Canberra, 1997.
Levin, K., Fransen, T., Schumer, C., Davis, C. and Boehm, S., 'What does "net-zero emissions" mean? 8 common questions, answered', *Insights, World Resources Institute*, 20 March 2023.
Nogrady, B., 'Is Fukushima wastewater release safe? What the science says', *Nature*, 618, 894–895, 2023 (https://doi.org/10.1038/d41586-023-02057-y)
Okaru, Valentina O. 'The basil convention: Controlling the movement of hazardous wastes to developing countries', *Fordham Environmental Law Review*, 4(2), 137–165, 1993.

Ozkan, M., Nayak, S. P., Ruiz, A. D. and Jiang, W., 'Current status and pillars of direct air capture technologies', *iScience,* 25(4): 1–23, 2022.

Ruth, M., Gabriel, S., Palmer, K., Burtraw, D., Paul, A. and Chen, Y. et al., 'Economic and energy impacts from participation in the regional greenhouse gas initiative: A case study of the state of Maryland', *Energy Policy*, 36(6), 2279–2289, 2008.

Schumer, C. and Lebling, K., 'How are countries counting on carbon removal to meet climate goals?', *Insights, World Resources Institute*, 16 March 2022.

Semenova, G., 'Environmental disasters as a factor of environmental pollution' *E3S Web of Conferences,* 217, 04007, 2020 (https://doi.org/10.1051/e3sconf/202021704007)

Wang, H., *Emissions Trading in China: An Opportunity for Renewable Energy? A Survey of Expert Opinions*, IISD Report, International Institute for Sustainable Development, April 2015. (https://www.iisd.org/system/files/publications/emissions-trading-in-china-renewable-energy.pdf)

Wright, I., 'Huge Cadia gold mine ordered to reduce polluting dust. Is it safe to live near a mine like this?', *The Conversation*, 22 June 2023 (https://theconversation.com/huge-cadia-gold-mine-ordered-to-reduce-polluting-dust-is-it-safe-to-live-near-a-mine-like-this-208111)

7 Consumer Demand and KN

Recall from Chapter 3, the distinction between a *good* and a *bad*. This distinction enables specific axioms pertaining to consumer behaviour to be developed. To recapitulate, a *good* is one where 'more' is preferred to 'less', and a *bad* depicts the reverse. Most consumer items such as food and clothing are *goods*, and we would usually prefer to have 'more' rather than 'less' of them. Pollution and acts of crime are clearly *bads*, and we would definitely prefer to experience 'less' rather than 'more' of these.

We focus on two axioms, namely those pertaining to comparison and transitivity. The *axiom of comparison* dictates that individuals can compare different consumption baskets – assemblages of actual or potential goods and services – and can display preference or indifference. For example, one individual might prefer to have a bowl of porridge along with a cup of tea for breakfast, whilst another might prefer a plate of bacon and eggs accompanied by a cup of coffee. The perception of good and bad is inherent in the display of preferences. One might also infer that there are different shades of good and bad in the declaration of preferences. The *axiom of transitivity* enables preferences to be placed in a consistent order. For example, suppose that an individual states that he/she prefers basket-A to basket-B, and further prefers basket-B to basket-C. Then, the transitivity axiom deems that this individual cannot prefer basket-C to basket-A. These axioms together provide a foundation for the development of the standard models of economic choice by consumers, who are assumed to be rational and who are seeking to maximise the enjoyment (utility) they derive from the consumption of goods and services. However, preferences need not be stationary – they can indeed change with improved information as we note below.

How would an awareness of the laws of thermodynamics and their implications for sustainability change consumer preferences and attitudes, and thus the standard instrument set deployed in microeconomics to identify optimal behaviour of rational consumers? As we have seen in Chapter 2, all processes carried out in an economy – production and consumption of goods and services – require materials and energy and inevitably result in waste products. The entropy generated by these processes is always greater than the order generated by them. For example, toothpaste and socks and television sets and many more of the products generated in an economy could be thought of as highly organised assemblages of (desirable) products. But achieving that state of desirability requires diffusion of materials and energy in the form of effluent discharge into water bodies, air emissions, release of waste heat and more. And after enjoyment of these products they are discarded, and even if they are not returned into KN sinks, the processes by which to capture, recycle or re-use them also require energy, which in turn is dissipated in the form of waste heat. In short, the laws of thermodynamics, when considered in the context of sustainability, suggest that less rather than more consumption is better, and thus the distinction between a *good* and a *bad* becomes tenuous. Furthermore, the stylistic example of baskets of goods mentioned above, wouldn't just include the 'goods themselves' – toothpaste and shoes and automobiles, etc. – but comprise alongside them the impact on KN caused by their production, use, and disposal, including the long-term harms (bads) this all causes, from public health impacts to climate change to biodiversity loss and beyond.

The axioms of consumer behaviour are central to the depiction of consistent consumer preferences. Consistency here refers to the trade-offs an individual is willing (and able) to make between desirable goods when preferences and incomes are given. But growing awareness concerning the sustainability of KN, for example biophysical limits on resource availability and waste absorption, may undermine that consistency. Complicating matters even further is the fact that consumers must also consider the timing of the benefits and the costs that are associated with the options available to them, as well as the geographic location and extent over which adverse impacts on KN may manifest themselves.

In this chapter, we focus on consumers' increased awareness about the importance of KN and the consequent changes in their preferences and substitution behaviour. We begin by summarising the following salient aspects of consumer demand theory as exposited in most standard texts:

- Conceptual premises of consumer demand, namely utility functions, indifference curves, budgets and substitution effects
- Factors that are influencing a demand function, such as prices, income, knowledge, and advertising
- Elasticity of demand and business profits

An important theme underlying the arguments developed here surrounds the notion that, using standard economic instruments – such as the concept of utility functions – one will arrive at very different insights if the KN context of human actions is considered as part of the decision-making process. It is noteworthy that behavioural economists (for example Kahneman et al. 1990) have questioned the basic premises of utility theory – especially that of rational economic behaviour. Such behaviour rests on the assumption that consumers are driven by utility maximisation in the context of self-interest and present gain. Our arguments, in this and subsequent chapters, lend credence to the notion of bounded rationality that behavioural economists exposit.

104 *Microeconomics and KN*

In order to explain the linkages between consumer demand and enhanced awareness of KN and sustainability issues (hereafter referred to as KN awareness), the following three items warrant consideration.

- First is the illustration of how increased KN awareness could be explained through the theory of indifference curves in conjunction with some observations from behavioural economics. This involves a concept termed the endowment effect and the disparity between willingness to pay (WTP) for a good or service, or for the reduction or avoidance of a bad, and willingness to accept (WTA) the loss of that good or service, or the incurrence of harm.
- It is possible to show that the market demand curve, which most decision makers in business study for the purposes of developing their strategies (for example, product development by business managers and regulation by social planners), can be easily influenced by the KN awareness of consumers as well as the KN attributes of the good in question.
- In just the same way as information on price and income elasticity of demand is important for both social and business decision-making, it is pertinent to argue that information on a new concept such as *the consumer elasticity of KN-friendliness* may prove useful for decisions involving sustainability considerations.

We will deal with each of these considerations in turn.

Utility Functions, Indifference Curves and Demand

The Standard Theory

Standard microeconomic theory assumes the existence of a rational, utility-maximising consumer whose choices are constrained by a fixed budget. The underlying utility function of a consumer is supposed to be a smooth, well-behaved function that displays diminishing marginal utility. When the consumer's consumption basket is limited to two goods, say A and B, such a function $\{U(Q_A, Q_B)\}$ will be a three-dimensional surface displaying the following properties:

$$\frac{\partial U}{\partial Q_A} > 0; \quad \frac{\partial U}{\partial Q_B} > 0; \quad \frac{\partial^2 U}{\partial Q_A^2} < 0; \quad \frac{\partial^2 U}{\partial Q_B^2} < 0 \tag{7.1}$$

These properties imply that the utility surface would resemble a hill containing both ascending and descending sections. Note that the descending section of the surface will correspond with decreasing utility, whilst the ascending section portrays the reverse. Given the assumption that consumers wish to maximise their utility, economists focus on the ascending section of the surface as illustrated in Figure 7.1. A horizontal cross-section of this surface will result in a contour describing how different combinations of A and B would yield the same level of utility. For example, the contour labelled CC in Figure 7.1 describes different combinations

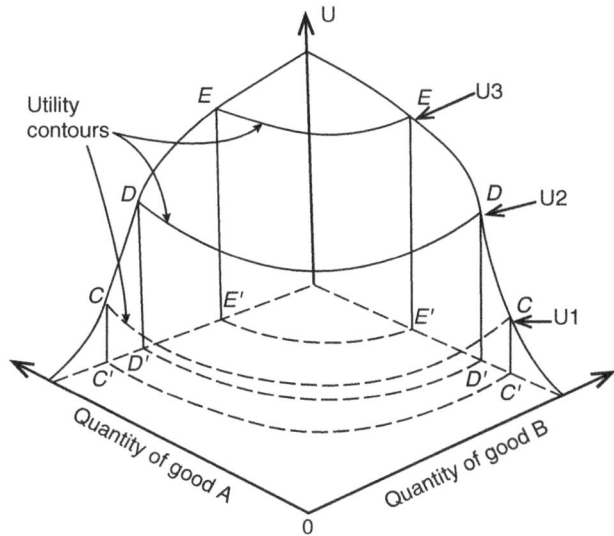

3-Dimensional Utility Surface
Adapted from: Hirshleifer, J. et al, Price Theory and Applications, Cambridge University Press, 2014.

Figure 7.1 Utility Surface for $\{U(Q_A, Q_B)\}$

of A and B yielding the same level of utility U_1. For this reason, each contour is named an indifference curve. That is, a consumer will be indifferent between various combinations of A and B as depicted by a single indifference curve (contour) because they provide the same level of utility. The indifference map is derived by taking a successive set of cross-sections (or contours) on the surface. And note that contours derived by taking horizontal cross-sections on the ascending portion of the surface are convex to the origin. Further, note that the indifference curves become smaller as the level utility gets higher – that is, as one reaches the higher end of the surface.

With reference to an indifference map, a consumer would prefer to be on the highest possible indifference curve. However, this preference is constrained by the consumer's resources – specifically his or her budget. Suppose that a consumer's budget is $Y for a consumption basket of two goods A and B, prices of which are respectively $$P_A$ and $$P_B$. The consumer's budget constraint is given by the inequality $\{(P_A Q_A + P_B Q_B) \leq Y\}$. The upper limit of this constraint is one where the inequality becomes an equation and is referred to as the budget line. As explained in most standard texts, a consumer maximises utility at the point where the budget line is tangent to the highest attainable indifference curve. Further, standard theory also illustrates how a price consumption curve can be derived for one of the two goods. For example, consider good B in Figure 7.2. By holding the price of good A constant and varying the price of good B, it is possible to trace a locus of tangencies between the budget

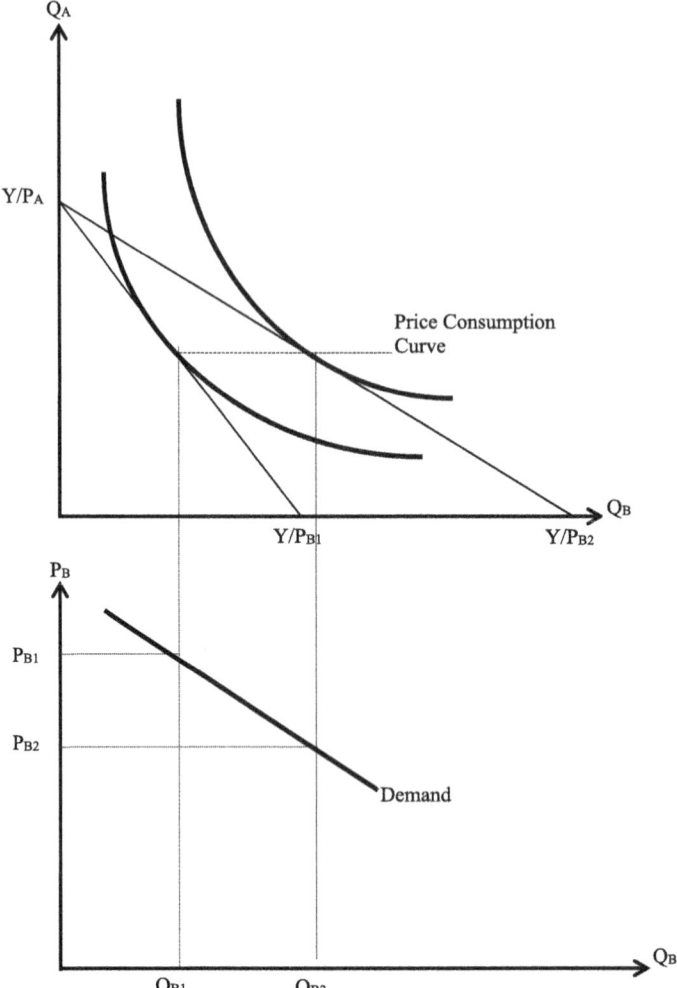

Figure 7.2 The Derivation of the Demand Curve

line and the highest feasible indifference curve. As illustrated in Figure 7.2, this locus of tangencies becomes the basis for deriving the demand curve for B.

Several permutations and analyses are possible from hereon. The position and shape of the budget line is influenced by changes in the prices of the goods and consumer's income. The latter will determine budget allocation and consequently shift the budget line. An increase in income will shift the budget line outwards. However, the resulting quantity of consumption would depend ultimately on the shape of the indifference curves and the elasticity of demand with reference to income. We consider these possibilities below. For the moment, we note the distinction between *normal goods, luxury goods* and *inferior goods*. With normal and luxury goods, an increase in income results in increased consumption, whilst the reverse unfolds for inferior goods.

In common language, the demand curve explains an individual's WTP for an extra unit of a good or WTA for the loss of that unit. At any consumer equilibrium, WTP = WTA. The WTP or WTA decreases as the quantity of consumption increases, due to the law of diminishing marginal utility. That is, the more a consumer already enjoys of a particular good, the less will be the contribution of an additional unit of that good to the consumer's enjoyment. The theory of indifference curves can explain this relationship mathematically and graphically.

Decision makers are ultimately interested in the nature and responsiveness of consumers' demand to pertinent parameters such as incomes, prices of the good or its complements and substitutes. In this regard, it is important to appreciate that the underlying utility function and the indifference curves themselves provide useful information for decision-making. For example, the slope of the indifference curve measures the marginal rate of substitution (MRS) and information concerning MRS can be useful for product development. However, from an empirical point of view, demand equations are derived usually from market and household surveys using econometric methods, and decision makers gain their information on substitution behaviour through the concept of cross-elasticity of demand, which we will consider later.

Even if we accept the basic assumptions that drive the applications of indifference curves, the theory that underlies them, as it is typically applied, needs modification to account for increased awareness concerning the importance of KN. Such modification is feasible by recourse to findings from the subdiscipline of behavioural economics. Of particular interest to us here is a behavioural attribute that some economists have labelled the endowment effect, which leads to the often-observed situation where *an individual's WTP for a good is consistently less than his or her WTA for the loss of that very good.* We will consider this next.

Adapting the Standard Theory for Effects Pertaining to KN

The disparity between WTP and WTA has been reported extensively (see, for example, Bateman et al. 1997; Franciosi et al. 1996; Kahneman et al. 1990; Knetsch 1989, 1995; Knetsch and Sinden 1984). In the literature, the cause of the disparity has been largely attributed to the *endowment effect* (Kahneman et al. 1990; Knetsch 1989). This term refers to consumption behaviour stemming from an individual's possession of a good. Generally, the effect is one where an individual is reluctant to part with the good after taking possession of the good. Once in one's possession, known and appreciated, sentimental or other attachments may lead consumers to request higher compensation to forego that good than what they were prepared to pay for it in the first place. It is due to this reluctance that WTA usually exceeds WTP. With a set of simple assumptions, it is possible to show how the endowment effect can be internalised into the theory of indifference curves. Then the derivation of demand curves within the framework of substitution behaviour, utility maximisation and a budget constraint demonstrate the disparity between WTP and WTA (Thampapillai 2000).

108 *Microeconomics and KN*

Important for us of course is the existence of the WTP-WTA disparity when a consumer substitutes between KN-Friendly (KNF) and KN-Unfriendly (KNUF) goods. The distinction between KNF and KNUF is necessarily relative. The production and consumption of KNF goods would inflict less damage on KN compared to KNUF goods. Larranaga and Valor (2022) examine the various attributes that consumers search for in differentiating between KNF and KNUF goods. Some of these attributes are indicated in Box 7.1. The context of absolute superiority could only exist if the production and consumption of KNF goods could enhance the size and quality of KN sinks and sources, and examples of these are indeed rare.

Box 7.1   KN-Friendly Attributes of Consumer Goods

Have you checked the labels of goods on your supermarket shelves? Many producers, no doubt, attempt to capture consumers' attention by highlighting the KN-friendly attributes of their goods. Some of these are as follows:

- Dolphin-safe tuna
- Teabags made of unbleached paper
- Shoes from recycled raw material
- Biodegradable detergents
- Organic farm produce
- Recycled paper
- Gin and cheeses made with 100 per cent renewable energy (often through the use of by-products from the agricultural processes on which they rely).

And the demonstrations of KN-friendly attributes are not confined to the supermarket shelves.

- Several automobile manufacturers have achieved significant resource efficiency in terms of water and fuel use and market themselves for having reduced the lifecycle costs of their products.
- Some electricity retailers now offer green tariffs to consumers in order to increase the renewable energy content of electricity. The offer of these green tariffs and the capability to generate green power is seen as a means of attracting KN-friendly consumers.

Pertinent Assumptions

Here we suppose that a consumption basket can be regarded as an endowment because a consumption basket is akin to a lifestyle to which individuals are accustomed. To illustrate this, consider varying endowments involving two goods A and B, where A is a KNUF good and B a KNF good. For example, when a person has become accustomed to consuming six units of A and four units of B on a regular basis, this consumption pattern or basket – which has become a specific lifestyle – can be regarded as similar to an endowment. That is, the person would be reluctant to change his/her accustomed lifestyle. Samuelson and Zeckhauser (1988) describe such reluctance as status quo bias.

As in standard theory, an individual displays a specific set of indifference curves, but the shape of these curves is governed by the individual's endowments. Specifically, the shape of the indifference curves will be such that the individual places a higher value on the loss of a good than on the gain of the other with reference to the endowment. To illustrate, consider Figure 7.3, where three different endowment positions **a**, **b** and **c** are shown.

For consumption behaviour involving the endowment at **a**, an individual displays a set of indifference curves labelled (I_1, I_2,...) as opposed to the set of indifference curves labelled (II_1, II_2,...), which describe consumption behaviour for the endowment at **b**. Note that the endowment at **a** is reflected by a greater consumption of A relative to B, while the reverse of this is observed for the endowment at **b**. That is, the shape of the indifference curves (I_1, I_2,...) is such that the loss of A commands a higher value than the gain of B, and the shape of indifference curves (II_1, II_2,...) is such that the loss of B commands a higher value than the gain of A. This distinction in valuation can be explained by differences in the marginal rate of substitution (MRS) and the ratio of marginal utilities (MU). For example, for the indifference curves (I_1, I_2,...):

$$\{(MRS_{AB})=(MU_A)/(MU_B)\} > \{(MRS_{BA})=(MU_B)/(MU_A)\}. \qquad (7.2)$$

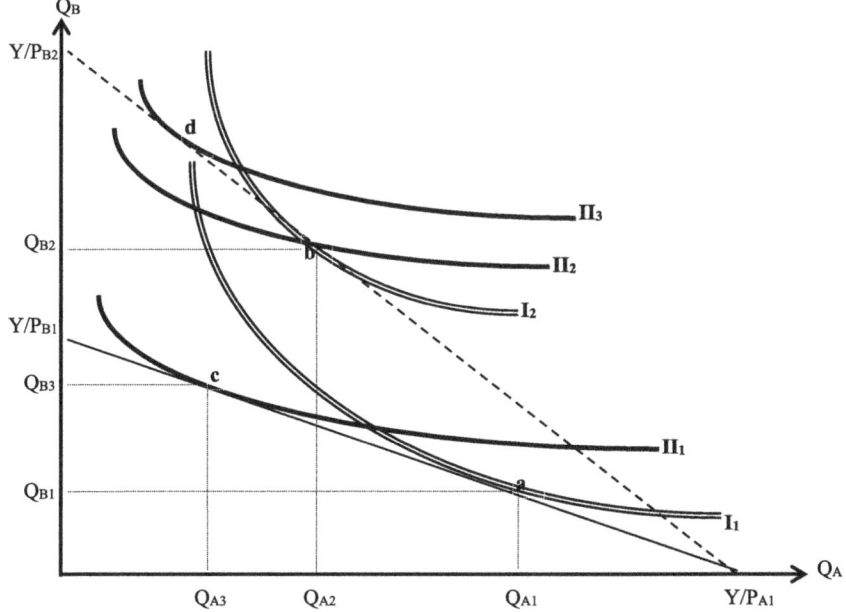

Figure 7.3 Varying Endowments and Indifference Curves

Note that MRS_{AB} measures the value of giving up one unit of A in exchange for gaining one unit of B. Similarly, MRS_{BA} measures the value of giving up one unit of B in exchange for gaining one unit of A. The inequality in (7.2) is reversed for indifference curves ($II_1, II_2,...$).

To generalise, for two goods A and B, should an individual's endowment contain more of A than of B, then the individual's underlying utility function is such that $\{(MRS_{AB})\} > \{(MRS_{BA})\}$.

The premise here is that the shape of an individual's underlying utility function, and hence the indifference curves, change when the individual's endowment changes. That is, should an individual be prompted to shift from **a** to **b** in Figure 7.3, then the individual's indifference curves will also change from ($I_1, I_2,...$) to ($II_1, II_2,...$). From a policymaker's perspective, it would be useful to explore the effects of shifting from KNUF to KNF goods.

The Elicitation of Demand Curves

Consider an individual who allocates a fixed budget of $Y towards the consumption of two goods A and B, the prices of which are respectively $$P_{A1}$ and $$P_{B1}$. We continue with the supposition that A is a KNUF good, whilst B is a KNF good. Assume that utility maximisation subject to the budget constraint occurs at point **a** in Figure 7.3; that is, the point of tangency of the indifference curve labelled I_1 with the budget line that satisfies $\{(P_{A1} * A) + (P_{B1} * B) = Y\}$. The consumption basket (or endowment) at this point is $\{(A = Q_{A1}), (B = Q_{B1})\}$. Suppose now that the price of B falls to $$P_{B2}$, resulting in a new budget line that satisfies $\{(P_{A1} * A) + (P_{B2} * B) = Y\}$. The utility-maximising consumer now changes his or her endowment to point **b**. The new endowment is defined by $\{(A = Q_{A2}), (B = Q_{B2})\}$.

The new endowment $\{(A = Q_{A2}), (B = Q_{B2})\}$ results in the consumer forming a new set of indifference curves, namely ($II_1, II_2,...$). In fact, the consumer may even further increase the consumption of B to the point of tangency between II_3 and the new budget line, namely point **d**. Should the price of B now revert to $$P_{B1}$, the consumer will alter his or her consumption behaviour to maximise utility at point **c**, which is the point of tangency of the indifference curve II_1 with the original budget line, namely the one that satisfies $\{(P_{A1} * A) + (P_{B1} * B) = Y\}$. The revised endowment in Figure 7.3 is $\{(A = Q_{A3}), (B = Q_{B3})\}$. This new endowment will result in the formation of a new set of indifference curves, and the process of changes in the utility function due to changing endowments will continue in this way.

Consider the consumer's movement from **a** to **b** to **c**. The endowment effect results in the demonstration of two price consumption paths: **a–b–d** for the price change of B from $$P_{B1}$ to $$P_{B2}$ and **d–c** for the price change of B from $$P_{B2}$ to $$P_{B1}$. The price consumption path **a–b–d** depicts a demand curve based on the WTP for B, because the path is based on the individual's acquisition of B. The path **d–c** depicts the demand curve based on WTA for B, because it is based on the individual's sacrifice of B. These are illustrated in Figure 7.4.

It is possible to conceptualise that the distinction between the two demand curves (and that between the shapes of the underlying indifference curves) would

Consumer Demand and KN   111

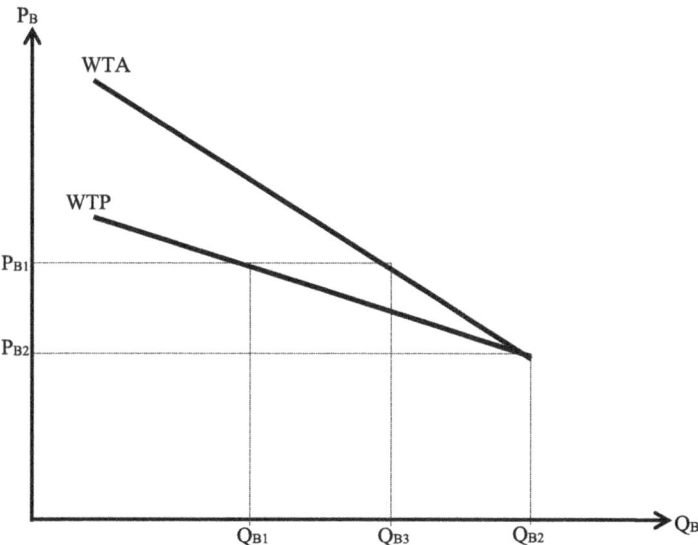

Figure 7.4 Two Distinct Demand Curves

diminish when the two goods A and B became increasingly similar and vice versa. So, for example, if a business manager is contemplating capturing a larger share of the market through product development, it is useful for the manager to consider:

- Product development strategies that would clearly distinguish the good from those offered by rival firms
- Marketing and advertising strategies that would clearly inform the consumer of the distinguishing attributes of the good

In the context of a society where KN awareness is on the rise, pursuing product development involving superior KN attributes and marketing them to inform consumers would no doubt be a sensible business decision. It may also be pertinent for businesses to initially lower the product price of a KNF good relative to the price of the rival KNUF good. This is because the sequence of events that follows a price reduction for a KNF good (as shown in Figure 7.3) indicates that a consumer will eventually settle on consuming more of the KNF good, even after the original price is restored. Similar reasoning can be given to justify government intervention with price incentives to encourage the consumption of KNF goods.

A further observation is that subsequent to the shift towards the increased consumption of the KNF good, it is likely that the KNUF good will become an inferior good. This is illustrated in Figure 7.5 for the shift from point **d** to point **c** for the fall in price of A (KNUF) relative to that of B (KNEF). The fall in the price of A, relative to the context at point **d**, takes the consumer initially to point **e**, indicating a reduction in the consumption of A (reduction from Q_{A21} to Q_{A31}), despite the fall in relative price. That is, the income effect displays a contraction in the consumption

112 *Microeconomics and KN*

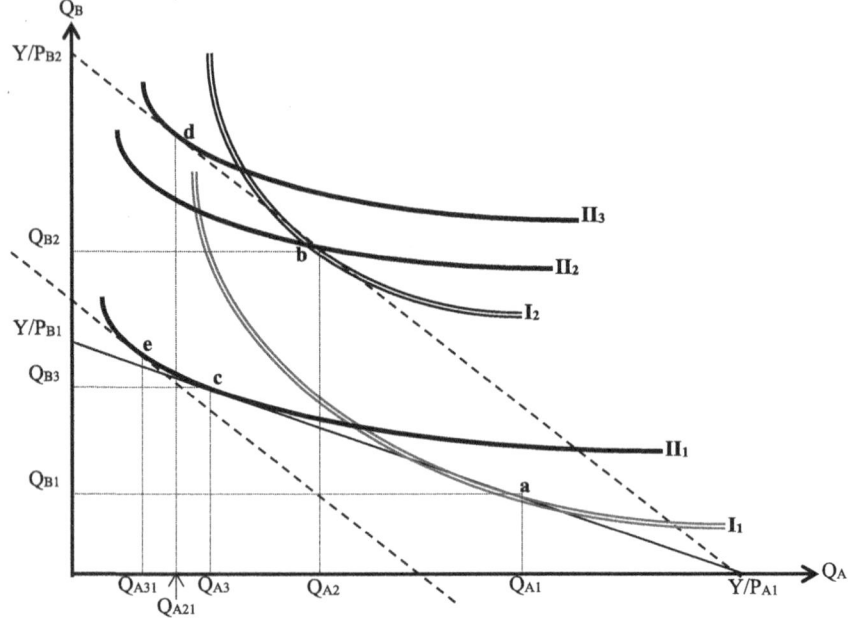

Figure 7.5 Income Effect and Substitution Effect

of A, implying that A has acquired the properties of inferior goods. It can be argued that the shift in consumption from point **e** to point **c** is due to the substitution effect.

The Market Demand Curve and KN Effects

Following the discussion presented above, it is possible for a decision maker to carry out WTP and WTA surveys to infer the merits of pursuing a particular strategy (such as product development or portfolio diversification). We can now show the relationship between the above discussion and the standard demand equation, which is often estimated from market surveys by econometric methods. These techniques are usually not quite capable of capturing the distinction between WTP and WTA, because the data for the analysis are often taken from actual recorded market transactions.

For convenience, suppose that the demand equation is linear, as follows.

$$Q_D = \beta_1 P + \beta_2 P_A + \beta_3 I + \beta_4 A + \beta_5 T + \beta_6 N \tag{7.3}$$

where

Q_D represents the quantity demanded
P is the price of the good
P_A is the price of the alternative good

I is income
A is advertising
T stands for tastes or attitudes
N is population
β coefficients measure the responsiveness of Q_D to each variable.

Note that there can be several more variables that would influence demand than what is considered below. However, for purely illustrative purposes we limit these to six variables, which are generally considered in most economics texts. As explained in most standard texts, changes in P explain the movement along a given demand curve, while the remaining variables in Equation (7.3) explain shifts in the demand curve. For example, if the price of an alternative good falls, then the demand for the good considered in Equation (7.3) is likely to be lower at any given price; or if consumers' disposable incomes rise, then the demand for it will be higher.

If, for example, a business manager or policymaker is exploring the potential for promoting a good with favourable KN attributes, then in the first instance the survey data on tastes and attitudes could focus on the consumers' KN awareness. That is, the demand curve for a KNF good is most likely to shift to the right as consumers become more aware of its KN-friendly attributes. Further, if the business manager expects that the KNF good has to be sold at a higher price, then he or she will want to learn more about the possible changes in incomes and attitudes of the clientele. This is because increases in income are often associated with higher demand for better-quality goods and improved attitudes towards KN.

Elasticity of Demand and KN

Elasticity of demand measures the responsiveness of quantity demanded (Q_D) to specific variables such as price (P) and income (I). It is defined as the *percentage change in quantity demanded in response to a 1 per cent change in the selected variable.*

Price elasticity of demand explains the responsiveness of demand for the good with respect to the good's own price, while *cross-elasticity of demand* considers the responsiveness with respect to the price of an alternative good. *Income elasticity of demand* explains the responsiveness of demand for the good with respect to changes in consumers' income. If price, income, and cross-elasticity of demand are denoted respectively as η_P, η_I and η_{PA}, then these can be defined with reference to Equation (7.3) as follows:

$$\eta_P = \frac{\partial Q_D}{\partial P} \cdot \frac{P}{Q} \tag{7.4}$$

$$\eta_I = \frac{\partial Q_D}{\partial I} \cdot \frac{I}{Q} \tag{7.5}$$

$$\eta_{PA} = \frac{\partial Q_D}{\partial P_A} \cdot \frac{P_A}{Q} \tag{7.6}$$

114 *Microeconomics and KN*

Business managers are specifically interested in the relationship between price elasticity of demand and total revenue. This is illustrated in Figure 7.6 for a linear demand curve, and we observe that total revenue is at a maximum when the elasticity of demand equals one. So if the observed elasticity of demand is less than one, managers will be inclined to raise the price of the good, and they will be inclined to do the reverse should the observed elasticity be larger than one. Cross-price

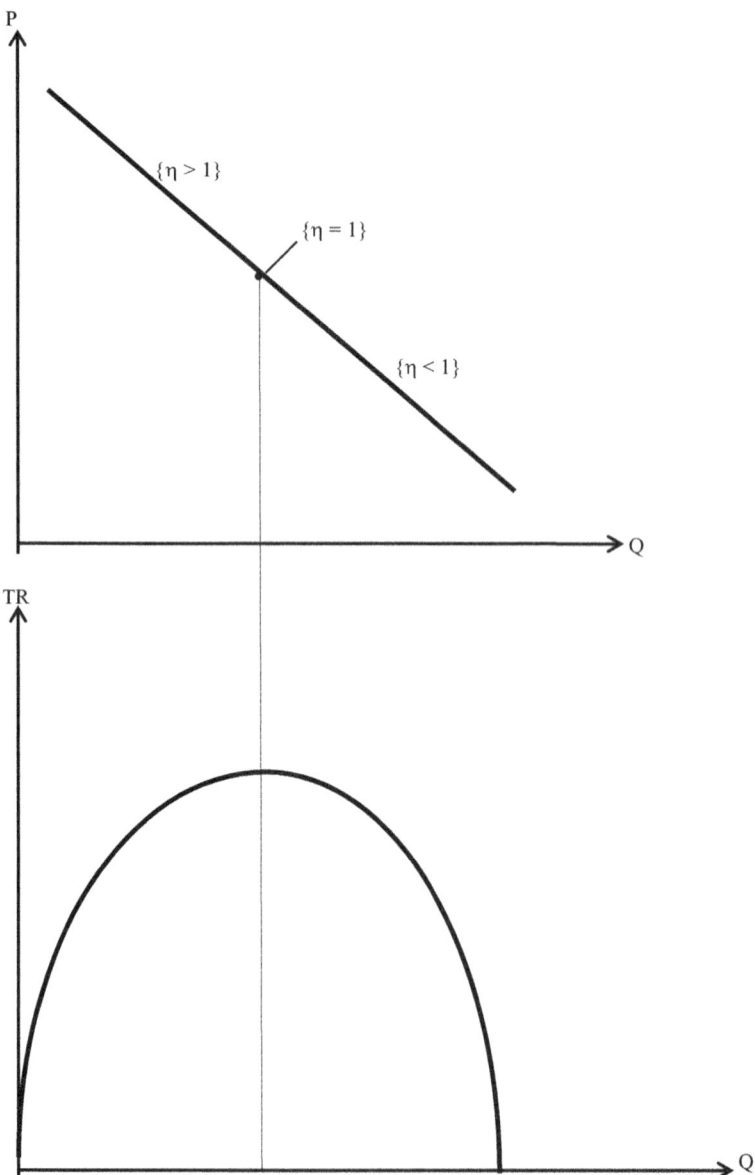

Figure 7.6 The Relationship between Price Elasticity of Demand and Revenue

elasticity information helps managers with the effects stemming from competition and expansion (or otherwise) of alternative or rival goods.

The challenge for us is to relate the elasticity principle to KN issues. In the previous section, we noted that the KNUF good in Figure 7.5 displayed the feature of inferior goods. That is, the income effect, namely an increase income, due to the increase in price from P_{B2} to P_{B1}, resulted in a contraction in the quantity of the KNUF good. In general, the income elasticity of demand (η_I) is negative for inferior goods, and positive for normal and luxury goods. In fact, η_I is in excess of 1 for luxury goods. A common observation is that as personal incomes increase, consumers seek KNF goods such as energy-saving devices and electricity from renewable sources. Increases in income are also associated with residents seeking KNF locations such as housing estates further away from sources of pollution. From a policy perspective, it is important to render KNF items cost-effective so that they cease to be luxury goods and hence their adoption would be widespread. To some extent such cost effectiveness and broadened adoption has happened with KNF items such as solar panels and electricity from renewable sources.

In this context, it would be useful to develop some elasticity concepts based on the KN attributes of a good. This is because, as indicated, the elasticity concept provides useful information for a business manager for pricing and revenue considerations. So, if a manager is contemplating offering a new product on the market and its main feature is going to be its so-called *KN-friendliness*, then a concept such as the ***elasticity of KN-friendliness*** (η_{KNF}) could become pertinent. That is, the higher the value of η_{KNF}, the higher the potential for the manager to capture a share of the market.

An example, though a hypothetical one, may help explain the relevance of the η_{KNF}. In almost all households (at least in the developed world), items of crockery are ceramic goods. The manufacture of these goods involves depletive extraction processes (for example, sand-mining) and high energy-consuming production processes (for example, kiln-drying or glass-blowing). We should of course note that the manager's proposed crockery goods are those used on a daily basis and not those that possess an intrinsically high value due to tradition, such as the items that are stored away in household cabinets for special occasions, for example, branded names like Royal Copenhagen, Rosenthal or Royal Doulton. Suppose that the manager's substitute goods are cups, saucers and plates made from the shell of the humble coconut. The manager thinks that a production process can be improvised so that many of the attributes of the new good do not differ from those of the ceramic good; for example, appearance, and the taste of the food or beverage when consumed using the new good. The main distinguishing KN-friendly attribute of the new good, however, is that it originates from a sustainable source. The business manager could of course be guided by other considerations. Most coconut plantations are found in developing countries, where labour is plentiful and cheap and investment in plant and structures may not be as high as in developed countries. However, the manager also needs to have some knowledge of whether he or she would be able to capture a share of the ceramic goods market. When KN-friendliness is the significant attribute of the new good, the η_{KNF} could prove to be a useful concept.

116 *Microeconomics and KN*

In the first instance, it is possible to gauge the size of the η_{KNF} by recourse to the cross-elasticity of demand. But here, the more influential parameter would be the price of the new good and not its KNF attributes. Suppose that we are able to construct an index of KN-friendliness (IKNF) for goods through a study of lifecycle analysis that explores the life history of a good from its cradle to its ultimate demise. For example, this can involve aggregating the sum of toxic emissions in the lifecycle of the good and using the inverse of this aggregate as an index. So, the η_{KNF} can be defined as the percentage change in quantity demanded in response to a 1 per cent change in IKNF. That is:

$$\eta_{KNF} = \frac{\delta Q}{\delta IKNF} \frac{IKNF}{Q} \qquad (7.7)$$

As we will observe later, much work has been done on the formulation of KN indices – especially pollution indices – and in some cases methods have been proposed to translate these indices into abatement costs.

Concluding Remarks

In this chapter, we have attempted to show how some revisions to the standard theory of consumer demand may assist decision makers in formulating strategies that are friendly to KN. The first was the deviation between WTP and WTA. If this deviation is marked in the context of a KNF good that is proposed by a business manager, then there is room for the manager to perform further market surveys to study the potential for introducing the good. The manager could also ascertain the potential for the good through an understanding of the KNEF. The higher the KNEF, the better the market prospect, and the manager could then examine the cross-price elasticity. Once the decision to produce the KNF good has been made, the manager can promote demand for it through advertisements and various marketing strategies that are aimed at improving the KN awareness of consumers. The role of the endowment effect illustrated above suggests that an initial lowering of the price of the KNF good could prompt consumer towards consuming more of this good than less of it in the long run. It is also useful to note that enhanced KN awareness has driven producers to attempt closed-loop production methods, where few contaminants enter KN sinks. We will consider this strategy of improving KN-friendliness in the next chapter.

Box 7.2 The Elasticity of KN-Friendliness

Monsanto, the major multinational firm that specialises in the sale of agricultural commodities, had developed in the 1990s a new 'KN-friendly' product – the New Leaf Potato. Of course, there is nothing new about the humble potato that millions of people across the world chew on for their daily dose of carbohydrate. Instead, the New Leaf Potato is a bio-engineered product. This new species of potato can defend itself against species of beetles that prey on the leaf of the potato plant. Therefore, this new product could potentially remove the need for aerial spraying of chemicals to control the potato beetle, and in turn prevent the accumulation of chemical residues in soils and in waterways. But the product had a relatively short life of two to three years because of opposition to genetically modified products

Let us now consider a hypothetical example. Suppose that farmer Keller from Germany decides to buy a newly developed insect-resistant variety of seed potatoes from a certified source. But will her consumers be impressed with the fact there will be less chemical spraying on her property? Given the level of KN awareness among German consumers, the answer is probably yes. Ms Keller's promotional campaign states that her potatoes will not only help protect KN but will also reduce the national import bill. The control chemical is currently imported into Germany.

At the moment, Ms Keller sells her produce directly to major German retailers such as Aldi and Karstadt. Having reviewed Ms Keller's material about the enhanced KN attribute of her good, all retailers have decided to increase their purchases from Ms Keller. This means that Ms Keller will enjoy a 10 per cent increase in sales. Suppose that we nominate the cost savings in chemical spraying to be a measure of IKNF. If the need for chemical spraying is eliminated, the change in IKNF is 100 per cent. In this hypothetical example, the EKNF is equal to 0.1. This information would prompt Ms Keller to review his pricing strategy.

There is one important caveat to this hypothetical example. Genetically modified organisms do have their own impact on KN. Pollen can be blown away from the field and thus affect the genetic make-up of wild species, for example. Incidentally, Germans are quite weary of these kinds of KN impacts as well, and so farmer Keller would need to not only concern herself with the elimination of chemicals, which may be straightforward to quantify in economic terms, but also the genetic impacts on the larger ecosystem, which are much more difficult to assess.

Review Questions

1 Do you agree with the basic definition of a 'good' as being one where more is preferred to less?
2 Critically evaluate the theoretical premise presented in this chapter to explain the distinction between WTA and WTP.
3 Do you agree with the statement that firms can succeed in promoting goods with KN-friendly attributes if the promotion includes a price reduction?
4 Provide a few examples where the estimation of the elasticity of KN-friendliness (EKNF) would prove useful. How could this EKNF be quantified?

References

Bateman, I. J., Munro, A., Rhodes, B., Starmer, C. and Sugden, R., 'A test of the theory of reference-dependent preferences', *Quarterly Journal of Economics*, 112(2): 479–505, 1997.

Franciosi, R., Kujal, P., Michelitsch, R., Smith, V. and Deng, G., 'Experimental tests of the endowment effect', *Journal of Economic Behaviour and Organisation*, 30(2): 213–226, 1996.

Kahneman, D., Knetsch, J. L. and Thaler, R. H., 'Experimental tests of the endowment effect and the Coase theorem', *The Journal of Political Economy*, 98(6), 1325–1348, 1990.

Knetsch, J. L., 'The endowment effect and evidence of nonreversible indifference curves', *American Economic Review*, 79(5): 1277–1284, 1989.

Knetsch, J. L., 'Asymmetric valuation of gains and losses and preference order assumptions', *Economic Inquiry*, 38(1): 138–141, 1995.

Knetsch, J. L. and Sinden, J. A., 'Willingness to pay and compensation demanded: Experimental evidence of an unexpected disparity in measures of value', *Quarterly Journal of Economics*, 99(3): 507–521, 1984.

Larranaga, A. and Valor, C., 'Consumers' categorization of eco-friendly consumer goods: An integrative review and research agenda', *Sustainable Production and Consumption*, 34: 518–527, 2022.

Samuelson, P. A. and Zeckhauser, R., 'Status quo bias in decision making', *Journal of Risk Uncertainty*, 1(1): 7–59, 1988.

Thampapillai, D. J., 'Willingness to pay and willingness to accept: A simple conceptual exposition', *Applied Economics Letters*, 7(8): 509–511, 2000.

8 Production, Costs Supply and KN

In the previous chapter, we illustrated how an individual's KN awareness can prompt changes in the display of consumer demand. In this chapter, we illustrate how producers can be prompted to re-examine their own production and supply capabilities in the context of realities concerning KN. The laws of thermodynamics as well as limited ecological resilience have a real effect on production capabilities. Therefore, there is a clear need to revise the standard theory of the firm in cognisance of these constraints. Such revision is the main purpose of this chapter.

The literature on production and supply within microeconomics commences with the description of a firm's utilisation of factor inputs and the resulting production function that displays diminishing marginal returns. In the short run, the factors utilised are usually classified into two types, namely fixed and variable inputs. In the long run however, all inputs are deemed variable.

Fixed inputs are generally durable items collectively labelled as capital and are independent of the quantity of production. For example, a farm's acquisition of durable inputs such as machinery and storage sheds would be referred as the farm's capital, and we include such items in the category of manufactured capital (KM). As the reader would appreciate, we have aimed, from the beginning of this text, to dentify items belonging to KM from natural endowments such as lakes, rivers and air sheds. We have argued such endowments constitute an important form of capital and have consistently labelled them as KN.

Variable inputs, as the name implies, change with the quantity of production, and usually include labour (L) and raw materials. Some of the raw materials, such as water, topsoil and energy are in fact drawn from KN. Further, emissions and various residuals emanating from the production process find their way into a range of KN sinks.

Nevertheless, most texts (for example, Chiang and Wainwright 2007; Hirshleifer et al. 2014) describe the production function as a two-factor model involving KM and L. In such functions, both KM and L are both regarded as variable inputs. The production function is also used to demonstrate the concept of isoquants and the cost-minimising combination of factor inputs. We discuss these below. The production function further serves as a basis for demonstrating the salient features of a firm's cost structure. Recall our brief review in Chapter 3 of specific relationships between short-run average costs and marginal costs, with the latter providing the basis for the supply curve in the short run. Finally, the production function and isoquants also provide the basis for explaining the role of technology and economies of scale.

There are many ways by which insights from thermodynamics may enter the analysis of production processes. One of them is briefly described in Box 8.1. Others can be found, for example, in Ruth (2018). The thesis we advance here is that nature, namely KN, is an important factor of production. While this may be readily apparent in the case of the farm mentioned above, which cannot produce its output without recourse to land, clean air and water, and somewhat reliable climate conditions, KN is, of course, an important factor of production for all kinds of firms. If a firm recognises this additional factor of production, then a sequence of implications – different from that suggested in standard texts – follows. First, the underlying premises and shape of the production function itself is altered. This leads to the isoquants being altered as well, which in turn has implications for the selection of factor inputs. Finally, the cost curves may also take on a different shape and this in turn has implications for a producer's supply response and perceptions concerning the role of technology and scale effects. We consider each of these in the sections that follow.

Box 8.1   Thermodynamic Constraints on Production Processes

One way in which the laws of thermodynamics express themselves is in the form of minimum material and energy requirements that must be met for a process to occur. For example, it takes a minimum of energy to sort the minerals in an ore from rock, to melt and refine the output, and to process it into a semi-finished or finished product, such as an ingot of copper or bar of steel. Similarly, for a given ore grade – the percentage of mineral in the ore – one needs a certain amount of ore to start with to yield a ton of the mineral. These minimum energy and material requirements are binding at any given point in time, and they also define the maximum efficiencies that may be achieved through refinements of technologies over time. While improvements in efficiencies may be relatively easy in the early stages of resource extraction and processing, they become increasingly difficult as both ore grades decline and technologies approach their thermodynamic limits. This decline in the potential for further progress has particularly far-reaching consequences in the minerals and metals processing industries, which are very mature, characterised by highly depleted reserves, yet are industries whose output is at the core of many infrastructures on which a growing and increasingly urbanised population depends – from the copper needed to build electric motors to the steel required for railroads, ships, bridges, skyscrapers and much more. (For empirical examples for five metals industries see Ruth 1998.)

An exposition of the altered theory of the firm has important implications for business management. Business managers, when cognisant of such relationships, are likely to voluntarily protect and conserve KN. Hence, there will be less need to enforce the systems of incentives and disincentives as well as regulatory measures that were discussed in Chapter 6. That is, when KN is acknowledged as an explicit argument in the production function, then the benefits of protecting KN become clearly visible. The foregoing statements are indeed correct if one takes an atomistic view of an individual representative firm. However, there is also the possibility of a 'Tragedy of the Commons' – even if an individual firm appreciates the contributions of KN to its operation, it may still overexploit KN, if there is the risk that other firms do so first. Thus, the arguments made in this chapter assume that a property rights framework and other legal mechanisms are firmly in place to prevent the tragedy of the commons.

The Production Function and KN

As indicated, the standard theory of production in economics relates output (Q) to the utilisation of labour (L) and manufactured capital (KM). That is:

$$Q = f(KM, L) \tag{8.1}$$

This theory, which is based on the law of diminishing marginal returns, identifies three zones of production, as illustrated in Figure 8.1. The first zone commences from the nil utilisation of factors, and in this zone an increase in the utilisation of factors results in increases in output at an increasing rate. In the second zone, the increase in factor usage results in increases of output, but at a decreasing rate. Finally, the third zone is one where increases in factor utilisation results in a decrease in output.

Hence, for the function in Equation (8.1), the following relations would hold for the second zone:

$$\frac{\partial Q}{\partial KM} > 0; \frac{\partial Q}{\partial L} > 0; \frac{\partial^2 Q}{\partial KM^2} < 0; \frac{\partial^2 Q}{\partial L^2} < 0; \tag{8.2}$$

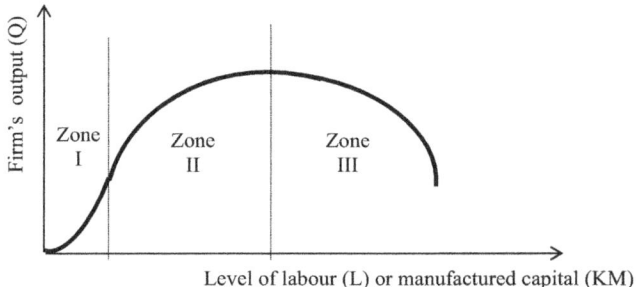

Figure 8.1 The Standard Production Function and the Zones of Production

The first and third zones are deemed irrelevant for decision-making purposes because they display the reverse of the characteristics identified in Equation (8.2). Economic theory describes the second zone of production as the rational zone of production, since it is in this zone that producers can choose profit-maximising allocation of factors. However, this theoretical framework changes when we incorporate KN as a factor of production. We illustrate these changes below.

Without any doubt, KN is a factor of production in every production context. Whether the site is a factory or an office building, indoor air quality and drinking water are clear factors of production. Similarly, soil, water and air are important ingredients of agricultural and related industries. Further, from a larger ecosystem perspective, KM and L are also dependent on KN. L cannot exist without clean water, air, or food and similarly, KM requires minerals and energy – all of which are drawn from KN.

Prior to conceptualising the characteristics of a production function involving KM, L and KN, we need to explain how one would define and measure KN.[1] As an example, consider the Industrial Park at Kalundborg in Denmark. Activities in the industrial park include coal-fired power generation, refining of crude oil, and manufacture of plasterboards and bio-chemicals. The park is located on the shores of Lake Tissø, which serves as a source of water for the park, for example, as a coolant in power generation, as well as a receptacle for effluent discharges from industrial operations. Lake Tissø is one of the KN items that contribute to the generation of output from the industrial part. There are, of course, several other items of KN that are interconnected with the lake; for example, tree density on the lake shores and in adjoining areas, micro- and macro-organisms in the lake and so on. For illustrative purposes, we shall confine our consideration to Lake Tissø as the only item of KN. We could state that the maximum amount of KN is available if at least the following criteria are satisfied:

- Lake Tissø is free of any contaminants.
- A stable ecosystem and biological population are present as one would expect, given the location and topography of the lake.
- Lake Tissø is at its maximum capacity based on groundwater hydrological characteristics.

Conversely, when untreated effluents degrade the quality of water and species are lost, and the extraction of water exceeds the rate of groundwater recharge and rainfall, we would conclude that the quantity of KN is declining. Therefore, the size measurement of KN must be a composite measure that accounts for the multiple attributes of the lake, including the quality and quantity of water. This measure usually takes the form of an index.

In a production function, we hope to explain the relationship between the size index of KN and the output generated from the Industrial Park. Note that an increase in the size of the index means that overall quality and/or quantity of the lake water is increasing. A decrease in the size of the index implies the reverse; that is, that the lake is degrading. When an increase in KN occurs, we would expect the

Production, Costs Supply and KN   123

magnitude of output from the Industrial Park to increase. However, this increase is not instantaneous. It occurs over time, as the historical evidence from the Industrial Park at Kalundborg confirms. As Valentine (2016) explains, the Industrial Park has developed a symbiotic relationship between several large industries on the site and has been able to significantly reduce the use of water from Lake Tissø by internally recycling and reusing water in industrial processes, thus effectively preventing the degradation of KN. The definition and measurement of KN usually follow the development of an index that attempts to account for as many attributes as possible. However, we must also note that decrease in KN (implying the utilisation of KN) can also lead to an increase in output in the short run, as we illustrate conceptually below.

Production and Assimilative Capacity

This and the subsequent sections closely follow the narrative provided in Thampapillai (2014, 2016).

Consider now the following production function:

$$Q = g\left(KN, \overline{KM}, \overline{L}\right) \qquad (8.3)$$

This function is illustrated in Figure 8.2. The illustration is based on the premise that some metric is available for the quantification of KN, as explained above, and that KM and L are held constant at some level, namely $\overline{KM}$ and $\overline{L}$. To reiterate, we assume that all components and attributes of KN have been aggregated into a single numerical scale, namely an index, and are represented by KN along the horizontal axis. The increase (accumulation) of KN runs from left to right, while its utilisation (depletion) is represented from right to left. Note that the

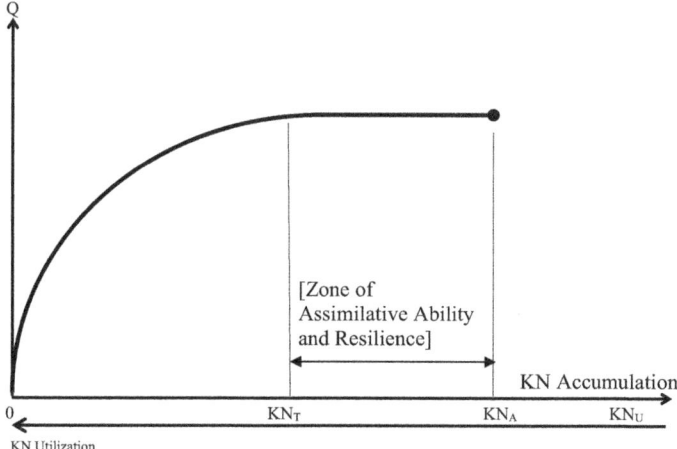

Figure 8.2 The Production Function Relating to KN

accumulation of KN will involve a much longer time compared to its depletion. For example, a firm that specialises in forest products can fell a tree and make relevant products within a few days. However, it would take several years for the tree to grow back to the stature it held prior to its felling. We expand on this premise below.

When one attempts to conceptualise the relationship between Q and KN (while holding KM and L constant), we posit that the production function will assume a specific shape, which is distinct from the standard production function shown in Figure 8.1. This distinctiveness has much to do with the multiple attributes of KN and is illustrated in Figure 8.2. Further, as shown, the function has a fixed domain. It is for this reason that the curve describing the changes in Q has a bold circle at the end of it. The fixed domain exists because there is a fixed upper limit for KN. Laws of nature and the level of economic activity dictate this upper limit. For example, in the case of air quality, the law of nature dictates that we cannot have more than 20 per cent oxygen in the air. Similarly, it is not possible (at given temperatures and pressures) to exceed a certain level of dissolved oxygen in water. In the primitive state, where (Q = 0), KN has maximum capacity, that is denoted by KN_U. However, this upper limit can decline because of economic activity, leading to higher levels of Q. For the function displayed in Figure 8.2, the upper limit is defined as KN_A.

Note that, in Figure 8.2, we consider the accumulation as well as the depletion of KN. For the moment ignore the issue of time and consider an increase in KN, say from the origin, while holding KM and L constant. There will be an increase of Q. But beyond a certain threshold level of KN, denoted by KN_T in Figure 8.2, increases in KN have no further effect on the size of Q. We can explain this in the reverse order as well. That is, if we deplete KN from its upper limit (KN_A) – again while holding KM and L constant – Q remains unaffected until KN reaches KN_T. This proposition can be supported by the fact that KN has the characteristics of *assimilative ability* and *resilience* – concepts that we introduced in Chapter 2. That is, despite being depleted, KN can continue to provide its services – primarily source and sink functions – notwithstanding the depletion. For example, consider the Kalundborg illustration above. Should some pollutants contaminate Lake Tissø, the pollution would deplete the lake by a certain quantum. Yet, the lake could continue to support the industrial park up to a threshold level of contamination because the flora and fauna of the lake could assimilate the pollutants and thereby enable the lake to be resilient and fulfil its role. But such resilience would prevail only up to a certain level of depletion or contamination. When the size of KN drops below a critical level – KN_T in Figure 8.2 – Q begins to fall. So, for the production function displayed in Figure 8.2, the domain ($KN_T \leftrightarrow KN_A$) can be described as the zone of assimilative ability and resilience, and this domain defines the *assimilative capacity* of KN. In the primitive case, where (Q = 0), the domain of assimilative ability and resilience is represented by $\{0 \leftrightarrow KN_U\}$. The state (Q = 0) is consistent with the lowest level of entropy for the system representing KN. Unlike in the standard function in Figure 8.1, the rational zone of production in Figure 8.2 will be the zone of assimilative ability and resilience. The reasons for this are outlined below.

Desired Output and the Generation of Entropy

In Figure 8.3, we show a family of curves for the production function (Q), where $\overline{KM}$ and $\overline{L}$ are held constant – but at different levels; for example, $(\overline{KM}_3, \overline{L}_3)$ < $(\overline{KM}_4, \overline{L}_4)$ and, accordingly, $Q_4 > Q_3$. That is, as the level of $(\overline{KM}, \overline{L})$ gets higher, the size of Q becomes larger. Further, because raising the intensity of KM and L inevitably involves the enhanced utilisation of KN, the maximum upper limit of KN could recede to the left. As a result, the zone of assimilative ability and resilience becomes smaller.

As illustrated, the enhanced utilisation of KN for achieving higher levels of Q also follows the law of diminishing marginal returns. It is possible to narrate the concept of entropy, which we considered in Chapter 2, as the primary driver of diminishing marginal returns. That is, as KN gets utilised and its upper limit is reduced, the level of entropy of the remaining stocks of KN is raised. This in turn warrants the utilisation of higher quantities of KN for similar levels of increases in Q.

Another feature in Figure 8.3 is that the gradient of the production function, δ, for a given value of Q is shown to be steeper at higher levels of $(\overline{KM}, \overline{L})$ than at lower levels. This implies that when KN deteriorates to a level below its threshold level (KN_T), then the fall in output is much faster at higher $(\overline{KM}, \overline{L})$ levels than at lower levels. Hence this gradient can be also regarded as a gradient of degradation. The explanation afforded thus far enables the argument that, at higher intensities of $(\overline{KM}, \overline{L})$, we expect KN to become increasingly fragile. We can easily relate this feature of increasing fragility of KN to a variety of business contexts, regardless of

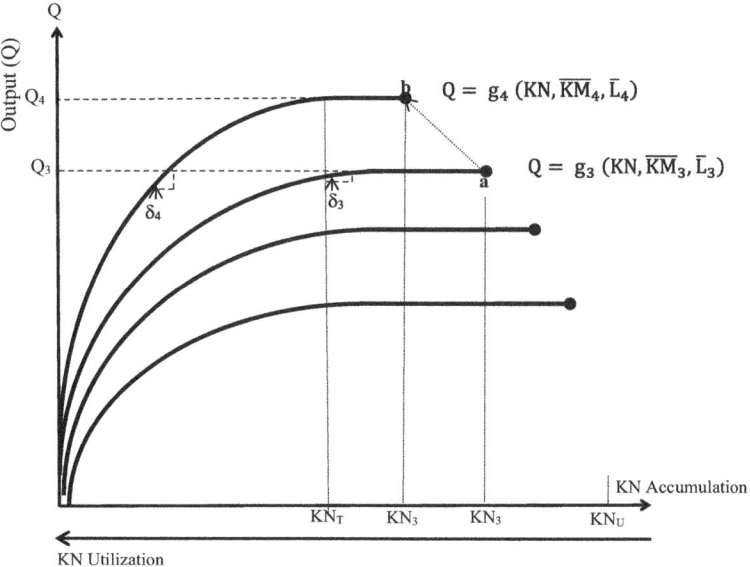

Figure 8.3 Family of Production Functions and Increasing Fragility of KN

126 *Microeconomics and KN*

whether it is a farm-firm or a factory floor with a production line. In the example of Lake Tissø, the higher the intensity of industrial plants on the lake shores, the less the intensity of the ecosystem supporting the lake, and therefore the higher the vulnerability of the lake.

The essence of the conceptualisation presented in Figure 8.3 is that KN becomes increasingly fragile, with higher rates of entropy when KN utilisation and output increase. For example, suppose that the present position of a firm is point **a** in Figure 8.3 and that an increase in desired output Q will take this firm to point **b**. The output increase $(Q_4 - Q_3)$ is associated with:

- Contraction in the domain of assimilative ability by $\{KN_3 - KN_4\}$
- Display of diminishing marginal returns with respect to KN—which we have argued is due to increased generation of entropy
- Increase in the gradient of degradation by $\{\delta_4 - \delta_3\}$

Note that $\{\Delta k = (KN_3 - KN_4)\}$ is the amount of KN that gets utilised or depleted when output increases from Q_3 to Q_4. We argue that further increases in Q would warrant Δk to increase due to entropy.

Transfer of Entropy and Time

It is possible to show how the entropic effects on KN in the form of diminishing marginal returns can be transferred to KM and L. This is illustrated in Figure 8.4.

For illustrative purposes, we assume that KM and L can be measured as a composite factor, and we denote this by K. Such aggregation is feasible if L is measured in KM-equivalent terms. We further suppose that the utilisation of KN can be also measured in the same metric and denote it by k. The left-hand panel of Figure 8.4

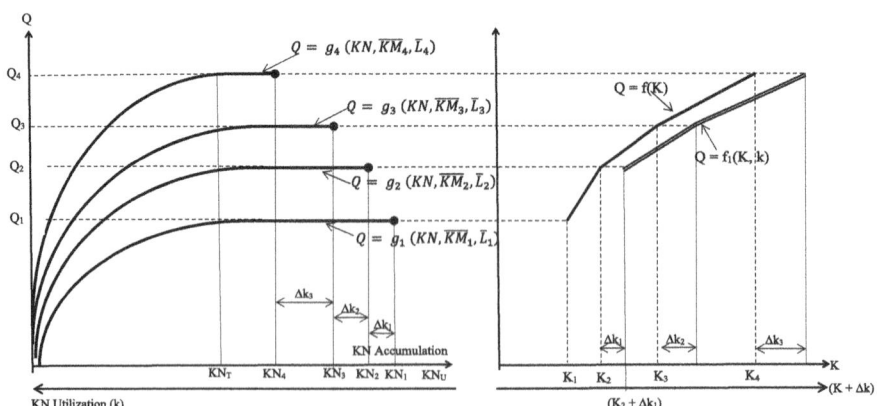

Figure 8.4 The Display of Entropy and Diminishing Marginal Returns

displays the family of curves for the function that we have considered thus far in Figure 8.3. The right-hand panel illustrates the elicitation of two types of functions, both displaying diminishing marginal returns. But the primary source for this display is evident in the left-hand panel through the law of entropy considered in Chapter 2 and the following two propositions are pertinent.

Proposition-1: The utilisation of KN and the associated reduction in its upper limit leads to the increase in entropy of the remaining stocks of KN.

Proposition-2: The increase in entropy of the remaining stocks of KN warrants the utilisation of higher quantities of KN for further increases in Q of similar magnitudes. That is, $\Delta k_1 < \Delta k_2 < \Delta k_3$ and so on.

Schneider and Kay (1994) provide support for the above two propositions. They develop a thermodynamic concept of energy gradient dissipation in the context of biological systems. They then empirically illustrate that complex mature ecosystems contain much lower entropy compared to systems that have been cleared. The corollary of this evidence is that KN utilisation leaves behind an ecological system that has high entropy and hence is of lower quality (as indicated, for example, by reduced species diversity, water retention, carbon sequestrations and other ecosystem functions). As a result, successive increases in Y (such as the progression from Y_2 to Y_3 to Y_4) warrant successively greater utilisation of KN.

Also note that, when the utilisation of KN is ignored in the right-hand panel of Figure 8.4, the role of KM and L is overstated. For example, output level Y_3 could be attributed to the role of (KM–L) as K_3. However, as shown, when the utilisation of KN is also considered, the output level Y_3 is truly due to the influence of $(K_3 + \Delta k_2)$.

We are now able to introduce the time dimension. In Figure 8.5, we have reproduced the two functions from Figure 8.3 pertaining to $(\overline{KM_3}, \overline{L_3})$ and $(\overline{KM_4}, \overline{L_4})$.

The time taken for the firm to move from point **a** to point **b** is relatively short (T_0 to T_1) compared to the time taken for the firm to move from point **a** to point **c** (T_0 to T_T). The former involves utilising KN, while the latter involves augmenting KN. The move (**a → b**) will give higher income within a short time, but at the cost of higher entropy and fragility. Alternatively, the move (**a → c**) will not add to income, but over a longer time will lower entropy and enhance stability of the natural system. Recall the earlier example of felling a tree and making a product, all within a week. However, it could take many years for a seedling to grow into a tree of the same dimensions as the tree that was felled.

If business managers based their decisions on this type of production function (that is, as shown in Figures 8.2–8.5), then we would expect to observe changes in their business strategies. Increasing the intensities of KM–L to achieve higher levels of output may not always prove to be prudent because, with the increasing fragility of KN, a manager can easily lose his or her production capability. These pressures become more explicit when we consider substitution behaviour within the framework of isoquants.

128 *Microeconomics and KN*

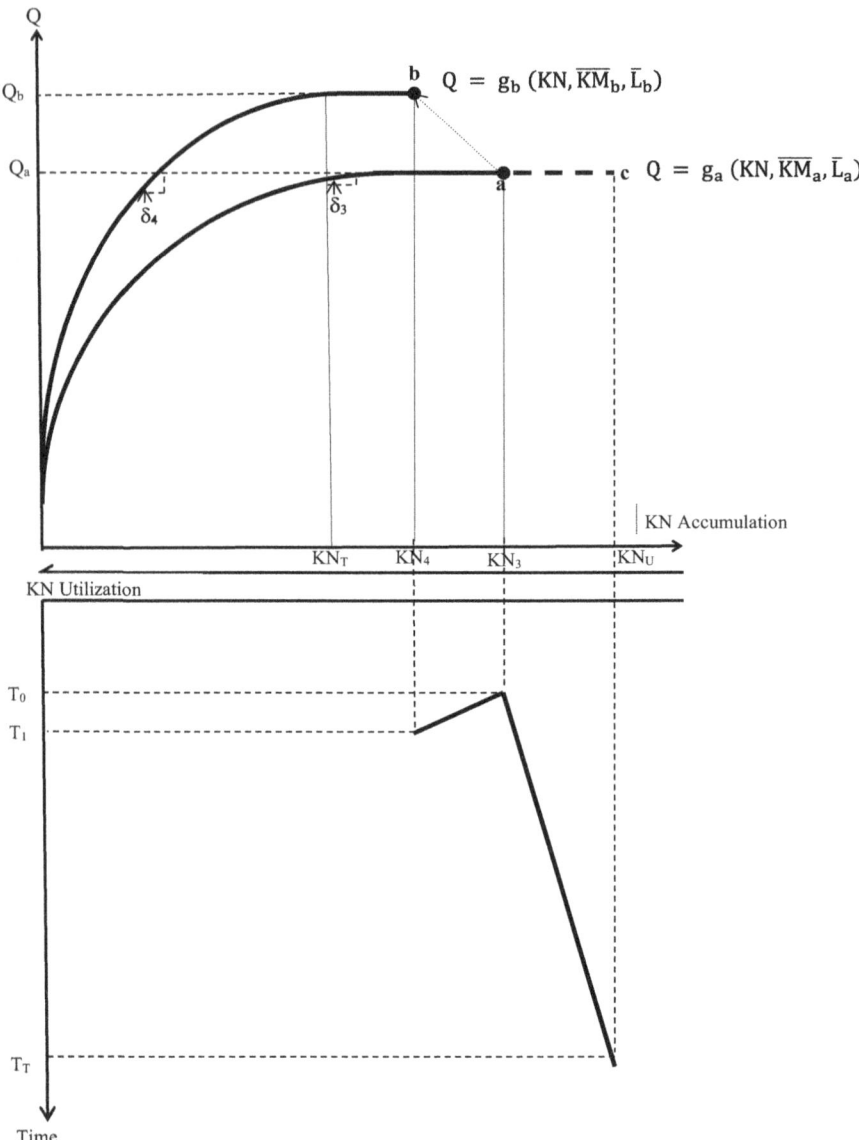

Figure 8.5 Utilisation Vs Augmentation of KN and Time

Isoquants, Substitution and Input Mixes

An isoquant is a contour resulting from a horizontal cross-section taken out of a multidimensional production surface – very much the same way as we explained the concept of an indifference curve in Chapter 7. For purposes of illustrative convenience, we shall confine ourselves to a two-dimensional surface.

Production, Costs Supply and KN   129

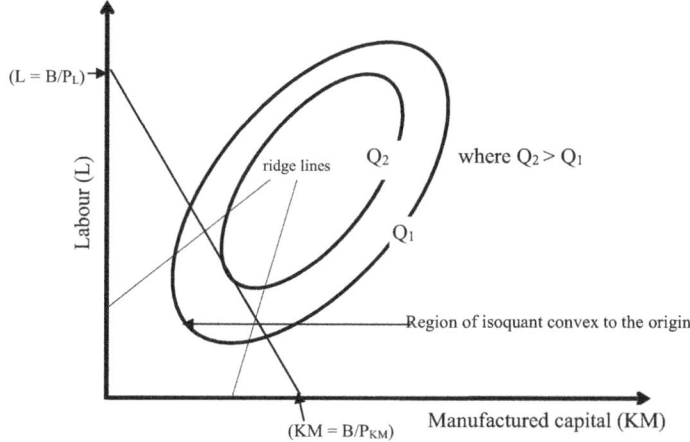

Figure 8.6 Isoquants in Standard Theory

In Figure 8.6, we present the isoquants that correspond to Equation (8.1) (Q = f (KM, L)).

Each isoquant describes a locus of points along which output remains constant for various mixes of the inputs. The region of an isoquant that is convex to the origin corresponds to zones I and II in Figure 8.1. In this region, there is positive substitutability between inputs. The region of the isoquant that is concave to the origin corresponds to zone III in Figure 8.1. In this region, we need to increase both L and KM together to maintain a certain level of output. The ridge lines are usually constructed to separate the zones in terms of substitutability.

A firm would confine its decision-making to the region of the isoquant that is convex to the origin because it would not want to use larger quantities of inputs necessary to produce a particular level of output. A manager who is restricted to a fixed budget of $B, and facing input prices P_{KM} and P_L, would optimise the selection of inputs at the point where the budget line (defined by $[P_{KM} * KM] + [P_L * L] = B$) is tangent to the highest attainable isoquant. This is illustrated in Figure 8.6. A familiar result in microeconomics is that, at the point of tangency, the marginal rate of substitution (MRS) between inputs can be defined by the ratio of marginal products, which in turn equals the ratio of input prices. That is:

$$\text{MRS}_{L,KM} = \left(MP_L / MP_{KM}\right) = \left(P_L / P_{KM}\right) \tag{8.4}$$

As the firm's available budget increases or the price of one or more inputs drops, the manager can move to a higher isoquant and thereby trace an expansion path.

The above explanation would change when KN is also recognised as a factor of production. Because we want to keep our illustration to a two-dimensional level, we shall regard KM and L as a composite factor input (K), as we did in the previous section. Equation (8.3) will be modified to read as $\{Q = g_1[K, KN]\}$. The isoquants that describe this production function are shown in Figure 8.7.

130 *Microeconomics and KN*

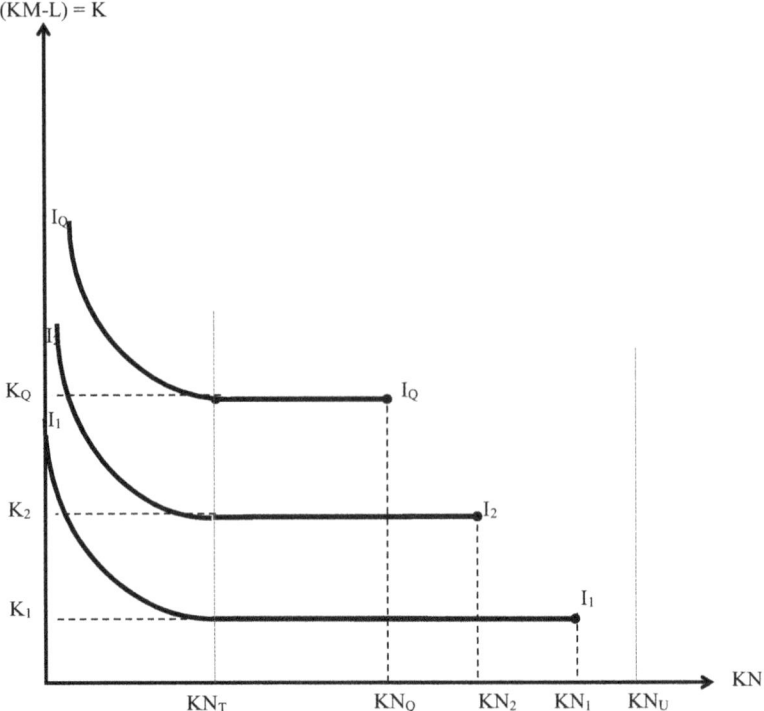

Figure 8.7 Isoquants for KN and KM-L (K)

Note that there is a correspondence between these isoquants and the production function shown in Figure 8.3. One can envisage the existence of a family of an infinite number of curves (of the type shown in Figure 8.3) in the KN–Y space. If we were to take a series of horizontal cross-sections on Figure 8.3, then we would obtain the isoquants of the type displayed in Figure 8.7.

Consider the isoquant labelled I_1I_1. Substitutability between KN and (KM–L) begins when the size of KN falls below its threshold level KN_T. That is, there is no substitutability between factors in the region of assimilative ability and resilience, which is defined by $\{KN_T < KN < KN_1\}$. Substitutability between KN and K also ceases at some extremely low level of KN, where output approaches zero. At this level, the isoquant becomes asymptotic to the vertical axis, implying that we would need infinite quantities of K to replace KN. For each isoquant, the region of substitutability ranges between 0 and KN_T. However, as we move to a higher isoquant, say I_2I_2, we note that the gradient of the isoquant (dK/dKN), namely the marginal rate of substitution between K and KN ($MRS_{K, KN}$), becomes steeper. That is, at higher levels of Y, larger quantities of K need to be given up for a small increase in KN.

In standard microeconomic theory, the ridge lines help distinguish the convex region of the isoquant from the rest. This then helps us understand the nature of substitution behaviour in the rational zone of production. It is not possible to directly extend this rationale to the conceptualisation that we have advanced here. This is because the region of convexity, which displays the possibility of substitution behaviour, is also the region of vulnerability. The rational zone of production is the region of assimilative ability and resilience, where no substitution possibilities exist. If we are to be guided by the rationality of retaining sustainability and stability, then ridge lines will consist of the vertical line drawn at KN_T and the line or curve that connects the upper limits of the successive production function.

In standard production theory, the analysis of isoquants also helps determine the input mix, as illustrated in Figure 8.6. The difficulty in extending this to the revised framework is the absence of a price for KN. However, it is certainly clear that producers should not choose a production strategy that takes them to the region $(0 \leftrightarrow KN_T)$. This is because they run the risk of losing their production capability. Alternatively, a producer may nominate (if he or she is able identify them), the upper limits (that is, KN_1, KN_2,..., KN_Q) as the basis for an expansion path, but needs to be cautious of the fact that the gap between the upper and lower thresholds (KN_T) becomes progressively narrower as the intensity of K gets higher.

One major implication of the framework we have considered here concerns the relationship between the government and business. If governments wished to force business managers to operate within the region of assimilative ability and higher stability, then they need to have strict regulations in place. This type of government intervention can have a positive spin-off. It forces industries to innovate and extend the productivity of KN by way of technology instead of by accumulating more K.

Consider again Denmark's Kalundborg. As indicated, activities in the industrial park include coal-fired power generation, refining of crude oil, and manufacture of plasterboards and bio-chemicals. The industrial park tried to close material cycles in ways where the unwanted by-products from some processes were used as inputs into others. Regardless of whether this was prompted by higher taxes or whether it was the result of efforts to minimise costs – the waste-generating firms save cost for not having to dispose of the waste, and the buyers get a waste product, which serves as an input into their production, at lower prices. Co-location of businesses such as the coal-fired power plant and the plaster manufacturer allowed also for transport costs to be reduced. Everyone, including KN, wins – at least in theory.

Box 8.2 How Does One Define Substituting between KN and KM–L?

Suppose that the premises of an industrial plant contain six buildings dispersed across a land area of several hectares. Further, suppose that two of these buildings need to be renovated and refurbished. The CEO can explore the following option: demolish the two buildings completely and extend the remaining buildings to accommodate the activities that are being conducted in the two buildings to be demolished and use the land area rendered vacant by the demolition to establish a wooded area of recreational value to the workers as well as to visitors to the plant. One might say that the CEO would have to spend much more on this strategy of investing in KN than on renovating the two buildings. However, it could also be said that the entire industrial estate was once a wooded area and that the erection of the buildings and construction of the industrial infrastructure has reduced the magnitude of KN. Hence, the establishment of a wooded area could be seen as a substitution between KM and KN. Investment in KN may lead to several positive externalities such as co-benefits in the form of improved scenic beauty and improved public health. One interesting question then is how to internalise these externalities so that higher levels of KN are achieved.

Sometimes the distinction between 'increasing KM' and 'increasing KN' can be elusive. Return to the earlier example of the Industrial Park on the shores of Lake Tissø in Denmark. Here again, if the CEO has to choose between allowing the installation of a mechanical plant to increase daily production as against planting trees, then he or she has an issue of substitution. However, this clarity diminishes with some activities that are designed to maintain the quality of KN. For example, suppose that the problem facing the CEO is to choose between setting up a plant that will increase the daily output of electricity and plasterboards as against upgrading the plant that treats effluent and uses water. Doing the latter will prevent the degradation of the lake. If the CEO chooses to upgrade the effluent treatment plant, then this can be seen as substituting KN for KM.

Note that not all cases of KN are found out of doors. Indoor air quality is equally important in a production context. Many firms have invested in air-cleaning devices to ensure that indoor air quality is maintained at an acceptable level. Hence, another example of substitution is one where a manager has to choose, say, between hiring a worker and upgrading the air-cleaning device. Recall that, within the conceptual framework we have presented here, increasing the accumulation of KM-L renders KN more fragile. The limit on substitutability when KN has been rendered fragile at high intensities of KM-L has implications for the theory of costs.

Source: Thampapillai and Ohlmer (2009)

Analysis of Costs

As indicated, the theory of costs has its foundations in the underlying production function. The total cost (TC) function, which comprises fixed costs (FC) and variable costs (VC), is usually described as one that displays a specific shape, reflecting varying returns to scale at different levels of production; (Figure 8.8). Most standard texts explain how this total cost curve emerges from the production function. Note that the horizontal axis in Figure 8.8 represents Q – which was also

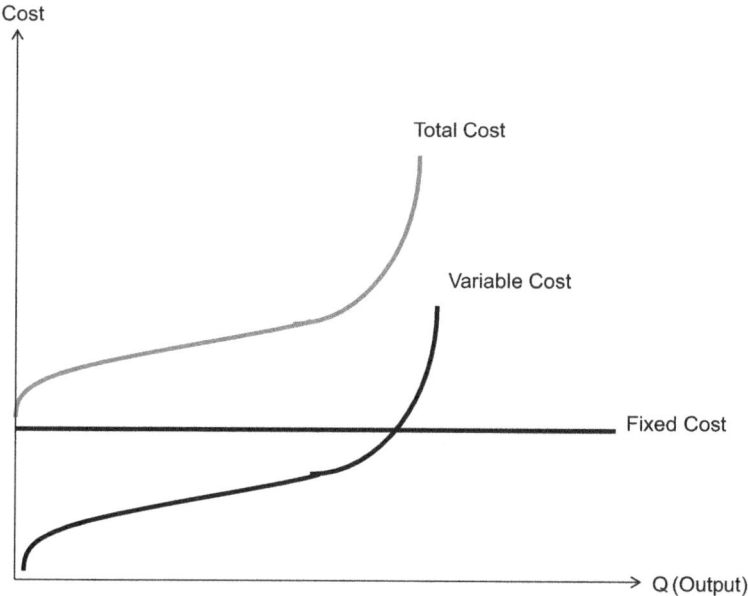

Figure 8.8 Cost Structure of a Firm

represented by the vertical axis in the production function described in Figure 8.1. We can intuitively argue that initial ascension of the TC and VC curves corresponds to zone I of the production function. The gentle ascension in the middle corresponds to Zone II. And the steep ascension at the end corresponds to Zone III.

In decision-making, apart from considering TC, managers are also interested in the shape and properties of average costs (AC) and marginal costs (MC). We briefly reviewed this in Chapter 3. To recapitulate, TC is the sum of all relevant costs during a production period. AC is the average cost per unit of output produced; that is, (TC/Q) where Q is the quantity of desired output. MC is the TC of producing one extra unit of the good and is usually denoted by (dTC/dQ); that is, it is the slope or gradient of the TC function.

For decision-making, managers would usually compare the TC with total revenue (TR), where TC is the opportunity cost associated with the use of all factors of production, usually assessed at market prices, and where TR is captures sales of desired products at market prices. Typically, not considered here are the many unpriced inputs used and outputs generated by the firm, such as oxygen in combustion processes, water for cooling, or emissions into the air. The firm then selects a production quantity that maximises the departure between TC and TR functions.

The relationship between AC and MC also plays an important part in decision-making. Following our earlier discussion in Chapter 3, as Q increases, both AC and MC would initially decrease, and then increase beyond a certain value of Q, which represents the shift from the first to the second zone of the production function. However, there is one important property of cost functions that should be noted. That is: *when costs are falling, AC is greater than MC, and when costs are increasing, MC is greater than AC.* This is illustrated in Figure 8.9. It would be irrational for a decision maker to produce in the region where AC is greater than MC, because in this region TR falls short of TC. Therefore, decision makers would choose their production strategies in the ascending region of the cost functions. It is for this reason that *the supply curve in microeconomics is often described as the ascending segment of the MC curve.*

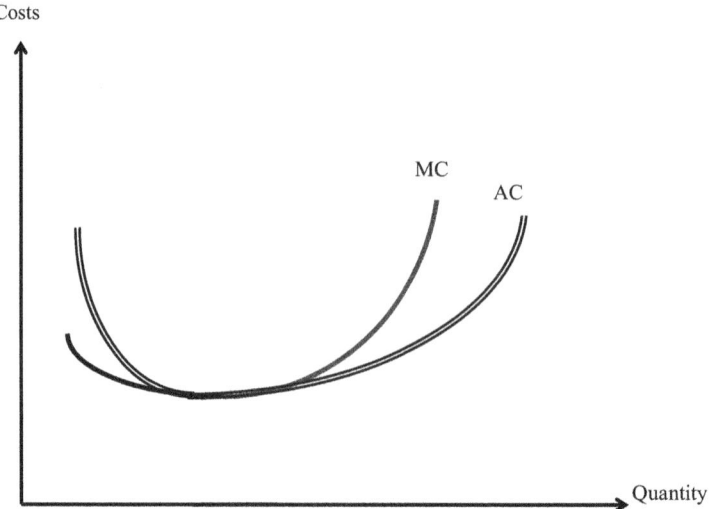

Figure 8.9 Average and Marginal Cost in Standard Theory

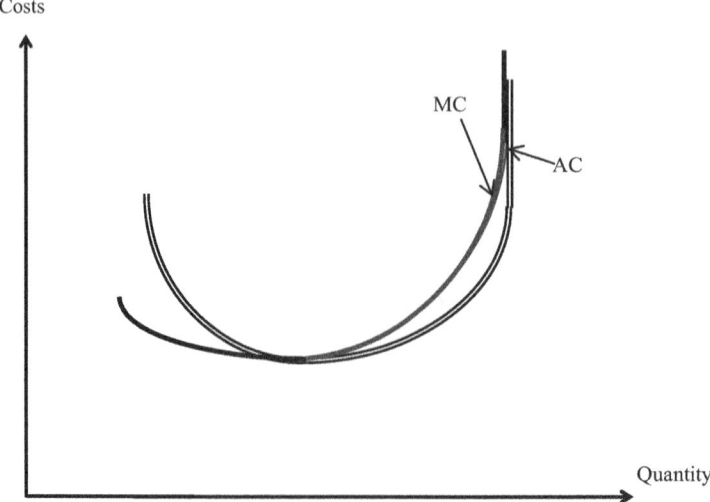

Figure 8.10 Average and Marginal Cost That Tends to Infinity

Consider now the implications of the conceptualisation we have introduced above with KN as an additional factor of production. If the business manager were to increase output by increasing the quantities of L and KM, he or she would be shifting upwards from one production schedule to another, as shown in Figure 8.3, whereby the fragility of KN progressively increases. At some level of K, output capability completely disappears due to the complete breakdown of KN. Hence, we can describe the cost curves in very much the same way as they are described in standard texts, but with one exception. This is that the manager's costs will tend towards infinity, at the point where KN breaks down completely. Therefore, the supply curve of a business will be upward sloping, as expected in theory, but turning vertical at some limiting point (see Figure 8.9). In other words, recognising KN as a factor of production makes the supply curve of most goods resemble that of an exhaustive non-renewable resource in natural resource economics.

Implications

In this chapter, we have considered three concepts that economists employ to illustrate how producers can formulate production strategies. These are: the production function, the isoquant and the marginal and average cost functions. This formulation is of course subject to the provision that other pertinent management variables, such as market potential and competitive advantage, have been considered. The production strategy will basically involve the choice of quantity of production, the allocation of resources, the choice of an expansion path, and response to prices and relevant market signals. In the absence of KN considerations, we can summarise the role of the three concepts as follows:

1 The production function assists with the selection of input quantities and identifies the range within which output decisions can be made.
2 The isoquants provide the same information as in (1) but also, at the same time, enable the producer to trace out an expansion path when budgetary conditions improve. This can be of assistance with medium- to long-term planning.
3 The cost curves assist with information on how the producer could respond to prices. Note that the upward-sloping segment of the marginal cost curve is the supply curve. So, in theory, if prices rise the producer can incur higher marginal costs and raise production. In practice, however, producers cannot vary production quantities to match every price variation. Yet the cost curves at least do provide information on the range over which supply responses can be varied. For example, if prices are expected to fall below a certain level, say where AC exceeds MC, then the producer must seriously consider plans to exit from the industry.

When KN is introduced as an additional input, we find that there are some important changes. First, it is not so straightforward to pursue an expansion path when the budgetary conditions improve. This is because higher levels of the KM-L factor intensities are associated with increased fragility of KN. This also implies that the benefits from the economies of scale that are observed with higher production quantities are overstated.

The main implication of the conceptualisation that we have presented here is that there is an additional limit to the production capability and the response capability that is induced by KN. Note that the producer's endowment with respect to KM and L also represents a limit on the production and response capabilities, at least in the short run. Therefore, it is pertinent for the manager to determine which limit is enforced first; that is, whether the limit enforced by KN emerges first or whether the limit enforced by KM–L emerges first. Should the limit enforced by KM–L emerge first, then, there is scope for expansion up to the limit imposed by KN. A cautionary note is in order owing to the influence of time. How fast do managers recognise that a KN limit is approached or has been reached? How fast can they respond to that information? The time-delayed nature of many KN-related responses add complexity to the decision-making process. For example, some metals accumulate only slowly in the tissue of animals, and concentrations are magnified along food chains, long before the public health impacts on humans can be, and are, detected. Similarly, emissions of carbon dioxide and other GHGs for centuries have resulted in their atmospheric concentrations being so high, that the global climate has been altered. Many of the consequences of global climate change are being felt now and will be felt for centuries to come. Changing production processes takes time, and the shift from carbon intensive to carbon free energy sources is slow.

It is customary to explain the role of technology in production through upward shifts in the production function for fixed levels of factor-utilisation. This gets manifested in the proximal bunching of the isoquants (compared to isoquants being spaced apart). When KN becomes a limiting factor, it is possible for the producer to explore, at least in principle, whether it is worthwhile to focus on developing a technology or production method that would alleviate the limitations imposed by KN. This can involve searching for a technology that extends the region of assimilative capacity on the production function or increases the range of substitutability on the isoquant. The appropriateness of seeking such a technological innovation would of course be determined by a wider range of factors, such as price expectations, competitiveness in the market and demand expectations. Should such variables signal that the innovation is appropriate then KN becomes an instrument that contributes to the manager's competitive advantage.

But how real is KN on the production line? As indicated above, we can observe direct relevance in agriculture and forestry. The quality of farm products derived from quality soils is bound to be superior to the quality of those drawn from contaminated soils. The performance of workers on worksites that have higher aesthetic qualities is also bound to be higher than that on sites that display poor aesthetic qualities. It is possible to show that output per worker in a steel plant in Australia, where all emissions are strictly filtered, is higher than in a similar plant in a developing country, where the emissions are not controlled, despite lower wages in the developing country. Hence, despite higher costs, it is possible for Australian steel to remain competitive.

However, in many instances, the producer does not have complete control over the deterioration of KN. There is always a residual risk that something goes wrong. For example, the cyanide spill in Hungary by Esmeralda, the Union Carbide accident in Bhopal India and several other disasters illustrate this fact. It is in this context that we need to examine the role of the government and regulatory bodies, and this was considered in Chapter 6. See Box 8.3 for an example of the fragility of KN.

Box 8.3 The Murray River – A Case of Fragile KN in Australia

A quintessential example to illustrate the fragility of KN in Australia is the Murray River (the third longest navigable river in the world). For a moment, suppose that there is just one isolated and very small farm property in the entire area that spans the Murray River Basin. This is indeed a purely hypothetical situation. In this abstract scenario, the intensity of KM–L would be so low that the Murray River would not be threatened with the problem of degradation and fragility. However, in reality, as reported in the Australian Broadcasting Company's (ABC) Landline programme (24 September 2000), the highly fragile state of the Murray River has been due to:

- Many farms and associated activities such as land clearing
- Several secondary industries such as canneries, wineries and fruit juice manufacturers

The high level of economic activity in the Murray Basin points to a very high level of intensity for KM–L. Because the Murray River is an explicit input in the production function of the various farms on the Basin, either in terms of providing water or acting as a receptacle for runoff and discharges, efforts have been taken to restore and stabilise the river. One such effort was pioneered around 1986 by Mr. John Moore, the director of one of Australia's largest fruit juice manufacturing companies in the town of Loxton in South Australia. Mr. Moore's initiative was to plant some 250,000 eucalypt trees on a land area of 100 hectares on the river basin, and then pump the wastes from the fruit juice factory, together with the effluent from the town, into this land area. Wastes, which would otherwise have ended up in the Murray, became useful nutrients for the eucalypts. In turn, the eucalypts were capable of reducing salinity in the river system. The project was of course initially successful because, as stated in the Landline programme: 'Eucalypts like river red gum and flooded gum are proved water users; they act like a set of lungs pushing water out of the ground and into the atmosphere at twice the rate of passive evaporation'. Unfortunately, around 1996 Mr. Moore's company went into receivership and the effluent from Loxton was not sufficient to maintain the trees. The eucalypts in the land area has become known as 'Moore's Woodlot' and had begun to die.

Source: http://www.abc.net.au/landline/stories/s184313.htm

Review Questions

1 Critically discuss the claim that entropy is the primary driver of diminishing marginal returns.
2 Consider the following sectors of the economy: tourism, manufacturing, agriculture, fisheries and mining. Illustrate how firms in each of these sectors can conceptualise the existence of a production function: $Y = g(KM, L, KN)$.
3 In the examples chosen for question 2, illustrate the substitutability between KM–L and KN.
4 In the same examples chosen above, describe how firms could utilise KN more efficiently.

Note

1 This does not mean that KM and L are devoid of measurement difficulties. Different items of KM, such as buildings and machinery are usually aggregated into a monetary value using constant prices. L is hardly homogenous. Although the number of persons employed by a firm is convenient measure, variations in skill and managerial capability are inevitable.

References

Chiang, A. C. and Wainwright, K., *Fundamental Methods of Mathematical Economics*, McGraw Hill, New York, 2007.
Hirshleifer, J., Glazer, A. and Hirshleifer, D., *Price Theory and Applications: Decisions, Markets, and Information*, Cambridge University Press, Cambridge, United Kingdom, 2014.
Ruth, M., 'Energy use and CO_2 Emissions in a dematerializing economy: Examples from five US metals sectors', *Resources Policy*, 24: 1–18, 1998.
Ruth, M., *Advanced Introduction to Ecological Economics*, Edward Elgar, Cheltenham, England, 2018.
Schneider, E. D. and Kay, J. J., 'Life as a manifestation of the second law of thermodynamics', *Mathematical and Computer Modelling*, 19(6–8): 25–48, 1994.
Thampapillai, D. J., 'Lessons from science: Need for a rethink of concepts in economics', *Economic and Political Weekly*, 49(37): 79–83, 2014.
Thampapillai, D. J., 'Ezra Mishan's cost of economic growth: Evidence from the entropy of environmental capital', *Singapore Economic Review*, 61(3): 1–10, 2016.
Thampapillai, D. J. and Ohlmer, B., 'Environmental economics and stewardship', in Staib, R. (ed), *Business Management and Environmental Stewardship*, Palgrave MacMillan, London, 2009.
Valentine, S., 'Kalundborg symbiosis: Fostering progressive innovation in environmental networks', *Journal of Cleaner Production*, 118: 65–77, 2016.

9 Market Organisation and KN

In the preceding two chapters, we considered the changes that can be expected when the laws of thermodynamics and the principles of sustainability are explicitly introduced into frameworks that explain the concepts of demand, production, costs and supply. These changes need to be extended towards explaining how markets and their organisation would or should change.

As explained in Chapter 3, the concept of a market stems from the interaction between demand and supply. A central tenet in microeconomics is that the perfectly competitive market equilibrium leads to the maximisation of net benefits by recourse to the fulfilment of basic functions of *what to produce, how much to produce* and *how to produce*. (We'll leave the question 'where to produce' to economic geographers and regional scientists but do want to acknowledge here the importance of space and location, especially in the context of globalisation and the relations between local and global market dynamics that have discernible bearing on social outcomes and KN quality.)

A further precondition for equilibrium is a well-functioning system of private property rights. Although the perfectly competitive equilibrium is seldom evident in practice, it serves as an important benchmark against which to compare, imperfectly competitive market organisations. Such comparisons then provide the basis for correcting imperfections. These corrections generally fall under the heading of 'shadow pricing' in the literature of Cost-Benefit Analysis; Mishan and Quah (2007), Sinden and Thampapillai (1995). As observed in Chapter 4, an additional key challenge stems from the absence of markets for goods and services that originate from KN – hereafter labelled as KN_{GS}. This is because many KN_{GS} are essential ingredients in the production and consumption of items that are regularly exchanged in markets.

The absence of markets for KN_{GS} is mainly because of their intangible nature. For example, Mäler and Vincent (2003) cite such absent markets as a prime cause of resource misallocation resulting in KN and its KN_{GS} becoming unsustainable. The standard approaches in environmental economics to deal with such issues are generally two-fold:

- Development of methods of valuation (and measurement) so that KN_{GS} can be included in market analyses and market failures can be minimised. We will deal with these methods of valuation in Chapter 16.

140   *Microeconomics and KN*

- Assignment of property rights for (selected) KN_{GS} to private entities in the expectation that a market capable of maximising net social benefits will eventuate. We have dealt with the subject of property rights in Chapters 3 and 4.

The corrective measures in environmental economics are 'market-centric', which implies that the perfectly competitive market equilibrium may also coincide with the incidence of sustainability. The main aim of this chapter is to argue that such coincidence is unlikely and that the benchmark of *perfect competition* (PC) should be revised to include *sustainability* as an explicit condition besides the standard ones. Recall the five conditions of PC from Chapter 3: *anonymity, homogeneity, perfect mobility, perfect information* and *full employment*. We now add a sixth condition – *sustainability* – and the revised benchmark should then be appropriately renamed *perfect competition and sustainability* (PCS).

The distinction between PC and PCS is considered in the next section. As we illustrate, the recognition of PCS as the benchmark results in radically different outcomes – especially with reference to the axiomatic distinction between *goods* and *bads*. This is followed by an illustration of how the standard expositions, when comparing PC and imperfect competition, can markedly vary when PCS is used as the benchmark instead. We consider three types of market imperfections, namely monopoly, oligopoly and monopolistic competition. As we illustrate below, the acknowledgement of PCS as the benchmark could prompt individuals and firms to adopt stewardship towards KN. Such stewardship could also include the search for technologies that would extend the domain of assimilative ability and resilience of KN. We conclude the chapter by arguing that the conditions of perfect information and full employment require an in-depth understanding of issues surrounding sustainability.

Perfect Competition and Sustainability

Following our discussion in Chapter 2, sustainability can be defined as the perpetual flow of KN_{GS}, irrespective of the utilisation of KN stocks that generate them. In terms of our discussion in Chapter 8, the region of sustainability would coincide with the region of assimilative ability and resilience. That is, some threshold level of KN stocks (sinks and sources) should be perpetually retained. We identified this threshold level in Chapter 8 as KN_T.

Our claim in this chapter is as follows. In just the same way as PC provides a benchmark for economic analyses, PCS, with the added condition of sustainability, provides an appropriate benchmark for the analytics of sustainable development. One could argue that the unique conditions of the PCS benchmark are such that the imbalances imposed on KN can be restored through either natural processes or explicit acts of stewardship. For example, the anonymity condition is such that any contaminating emissions should be small enough to not exceed the assimilative capacity and resilience of KN. At the same time, if the condition of perfect information encompasses KN, including the knowledge of the requirements to maintain the perpetual flow of KN_{GS}, then it is possible to argue that individuals will voluntarily control their extractions and emissions. For example, the much-celebrated work of Dasgupta and Heal (1974) and Solow (1974) reveals the feasibility of a permanent flow of consumption, despite the constraints of exhaustible resources. They of course allow for substitutions between exhaustible resources and reproducible resources. The important point is that, even if Dasgupta-Heal-Solow did not explicitly revise the PC benchmarks, their assumptions were near enough to the PCS benchmark that we address here. That is, while anonymity could ensure that the magnitudes of withdrawals are small, perfect information (and foresight) could prompt the choice of renewable KN over non-renewable KN. That said, the creation of 'reproducible resources' and their utilisation may, however, require non-renewable resources. If that is the case, then allowances will need to be made to ensure perpetual substitutability. Besides, as we noted in Chapters 5 and 6, renewable resources themselves have the potential to lose their renewability, i.e. when they are over-utilised.

If KN is perfectly sustainable, then the withdrawal of KN_{GS} from it would not entail an opportunity cost. As an example, consider the case of a small wine-maker who operates on a riverbank. The wine-maker draws water from the river and then discharges the residues of the wine making process into the river. The volume of withdrawal is such that it does not affect the hydrology and biology of the river. Further, suppose the river is resilient by being able to assimilate any residues and retain its ecosystem properties without the wine-maker having to treat the residues. Then the wine-maker does not incur an opportunity cost for using the river. The river displays a state of perfect sustainability. Alternatively, consider the case where the residues have to be treated prior to their discharge into the river because the river is unable to assimilate them without treatment. In this case, the river retains its properties and its sustainability, but has imposed an opportunity cost on the wine-maker. In the context of several small (anonymous) wine-makers along the river bank, the treatment of residues and, as a result, added marginal costs in order to maintain water quality, would become imperative. Besides, a set of river catchment management actions to assure adequate water flow in the river could also become imperative.

The distinction then between PC and PCS lies in the adoption of actions (if necessary) that would always render KN sustainable. The net result is that the size of net benefits in the context of PCS will be always smaller than those observed in the context of PC. Such a result could stem from the actions of both consumers and producers. Consumers could, in their awareness of sustainability as argued in Chapter 7, choose modest patterns of consumption, while producers would bear higher marginal costs to protect KN assets. Both sets of actions represent stewardship towards KN, prompted chiefly by perfect information that encompasses knowledge of the thermodynamic principles and of sustainability.

Consider Figure 9.1. The upper panel represents the perfectly competitive market for a specific commodity. The lower panel explains the formation of unwanted discharges and emissions – labelled hereafter as E – that emerge from the utilisation of KN for producing market output (Q). In other words, $[E = f_E(Q)]$ can be

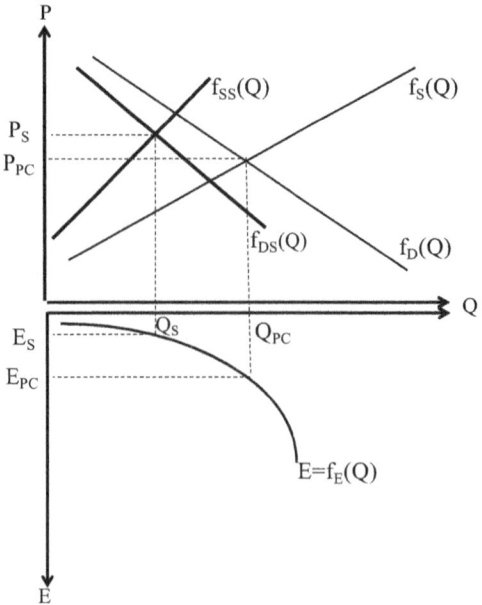

Figure 9.1 PC vs PCS

regarded as a KN damage function. Note that the vertical axis of the lower panel, is E. The horizontal axis of this lower panel, namely Q is jointly shared with that of the upper panel. E_S represents the volume of E that KN sinks can tolerate after allowances have been made for the utilisation of KN in all markets. The recognition of E_S differentiates the market equilibrium and sustainability conditions for PC and PCS, as illustrated in Figure 9.1:

$$\text{PC: Demand}\left(f_D(Q)\right) = \text{Supply } f_S(Q)) \rightarrow \left[\left(Q^*, P^*\right) \text{ and } \left(E = E^* > E_S\right)\right] \quad (9.1)$$

$$\text{PCS: Demand}\left(f_{DS}(Q)\right) = \text{Supply } f_{SS}(Q)) \rightarrow \left[\left(Q_S, P_S\right) \text{ and } \left(E = E_S\right)\right] \quad (9.2)$$

$$\left(\int_0^{Q^*} f_D(Q) \cdot dQ - \int_0^{Q^*} f_S(Q) dQ\right) > \left(\int_0^{Q_S} f_{DS}(Q) \cdot dQ - \int_0^{Q_S} f_{SS}(Q) dQ\right) \quad (9.3)$$

This result may seem very similar to that of internalising an externality. However, the binding configuration for PCS is (Q_S, E_S). Several permutations for $(f_{DS}(Q))$ and $(f_{SS}(Q))$ would be possible for achieving compliance with (Q_S, E_S). Perfect information could make producers adopt known technologies that limit market output to Q_S, while consumers armed with the same knowledge could also reduce their demand; and so on. The determination of (P_S) will rest on some specific production technology and consumption pattern within the bounds of (Q_S, E_S). That is, the same way as the invisible hand (discussed in Chapter 3) supposedly guides the emergence of (Q*, P*) in the context of PC, the invisible hand – in the context of PCS – now armed with perfect information about KN, the future, and the requirements of other markets for utilising KN sinks, will guide the emergence of (Q_s, E_s, P_s). The size of net benefits as identified in (9.3) above, would be smaller in the context of PCS than that of PC. However, as we illustrate below, with PCS, net benefits can continue indefinitely.

PC Versus PCS and Preference Relations

The preference relations for E and Q in the context of the binding configuration in PCS, namely (Q_s, E_s), can be explained as follows. Given the basic knowledge concerning E_s, two sets of values for E are identifiable. These are: $\left(e_1 \in (E \leq E_S)\right)$ and $\left(e_2 \in (E > E_S)\right)$.

The conditions of PCS would dictate that $(e_1 \succ e_2)$. This preference relation in turn will dictate the preference relation for two sets of market output quantities. That is, $\left(q_1 \in (Q \leq Q_S)\right)$ and $\left(q_2 \in (Q > Q_S)\right)$, where $(q_1 \succ q_2)$. Note here a reversal in the basic axiom of preference. That is, with PCS producing less of a good is preferred to producing more of it. This is due to the dis-amenities of resource use and waste generation stemming from production. Hence, the domains of utility maximisation would be smaller in the context of PCS than PC. Furthermore, in the event that such domains need to be extended, the innovation would primarily involve the search for technologies that would reduce the size of $(df_E(Q)/dQ)$ and shift $[E = f_E(Q)]$ to the right.

The market equilibrium in PCS will be such that a perpetual permanent net benefit (NB) with a present value (PV) as defined below is always feasible:

$$PV(NB) = \int_{t=0}^{\infty} \left(\int_0^{Q_S} f_{DS}(Q) \, dQ - \int_0^{Q_S} f_{SS}(Q) \, dQ \right) e^{-rt} dt \qquad (9.4)$$

In (9.4), r represents the social rate of discount. In the context of PCS, it is possible to argue that the discount rate (r) could tend to zero; then $\{PV(NB) \to \infty\}$.

One can further reinforce reversal in the axiom of preference explained above by distinguishing the equilibrium conditions that maximise net benefits in the context of PC from that of PCS. Recall from Chapter 3 (Figure 3.3) that PC results in a unique configuration of market equilibrium quantity and price, namely (Q*, P*) that yields maximum net benefits. That is, the area bound by the demand and supply curves is highest for (Q*, P*). A corollary of this result with PC is that the bigger the market the bigger is the net benefit. This observation, as we have mentioned before, sits comfortably with the traditional axiomatic differentiation between 'goods' and 'bads' in microeconomics:

Goods are those where *more is preferred to less*
Bads are those where *less is preferred to more*

Recall from Chapter 3 that 'goods' generate positive utility, whilst 'bads' such as pollution and crime do the reverse.

Consider now the PCS benchmark that includes sustainability as the explicit sixth condition besides the five conditions of PC. Then, the role of time is explicit in the benchmark as is evident in Equation (9.4). In this context, to find the highest net benefit, one would adopt the optimisation of an inter-temporal consumption function such as that illustrated by Dasgupta and Heal (1974). As a result, the criteria for the choice of markets would be those with a configuration of *quantity and price that maximises net benefits over an infinite time period*. Despite the illustration being theoretical and abstract, the basic axioms of economics would change quite drastically as follows:

Goods would be those where *less is preferred to more*
Bads would be those where *more is preferred to less*

This axiomatic change that unfolds from PCS, prompts moderation over excess and extravagance. It is the recognition of this premise that made Boulding (1945) deem:

> Any discovery which renders consumption less necessary to the pursuit of living is as much an economic gain as a discovery which improves our skills of production.

Kohr (1957) was even stronger in his condemnation of 'bigger is better' that emerges when sustainability is ignored:

> ... there seems to be only one cause behind all forms of social misery: bigness. Over simplified as this may seem, we shall find the idea more easily acceptable if we consider that bigness, or oversize, is really much more than just a social problem. It appears to be the one and only problem permeating all creation. Whenever something is wrong, something is too big.

Schumacher's (1973) influential book titled – *Small is Beautiful: A Study of Economics as if People Mattered* – synthesises and justifies conclusions such as those noted above. For the possibility of net benefits from markets to exist in perpetuity, the discount rate (r) in Equation (9.4) needs to be also as small as possible – close as possible to zero – to acknowledge the fact that the future is as important as the present. However, embracing the concept of 'small' can be tenuous in the contexts we consider below.

Monopoly

Monopoly is a market configuration characterised by only one seller, but with many buyers. The main sources of monopoly are: *exclusive control over inputs; network economies; patents and copyrights; government licenses or contracts;*

and economies of scale. In each instance, the monopolist is able invoke barriers to prevent the entry of rival firms into the market. Such barriers may be either strategic or legal. For example, if a steel-making firm acquires the mine that supplies iron ore to its steel-mill, then this firm gains monopoly power through exclusive control over an important input. Besides, if this firm can take control of a minor network of firms that contribute to steel-making, such as transport of materials and labour hire firms, then the steel maker has further strengthened his/her monopoly power. Legal sources of monopoly include government contracts and patents. For example, if the steel maker has won the contract to manufacture rail lines for a country's national railways – the firm's monopoly power is enhanced. The contract prevents rivals from entering the market for supplying rail lines. Patents, though common in the pharmaceutical industry, can be also found in other areas such as engineering and manufacturing design. That is, a firm holding a patent can affect the welfare of society through its monopoly pricing strategies as we demonstrate below.

One important barrier for entry of multiple firms into a monopoly market is the incidence of very high capital costs accompanied by low operating (variable) costs. Such incidence leads to the average costs (AC) declining over a large domain of quantity produced resulting in economies of scale. The direct implication of economies of scale is that a monopolist is capable of edging potential rivals out of the market by producing large quantities at a much lower cost. Recall from earlier narratives, in Chapters 3 and 8, on the relationship between marginal costs (MC) and AC. That is, if the AC is falling, then MC will remain below AC. The context where MC is perpetually below AC is referred to as a *Natural Monopoly*, examples of which are typically the provision of public utilities like telecommunications, water and electricity. In these cases, special pricing considerations are important. Leaving issues concerning natural monopoly to the end of this section, consider first how a monopolist sets price in general.

In PC, price is determined by the equilibrium between supply and demand in the market, and each firm is a price-taker. With a monopoly, the firm determines the price based on its perceived demand and aims to maximise profit. To illustrate, suppose that the monopoly firm faces the following linear demand function: $(P = \alpha - \beta Q)$. This firm will then estimate its potential total revenue (TR), namely $(P * Q)$ as:

$$TR = \left(\alpha Q - \beta Q^2\right). \tag{9.5}$$

Hence, the firm's marginal revenue (MR) will be:

$$MR = \left(dTR/dQ\right) = (\alpha - 2\beta Q). \tag{9.6}$$

The monopoly firm will maximise its profit by producing the quantity that corresponds with MR = MC.

To further illustrate suppose that this firm's total cost (TC) is given by:

$$TC = \gamma Q + [(1/2)\delta Q^2], \text{ and hence:} \tag{9.7}$$

MC = γ + βQ.   (9.8)

Based on (MR = MC), monopoly's profit maximising quantity (Q_M) is given by:

$Q_M = (α − γ)/(2β+δ)$.   (9.9)

Consider Figure 9.2, where we illustrate the standard textbook features of profit maximisation by a monopoly. Two features are noteworthy. First, the net social benefit in the absence of monopoly would be defined by the area ABC, whilst in the presence of monopoly the area would be smaller, namely AIJB. The loss in net social benefit due to monopoly is area IJC and is often referred to as a deadweight loss. The second feature is the size of profit. In Figure 9.2, we identify two sets of prices and costs, as follows:

- P_{PC} – Market price that would prevail under PC and coincides with P = MC;
- PM – Monopoly market price based on MR = MC;
- ATC_{PC} – ATC that corresponds with quantity (Q_{PC}) that is produced under PC;
- ATC_M – ATC that corresponds with quantity (Q_M) that is produced under monopoly.

If forced to adopt (Q_{PC}, P_{PC}), then the monopolist's profit would be: $\{(P_{PC} − ATC_{PC}) * Q_{PC}\}$. Monopoly profits based on (Q_M, P_M) would be $\{(P_M − ATC_M) * Q_M\}$. It is not difficult to appreciate that $\{(P_M − ATC_M) * Q_M\} > \{(P_{PC} − ATC_{PC}) * Q_{PC}\}$.

The comparison of PC with monopoly in standard texts (Frank 2010; Pindyck and Rubinfeld 2007) highlights the case of deadweight loss and the higher profits that unfold with a monopoly relative to PC. Policy intervention is largely aimed

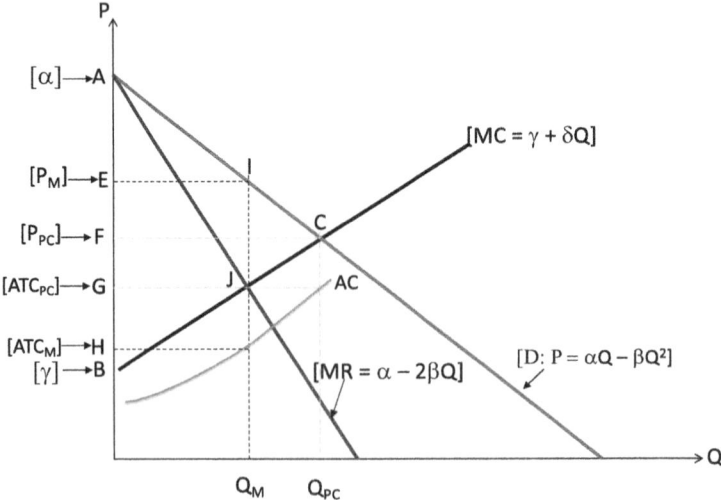

Figure 9.2 Pricing and Welfare Losses Due to Monopoly

at moving the market closer to a PC context. Standard texts often relegate KN issues to sections dealing with public goods and externalities. But if KN issues are also considered alongside the market outcomes due to monopoly, then a different approach to policy is likely to unfold.

We now introduce the KN damage function, namely $[E = f_E(Q)]$ in Figure 9.3. Consider again the equilibrium in the context of a monopoly as determined by the equality between marginal revenue and marginal cost, (Q_M, P_M). This equilibrium is likely to result in a lower level of emission E_M compared to E_{PC} that would prevail in PC. This inference can be misleading on at least two counts. First, following the foregoing argument, the actual comparisons should be made with PCS and not PC. Second, even the comparison with PC could be misleading, because it is unrealistic to assume that the monopolist will employ the same marginal cost, $\{f_S(Q) = MC\}$ as the aggregate of a collection of small anonymous firms. If endowed with a greater degree of economies of scale, then the monopolist's MC (labelled as MC_M in Figure 9.3) could be lower. Then the quantity-emission outcome could be higher and the so-called deadweight loss smaller compared to the narrative in standard text books. There is also the possibility that, for a monopolist with economies of scale, the generation of desired output may be higher while emissions may be kept more at bay than may be the case with the aggregate of small firms. The implications for the so-called deadweight loss are then less clear-cut.

Should the economies of scale be more significant, it is plausible for Q_{M1} to approach Q_{PC} or even exceed it, and likewise for E_{M1} and E_{PC}. This possibility can

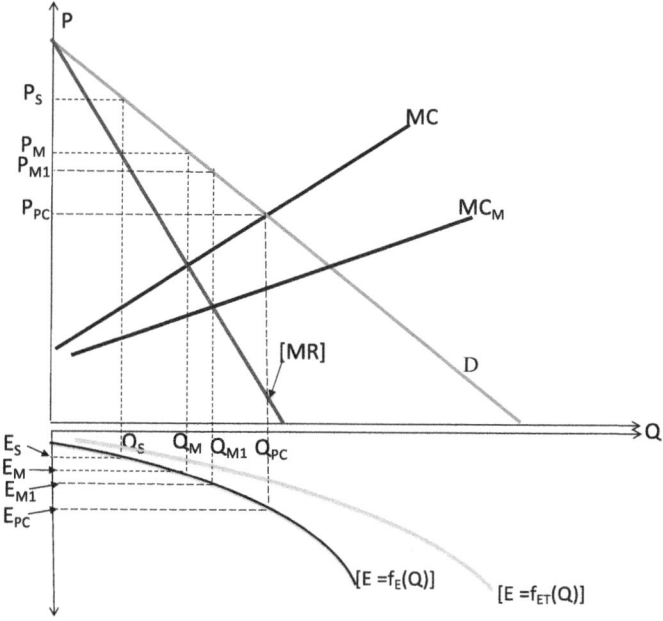

Figure 9.3 PC and PCS vs Monopoly

be illustrated as follows. Reconsider the demand and supply functions considered above, namely: $(P = \alpha - \beta Q)$ and $(P = \gamma + \delta Q)$. Suppose that the initial context is that of PC and transformation to a monopoly is in place. Such transformation includes a change in the supply schedule (because of economies of scale), namely $\left\{ P = \overline{\gamma} + \left[\dfrac{\delta}{\eta} \right] Q \right\}$ and $\overline{\gamma} < \gamma$, $(\eta > 1)$. That is, the gradient of the supply schedule becomes smaller than δ because δ is now divided by a positive number in excess of 1. Hence in Figure 9.3, for Q_{MI} to exceed Q^* the following inequality should be feasible:

$$[(\alpha - \overline{\gamma})(\beta + \delta)] > \left[(\alpha - \gamma)\left(2\beta + \dfrac{\delta}{\eta} \right) \right] \qquad (9.10)$$

The inequality stated in (9.10) is feasible, as long as ($\overline{\gamma} < \gamma$) and ($\eta > 1$). Hence, the possibility of E_{MI} exceeding E_{PC} is also feasible.

The comparison with PCS, however, could present a different picture. Note that the binding configuration in PCS is (Q_S, E_S). Therefore, transforming the PCS context into an unregulated monopoly, which displays economies of scale, raises the possibility of higher levels of (Q, E) relative to (Q_S, E_S). That is, there is a stronger case for the inequality expressed in (9.10) in the context of PCS than PC.

As indicated, the recognition of sustainability as a condition for the operation of the economy requires an understanding of (Q – E) space for KN and KN_{GS}, besides the usual (Q – P) space that deals with market analysis. Such consideration would also prompt the various permutations of feasible behaviour by monopolies, as illustrated above. Additionally, the nature and extent of government intervention need to be guided by the need to achieve compliance with (Q_S, E_S). As indicated in Chapter 5, the intervention could take the form of taxes and/or imposition of standards. Such intervention would reduce monopoly profits because of higher marginal costs associated with treating emissions ($E_{MI} - E_S$). Also, note that a monopolist who seeks to sustain profits over an infinite time horizon could also choose to undertake such treatment voluntarily. For example, the monopolist could search for, invest in, and implement, technologies that render $[E = f_E(Q)]$ in Figure 9.3 to be more cost-effective. That is, the monopolist's action would shift this function towards the right as illustrated by $[E = f_{ET}(Q)]$ in Figure 9.3. Such technologies, though not limited to monopolists, include measures such as biomimicry and closed-loop production systems. We consider these in Chapter 19.

Prior to concluding our narrative on monopoly, we need to briefly revisit the case of natural monopoly. As indicated earlier, this is a context where MC tends to always remain below AC, primarily because of very high capital costs compared to operating costs. Again, as indicated earlier, the contexts are usually the supply of utilities such as water, electricity, public transport and telecommunications. Consequently, as illustrated in Figure 9.4, equating price with MC will result in a loss. That is, $[TR = (P_{MC} * Q_{MC})] < [TC = (P_{AC} * Q_{MC})]$. Alternatively, if left to a private monopoly, the resulting price, P_M based on (MC=MR), could render this market inaccessible to poorer sections of the community. A popular option is for

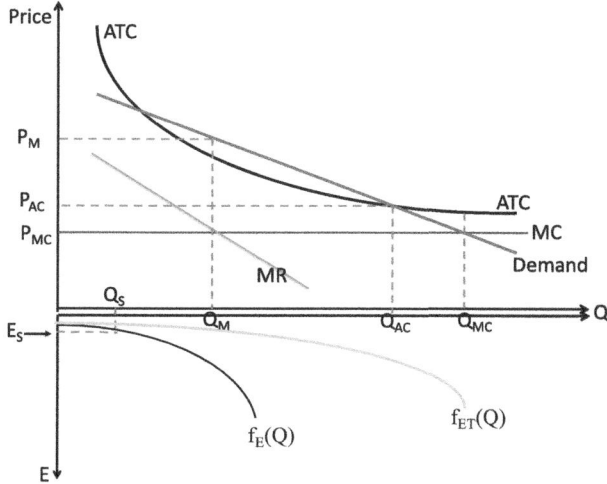

Figure 9.4 Natural Monopoly

a government agency to manage the market with average cost pricing. That is, (P = P_{AC}), yielding zero economic profit, but with some safeguards such as a conservation tax to act as a buffer.

Attempts to privatise the managing agency and run the market as a regulated monopoly have not always been successful. The trend has been to corporatise the government agency and make it behave like a private sector entity – but with limits on pricing strategies. Besides pricing, considerations pertaining to the regulation of E remain the same as we considered above. Achieving (Q_S, E_S) is more difficult with a natural monopoly than with a regular monopoly. This is because of the essential nature of the output with reference to items such as water and electricity, and hence the large volume of Q required to meet public demand. Nevertheless, for utilities such as electricity and public transport, transitioning away from the utilisation of GHGs has become a priority. Water supply decisions need to ensure that the hydrologic balance of the water resource system is maintained and conservation measures such as rainwater tanks and wastewater reuse have become important. Such measures are tantamount to shifting the KN damage function to the right; that is, as illustrated in Figure 9.4, a move from [E = $f_E(Q)$] towards [E = $f_{ET}(Q)$].

Oligopoly

Oligopoly is a market configuration where a few *dominant* and *interdependent* firms control both production and pricing decisions. They sell similar, but branded products. Examples include large supermarket chains, petrol retailers such as Shell, Caltex, and BP, print and television media outlets, pharmaceutical firms, major trading banks, and branded restaurants.

Most economics texts explain *dominance* by recourse to the concept of concentration ratio, namely share of the market that a firm controls. A monopolist, for example, has a 100 per cent concentration ratio with reference to the product it sells. At the other extreme, with PC and PCS, each participating firm would have the same and infinitesimally small concentration ratio. With oligopolies, the concentration ratio can be as high as 70–80 per cent and the number of dominant firms range from two to usually six or seven. A duopoly is found where two firms dominate the market as with the domestic airline industry in Australia. Some firms attempt to enhance their dominance through mergers and acquisitions as well as creating subsidiaries under different trading names. But such attempts need to survive the scrutiny of government legislations in the form of competition policy. Governments in general do attempt to increase competition in markets by reducing the concentration ratio. Oligopolists in return erect barriers to entry. Such barriers are like those witnessed with monopoly, namely patents and economies of scale.

Different degrees of collusion describe the *interdependence* between oligopolistic firms. Strict collusion results in the formation of cartels where each firm in the grouping abides by the rules on production and pricing. The Organisation of Petroleum Exporting Countries (OPEC) is an example of a cartel that was formed in 1960. Each cartel has a designated leader to expedite the strategies agreed on. Usually, collusion is tacit with an acknowledged market leader who informally sets prices and production quotas to which other producers respond. Despite such leadership, oligopolists tend to compete with and outsmart each other in the expectation of garnering higher profits. Strategic collusion between firms can also be an additional barrier to entry of other firms into the market. Such collusion may also be a driver to force some firms to exit the market.

In determining pricing and supply decisions, an oligopolistic firm will include in its deliberations not only the response of consumers, but also its speculation about how rival firms would react. Therefore, describing an oligopolist's behaviour is more complicated than formulating models of monopoly or PC/PCS. The general approach to modelling an oligopolist's behaviour is based on game theory. Here the oligopolist will assign probabilities to potential outcomes of its production and pricing policy. The compilation of anticipated behavioural responses is likely to reveal that an oligopolist will face a kinked demand curve as illustrated in the upper panel of Figure 9.5. If this firm attempts to raise its product price above the kink, that is (Q_0, P_0), rival firms are likely to be tardy in following suit for fear of losing their current market share. Further, consumers could also switch producers in search of lower prices for similar, if not identical, but differently branded products. Hence the demand curve above the kink is expected to be relatively elastic. Alternatively, the lowering of the price below the kink could prompt consumers to increase their purchases, and thereby enable the firm to enjoy a temporary increase in its market share. This increase in market share is temporary because rival producers would now follow suit in lowering prices, and a price war could ensue.

But, what of outcomes in Q-E space? Would this firm be able to attract more consumers and increase its market share by displaying stewardship towards KN? Oligopolists will face challenges similar to those faced by monopolists to comply

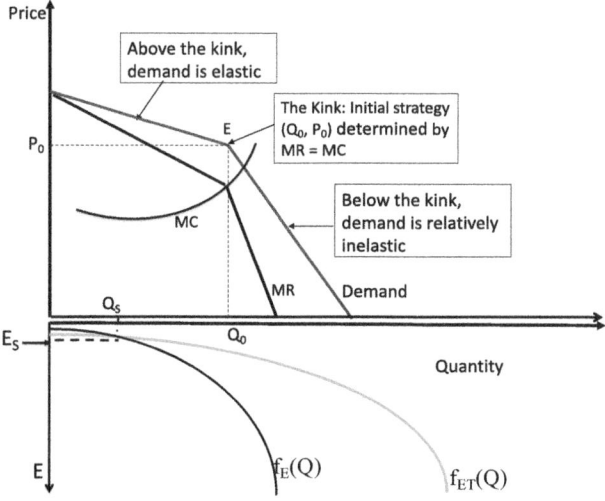

Figure 9.5 Oligopoly

with (Q_s, E_s) – that is the sustainable level of output and KN damage. As illustrated in Figure 9.5, they need to invest in cost-effective technology to shift the KN damage function outwards such as that portrayed by $[E=f_{ET}(Q)]$. An oligopoly that causes significant KN damage in Southeast Asia is the Palm Oil industry established on the waterlogged peat soils. The agricultural practice of draining the water before planting the palm seeds and exposing the peat soil to sunlight alone generates a significant volume of GHGs. As we indicate below, searching for a technology to reduce GHGs for this specific context is futile. It is best to leave waterlogged peat soils undisturbed. Further, oligopolists also tend to engage in token gestures of their stewardship towards KN. For example, major supermarket chains have replaced plastic shopping bags with ones made of paper – yet many of the items that consumers purchase are contained in plastic containers or wraps. As a further example, airlines sell so-called 'carbon-neutral' tickets, whilst some inflight services and activities associated with the sale could negate carbon neutrality.

Monopolistic Competition

Monopolistic competition is a market configuration that combines the features of monopoly and PC. Market imperfection stems from the fact that many firms compete by differentiating their products. Product differentiation removes the homogeneity and anonymity conditions of PC. But, as with PC, there are no barriers to entry or exit and the flow of information. Examples include restaurants – mainly in food courts, budget brands of household consumer items such as canned foods and cosmetics and hairdressers. Each firm in a monopolistically competitive market operates on the basis of its perception of residual demand. Such demand is the unsatisfied quantum of demand that a producer can still capture after a previous

152 *Microeconomics and KN*

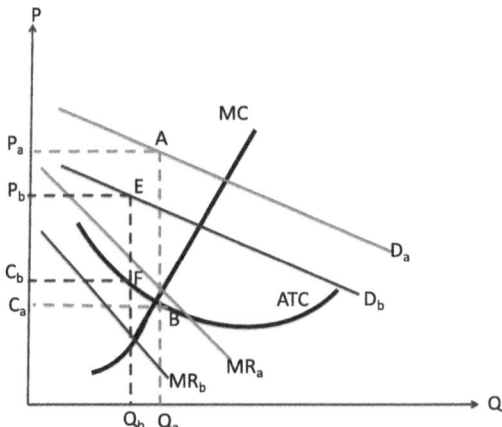

Figure 9.6 Monopolistic Competition

market transaction has been completed by another producer. To illustrate, consider a common household consumption item, namely rolled oats. A firm specialising in a superior brand with fancy packaging sells this item to its loyal set of consumers. Although brand loyalty could serve as a barrier to entry, the demand for rolled oats has not been exhausted. An unnamed producer could enter this market with a differentiated product – say with cheaper packaging and price compared to the superior brand. As a result, this new producer could capture the demand for rolled oats that still persists. As with a monopoly, the producer would aim to maximise profits by determining production quantity that complies with MR = MC.

We illustrate in Figure 9.6 a sequence of events that could unfold for single firm that enters a monopolistically competitive market. Suppose that this firm initially perceives a residual demand of (D_a) for its product and estimates the associated MR as (MR_a). Equating MR_a with MC, this firm would produce Q_a units of output and charge a price of P_a. The profit accruing to this firm would be: $(TR - TC) = [(P_a * Q_a) - (C_a * Q_a)]$ = Area $[P_a ABC_a]$. Because there are no constraints on information flows, other firms would learn of this profit and enter the market. Such entry would shift the demand faced by this firm down from D_a to D_b. And the revised MR would be MR_b. Based on (MR_b = MC) new quantity and price would be (Q_b, P_b). As illustrated in Figure 9.6, this firm's profit would reduce to area $[P_b EFC_b]$. The theory of monopolistic competition suggests that as long as positive profits can be discerned, other firms will continue to enter the market and drive the firm's economic profit to zero.

The pertinent question is: *How could monopolistically competitive firms display stewardship towards KN?* In fact, product differentiation enables monopolistic firms to differentiate their products by introducing KN-friendly attributes. We discussed such attributes in Chapter 7. That is, increasing competition with an improved menu of KN-friendly attributes could enable a cleaner industry to emerge. But it is likely that such measures could fall far short of meeting the requirements for ensuring the sustainability of KN.

Perfect Information and Full Employment of KN

Recall the narrative surrounding the rationale for Equation (9.4) above, leading to the sustainability of KN. Consumers and producers have to make well-informed choices that would result in the perpetual flow of net benefits. Perfect information is an essential condition for achieving this result. An important implication of this result is that *'smaller'* is better than *'bigger'*. That is, consumers armed with knowledge about sustainability, and acting in their self-interest on the basis of this knowledge, would consciously seek items that would safeguard KN including smaller quantities of purchase. For their part, producers too being perfectly informed, respond accordingly by producing smaller quantities of KN-friendly goods.

Standard economics texts such as Pindyck and Rubinfeld (2007), do consider the importance of perfect information. But the narrative is focused on how the absence of such information could cause markets to fail and not expand. The narrative is pursued under the heading of asymmetric information. That is, one party to a market transaction knows more about the items being transacted than the other party. Such information asymmetry is common amidst all the imperfect market organisations considered above. For example, monopolists and oligopolists may not disclose all characteristics of their production context in their quest to protect and/or expand their market share. For similar reasons, monopolistically competitive firms may falsely advertise their product features. The popular example cited in most texts pertains to used-car saleyards and the 'Market for Lemons'. That is, the seller knows which vehicles for sale are defective, whilst the customers are usually unaware. Most textbooks also differentiate this information asymmetry in terms of the timing of the imperfection with reference to the market transaction. 'Adverse Selection' occurs before the transaction is completed. For example, a producer may lure investors' commitment to their risky enterprise without full disclosure – the same way as in the Market for Lemons. Many of the transactions in the financial markets of the United States that led to the Global Financial Crisis are often cited as examples of Adverse Selection. 'Moral Hazard' occurs after the transaction has been completed. One might procure, for example, an insurance policy or a business loan on the strength of declaring compliance with the adoption of stringent safeguards – but flagrantly violate these safeguards after the procurement.

With regards to KN and the sustainable flow of KN_{GS}, perfect information is a tenuous premise. For example, consider the following claim: *'Achieving net-zero emissions by 2050 will prevent further global warming'*. There may be some element of truth with this statement – but it would be difficult to prove it. As per our discussion in earlier chapters, reducing emissions alone would not suffice. A confluence of several factors such as: changes in forest canopy density, hydrologic system balance, and mining would all contribute to GHG emissions and global warming. A much higher degree of uncertainty prevails with reference to KN-related claims than those pertaining to markets with asymmetric information considered above. Of significant concern is the recurrence of extreme weather events such as catastrophic fires and devastating floods with both increased frequency and

154 *Microeconomics and KN*

intensity. In these contexts, knowledge possessed by indigenous communities has gained momentum. As explained by McAllister et al. (2023): 'Indigenous knowledge ... includes long histories and guidance on how to live with, and as part of, nature'. The longest-living culture on earth is that of the Australian Aborigine and spans a period extending to 60,000 years. An important part of this culture is *Truth Telling* – which includes a narrative on how to live with nature.

Box 9.1   Perfect Information and Truth Telling

On the 14th of October 2023, Australia went to the polls on a referendum election. This referendum was called the Voice and included the following three features:

- Recognition of Aboriginal and Torres Strait Islander peoples, namely the indigenous people, in the Australian constitution as the original inhabitants of Australia
- Voice would advise the Australian Government on matters affecting the indigenous community
- *Truth Telling*

Australians were given a choice to vote either YES or NO. Nearly sixty per cent of Australians voted NO. More details about the Voice can be found at: https://www.anu.edu.au/about/strategic-planning/indigenous-voice-to-parliament

Our intention of bringing this event in the context of the section on Perfect Information is to argue that the entirety of the sixty-thousand-year history of the Australian original inhabitants was not revealed during the referendum campaign. That is, *Truth Telling* was incomplete. Quite understandably the narratives focused on the injustices done to the indigenous population during colonisation. However, an important feature of the indigenous culture, is stewardship towards KN, and this was not highlighted. In fact, several recruits to work in the national parks and wildlife service of the various Australian States and Territories are drawn from the indigenous community. Indigenous knowledge would no doubt be important in the context of combatting the increasing frequency and intensity extreme climate-related events.

Sadly, the majority of the Australian public failed to appreciate in full the importance of *Truth Telling*. In the midst of arguments and counterarguments, *Truth Telling* was dominated, understandably by injustices inflicted on indigenous communities. The indigenous population's respect and stewardship for KN did not gain any priority.

The condition of full employment with reference to KN in PCS requires clarification. As indicated above, a slight disturbance to the waterlogged peat soils in Southeast Asia can lead to the generation of significant volumes of GHGs. There are of course several other examples of sensitive KN assets such as old-growth forests, lakes and rivers, coastal estuaries and marine ecosystems, as well as subterranean ecosystems including groundwater resources. Disturbance of these systems through various economic activities could unleash a string of harmful externalities. In these contexts, the condition of full employment would be tantamount to the *preservation of the relevant ecosystem*. That is, to leave the ecosystems intact without harnessing them for any form of market development.

Recall our narrative in Chapter 8, especially that with reference to Figures 8.2 and 8.3. These figures were based on the function $\left\{Q = g\left(KN, \overline{KM}, \overline{L}\right)\right\}$, where KM and L are held constant whilst KN remains variable. We identified two features of this function, namely the domain of assimilative and resilience, and the gradient of KN degradation with respect to output (Q). These two features together identify the potential employability of KN. That is, it would not be sufficient to retain a domain of assimilative ability and resilience for KN to be productively employed. We need to also necessarily minimise the gradient of KN degradation so that the flow of KN_{GS} can continue without being hindered. Figure 8.3 illustrates that KN is more stable at lower levels of Q than at higher levels. *Truth Telling* of indigenous communities – the closest we can get to perfect information – would suggest that we must preserve important ecosystems.

Concluding Remarks

The implications of the foregoing simple analysis are at least threefold. First, in the sphere of applied economics and policy analysis, PC has served as an important benchmark – especially in the shadow pricing literature dealing with project appraisal and cost-benefit analysis. The main argument in this chapter is that PCS would prove to be a better benchmark, because it would facilitate the choice of decisions that would satisfy the criteria for sustainability. For example, consider the stimulus measures that were hurriedly mobilised in the wake of the (2008–2009) global financial crisis and the Covid-19 outbreak in 2019–2020. Very little attention was paid to the importance of sustainability. In fact, many of the measures introduced seriously compromise the possibility of sustainability. Had PCS been firmly registered as the benchmark in economics at the outset, the global financial crisis (for that matter, any such crisis) might not have surfaced. Even if it had, the response measures would have readily embraced modesty and the adoption of what environmental scientists label 'closed-loop' production systems; namely, those that do not pose additional burdens on KN sinks. The Industrial Park at Kalundborg in Denmark (discussed in Chapter 8) illustrates this concept.

Second, the recognition of PCS prompts the reassessment of comparisons with imperfect market organisations such as monopoly, oligopoly and monopolistic competition. As illustrated in the previous section, for example, within an unregulated context monopolies, if able to exploit economies of scale, can expand output beyond the limits dictated by PC and PCS and thus compromise the possibilities of sustainability.

Finally, and more importantly, the introduction of PCS as a benchmark enables the exposition of voluntary stewardship as a potential measure towards sustainability. The notion of stewardship is not readily found within the economics policy literature, barring perhaps Frank's (2010) foray into the economics of altruism. Voluntary stewardship inevitably ushers in modesty in favour of extravagance and compliance with the types of revisions suggested in Chapter 7. The distinction between extravagance and modesty is inevitably murky. Nevertheless, the recognition of PCS as the benchmark could propel economies towards some accepted norms of modesty.

Review Questions

1 Is it possible for the conditions of perfect competition to accommodate the existence of sustainability?
2 Critically evaluate the claim made in this chapter that the recognition of sustainability as an additional condition would lead to the axiom of preference being *less is preferred to more*.
3 Discuss how the PCS benchmark considered in this chapter could prompt voluntary stewardship towards KN.

References

Boulding, K., 'The Consumption Concept in Economic Theory', *American Economic Review*, 35(2): 1–14, 1945.
Dasgupta, P. and Heal, G. M., 'The optimal depletion of exhaustible resources', *The Review of Economic Studies*, 41(1): 3–28, 1974.
Frank, R. H., *Microeconomics and Behavior*, McGraw-Hill, New York, 2010.
Kohr, L., *The Breakdown of Nations*, Routledge & Kegan Paul, London, 1957.
McAllister, T., Macinnis-Ng, C. and Hikuroa, D. C. H., 'Indigenous knowledge offers solutions, but its use must be based on meaningful collaboration with Indigenous communities', *The Conversation*, 23 March 2023.
Mishan, E. J. and Quah, E., *Cost-Benefit Analysis*, Routledge, London, 2007.
Pindyck, R. S. and Rubinfeld, D. L., *Microeconomics*, Pearson-Prentice Hall, Singapore, 2007.
Schumacher, E. F., *Small Is Beautiful: A Study of Economics as If People Mattered*, Blond and Briggs, London, 1973.
Sinden, J. A. and Thampapillai, D. J., *Introduction to Benefit-Cost Analysis*, Longman, Melbourne, 1995.
Solow, R. M., 'Intergenerational equity and exhaustible resources', *Review of Economic Studies: Symposium on the Economics of Exhaustible Resources*, 1974.

Part III
Macroeconomics and KN

10 Some Important Concepts in Macroeconomics

In this and the chapters that follow, we explore how macroeconomic analyses need to change when we afford explicit recognition to the role of KN. We will begin first with a brief review of important concepts in macroeconomics as exposited in standard texts such as Frank and Bernanke (2009) and Mankiw (2004). This is the main focus of this chapter. An understanding of these concepts is a precursor to redefining the concepts in macroeconomics in the context of KN – the main focus of Chapters 11–14.

National Product

Macroeconomists regard an economy as a *single entity* that produces a *single (composite or aggregate)good*. This single good is termed national product (NP). It is defined as an aggregate that captures all facets of economic activity that occur in a nation. The price of NP is usually referred to as the price level. General increases in price level reflect a state of inflation in the economy, and hence some contemporary macroeconomists use inflation instead of price level as the price of NP. Although a nation could easily conform to a single-entity status, we need to appreciate that in many instances, KN does not have national boundaries. We address this issue in subsequent chapters. In this chapter, we focus on the notion of a nation producing a single good.

The size of national product is usually measured by one of three methods. These are either the sum of all:

1 *Real final expenditures; or*
2 *Real incomes accruing to factors of production; and*
3 *Real value-added during different stages of production.*

The rationale for these approaches rests on the circular flow model of the economy that was described in Chapter 2 (Figure 2.1). That is, with the circular flow, the sum of all incomes becomes expenditures and these in turn become incomes, and the process so continues. The underlying basis for transactions that yield both incomes and expenditures is value added in production. Hence, in the context of an

DOI: 10.4324/9781003408574-13

equilibrium, the size NP as identified by all three methods would be identical. The sum of all real final expenditures is referred to as *aggregate demand* and the sum of all incomes is referred to as *aggregate supply*. We qualify the meaning of the adjectives 'real' and 'final' below.

To illustrate the expenditure method, denote the size of the national product as Q_{NP} and suppose that this size is generated by N final expenditures involving the following quantities and real prices of goods: $(Q_1,\ldots, Q_N)$ and $(P_1,\ldots, P_N)$. Then

$$Q_{NP} = (P_1 * Q_1) + \ldots + (P_N * Q_N) \qquad (10.1)$$

Similarly, if the production of goods for the N final expenditures depends on K factors – mainly labour (L) and capital (KM), with quantities $(q_1,\ldots, q_K)$ and real prices (denoted by $W_1,\ldots, W_K$ to represent wages as well as returns), then:

$$Q_{NP} = (W_1 * q_1) + \ldots + (W_K * q_K) \qquad (10.2)$$

Note that NP is a physical quantity measure. Prices (or appropriate income measures) are used merely for the purpose of aggregating diverse items into a single unit. Further, real prices have to be used for this aggregation so that the inflationary effects on prices are absent and the resulting measurement adequately reflects the changes in the size of NP. Such adequate reflection also requires the aggregation to be performed in the context of final expenditures in order to avoid the duplication of transactions. The concepts of real values and final expenditures are now considered in turn below.

Real Values and Price Level Indexes

The prices that are observed in usual market exchanges are referred to as nominal values, which include the effects of inflation. Real prices, on the other hand, are ones in which the effects of inflation have been removed. It is usual to define real prices as follows:

$$\text{Real Price} = \left[\frac{\text{Nominal Price}}{\text{Price Level Index}}\right] \qquad (10.3)$$

Price level is usually quantified as an index of expenditures. The general formula for the price level index of a given year is as follows:

$$\text{Price Level Index} = \left[\frac{\text{Observed Expenditure in a given Year}}{\text{Expenditure at Base Year Prices}}\right] \qquad (10.4)$$

Note that using a real price is the same as using the price of the base year for aggregating the transactions of each year. To illustrate, suppose that we are dealing with a period of T years $(1,\ldots, T)$ and N goods $(1,\ldots, N)$ in each year. Using double subscripts for quantities and prices in Equation (10.1), we will get the following data set where P_{ij} and Q_{ij} denote respectively the price and quantity of the jth good in the ith year:

$$\begin{aligned} \text{Year 1}: & \ (Q_{11}, P_{11}),(Q_{12}, P_{12}), ...,(Q_{1N}, P_{1N}) \\ \text{Year 2}: & \ (Q_{21}, P_{21}),(Q_{22}, P_{22}), ...,(Q_{2N}, P_{2N}) \\ \text{Year T}: & \ (Q_{T1}, P_{T1}),(Q_{T2}, P_{T2}), ...,(Q_{TN}, P_{TN}) \end{aligned}$$
(10.5)

If we nominate Year 1 as the base year, then the size of national product for each year at base year (constant) prices will be:

$$\begin{aligned} \text{Year 1}: & \ Q_{NP1} = (P_{11}*Q_{11})+...+(P_{1N}*Q_{1N}) \\ \text{Year 2}: & \ Q_{NP2} = (P_{11}*Q_{21})+...+(P_{1N}*Q_{2N}) \\ \text{Year T}: & \ Q_{NPT} = (P_{11}*Q_{T1})+...+(P_{1N}*Q_{TN}) \end{aligned}$$
(10.6)

That is, the same set of prices ($P_{11}...P_{1N}$) is used for each year. Therefore, in most country statistics, one would observe sets of data presented at constant (real) prices as well as at nominal (current) market prices.

Comparisons of national products measured at nominal prices over time would invariably reveal a significant increase in national products. This is the case when prices increase much faster than quantities of output due to inflationary forces. However, such increases are typically modest when real prices are used. Hence, the measurement of national product at real prices gives a proper appreciation of the performance of an economy. For example, an examination of Indonesia's national income accounts reveals that the size of national product at nominal prices in 1993 was 70 times greater than that in 1973. However, at real prices, there is only a mere threefold increase over the same time period.

Final Expenditures

As indicated, changes in the size of national product are indicators of the performance of an economy. The use of final expenditures avoids the overstatement of performance. Consider the following example. A farmer has produced 200 kilograms of beans. A leading supermarket chain buys these beans from the farmer at $0.5 per kilogram. The same beans are then transported to a supermarket and sold to customers at $0.6 per kilogram. If all expenditures are counted, then we have $[(0.5*200) + (0.6*200)] = $220. However, in doing so, we have counted the same beans twice. Therefore, to avoid an overstatement of national product, only the expenditures pertaining to a good or service at the point of the final transaction are counted. With the example of the beans, if the supermarket transaction is deemed the final one, the contribution made by 200 kilograms of beans to the national product is $120.

Economic Performance and Goals of Macroeconomics

The goals of macroeconomics are often differentiated in terms of the time horizon. The main long-run goal of macroeconomics (as stated in most standard texts) is to achieve steady economic and employment growth as well as to maintain inflation at acceptable (target) levels. As explained below, maintenance of inflation at acceptable levels rests on the premise that continued economic growth is required.

Economic growth is a standard measure of performance for any economy. Simply put, it is the increase in the size of real per capita NP. The implicit assumption is that NP and employment move together in the same direction. The long-run goal requires NP to grow at rates that exceed the rate of population growth. It is only then that per capita NP can increase. In the short run, however, almost all economies display fluctuations in the size of NP – fluctuating between 'ups' and 'downs'. Some of these fluctuations, referred to as booms (accelerated expansions) and busts (sudden recessions) are often caused by disturbances or shocks arising from external and/or internal forces. For example, overconfidence in an economy leading to excessive investments could prompt a period of rapid expansion or a boom. In contrast, the reverse, say due to the discovery of poorly functioning investments, could prompt the withdrawal of investments and hence a bust. Such withdrawals could also happen when pandemics strike or when social unrest and terrorist activities occur. We consider some examples of these events below.

The aim of macroeconomics in the short run is to stabilise and protect economies from oscillations so that a steady (smooth) increase in the size of NP (and employment) will unfold in the long run. As illustrated in Chapters 12 and 13, chief stabilisation measures adopted by macroeconomists pertain to the instruments of fiscal and monetary policy. However, the effectiveness of such stabilisation will depend very much on the type and source of the disturbance and may also warrant

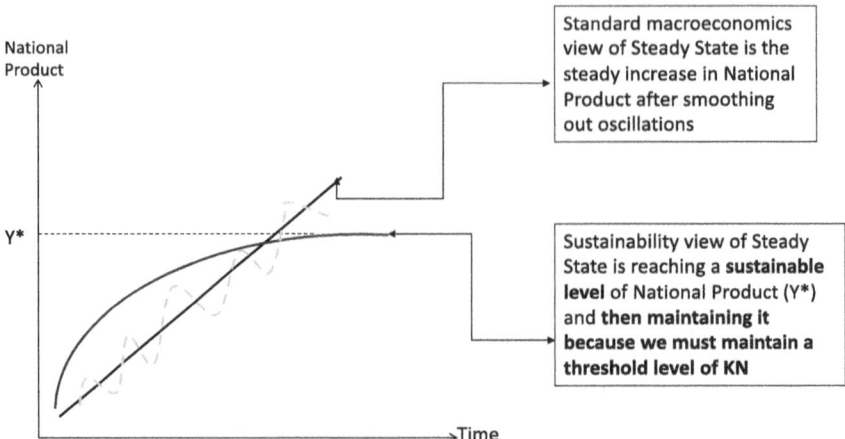

Figure 10.1 Standard View vs Sustainability View of Steady State

adoption of sound measures dealing with KN – not forgetting of course another important entity, namely social capital.

As indicated in the chapters that follow, the distinct departure in environmental economics pertains to the goal of sustained (steady) increases in NP. Environmental economists recognize that it is prudent to raise NP up to a specific level (e.g. to move people out of poverty and achieve a desirable standard of living) and then to expend efforts towards maintaining NP steady at that level so as to not deplete KN on which the economy and population rely. This is illustrated in Figure 10.1. In other words, a minimum threshold level of KN must be always maintained to sustain the economy. The reasons for this departure go back to the laws of thermodynamics and assimilative capacity and resilience as exposited in Chapter 2.

Measurement of NP

NP is usually measured at three levels. The simplest level is to count all final expenditures or factor incomes within the geographic bounds of a country during a specific time period, which is usually a year. This is referred to as *gross domestic product* (GDP). The expenditures that make up GDP are classified as follows:

1 Consumption (C)
2 Investment (I)
3 Government Expenditure (G)
4 Exports (X)
5 Imports (M)

The statisticians have various rules and guidelines for measuring these expenditures. For example, for I, only new investments that occur during a given year are counted. An item that entered a firm's inventory in the previous year would have been counted in the previous year and not in the following year, even if a sales transaction occurred in the following year. We shall, however, not deal with these rules and guidelines. A clear explanation of these is given, for example, in Jackson (1989). The definition of GDP with reference to expenditures is:

$$GDP = C + I + G + X - M \tag{10.7a}$$

The measurement of GDP with reference to incomes identifies three quantities, namely Compensation of Employees (CE), Gross Operating Surplus (GOS) and Net Taxes (T). As the term implies CE refers to wages and related payments received by L, and GOS pertains to income received by owners of KM. The third entity in the income accounts is government receiving T from various entities in the economy. Hence GDP in terms of the income accounts is:

$$GDP = CE + GOS + T \tag{10.7b}$$

A review of Equations (10.7a) and (10.7b) reveals some important relationships between the expenditure and income accounts. In the context of perfect trade, one would observe the identity (X ≡ M) and hence the trade balance (X − M) would be zero. Further in the context of a perfect fiscal system, the identity (T ≡ G) would prevail and consequently the fiscal balance (T − G) would be zero. Given perfect trade and fiscal conditions, the definition of GDP reduces to:

$$GDP = [C+I] = [CE+GOS] \tag{10.7c}$$

That is, given perfect trade and fiscal balances, expenditures pertaining to C and I would be exclusively distributed between incomes received by L and owners of KM.

The second level of measurement for NP, *gross national product* (GNP), considers the flows of income in and out of a country. This is defined as:

$$GNP = GDP - R \tag{10.8}$$

In Equation (10.8), R stands for 'net factor incomes remitted abroad'. For example, foreign nationals and firms located in Kenya would remit part of the incomes earned in Kenya abroad. At the same time, Kenyans and Kenyan firms located abroad would remit some part of their income earned abroad to Kenya. The term R is the difference between these two remittances. If Kenya's GNP is more than GDP, it means that remittances into Kenya by Kenyans living abroad are more than the outward remittances of expatriates in Kenya.

The third level of measurement is *net national product* (NNP). This recognises the wear and tear of the investments or capital stocks:

$$NNP = GNP - K_c \tag{10.9}$$

In Equation (10.9), K_c represents the depreciation allowance for capital stocks and is also sometimes referred to as capital consumption.

In gauging economic growth, NNP is a better measure than GNP, which in turn is a better measure than GDP. However, even many economists concede that economic growth should not be the sole criterion to measure an economy's performance. For example, when natural disasters strike, there is increased economic growth due to restoration activities. More generally, some aspect of growth is used to address the downsides of growth – for instance, when people buy water purifiers because previously pristine water sources have become polluted, or when households install burglar alarms when economic growth led to growing inequalities that stimulate crime.

The environmental economist argues that, because nature is capital, NNP should be refined further, by subtracting from it, the depreciation of KN. We will consider this in the next chapter.

The reverse of economic growth is a recession. Periods of growth and recession are also accompanied by changes in price level and employment. A sustained increase in the price level is inflation and the reverse of this is deflation. Periods of recession usually display unemployment and the reverse goes for periods of growth. Although the terms employment and unemployment usually refer to labour (L),

Some Important Concepts in Macroeconomics 165

they can also apply to capital stock (KM). For example, empty buildings and idle machinery represent unemployment of KM. Periods of accelerated growth can result in over-employment of L and KM. Yet another goal of macroeconomics is to keep inflation and unemployment under control within set limits. We present next a capsular review of the causes of these economic events (that is, growth, recession, employment, unemployment, inflation and deflation). We do so by examining the relationship between the size of NP and the price level or inflation.

The Relationship between NP and Inflation

An understanding of the relationship between inflation (π) and the size of NP (Q_{NP}) is useful for an illustration of some basic concepts in macroeconomics. The nature of this relationship depends on how NP is measured; namely, whether as the *sum of all real final expenditures* or as the *sum of all real incomes*. Note that inflation is usually measured as the rate of change in the price level index. The commonly used price level index in macroeconomics is the GDP deflator. Following the rationale of Equation (10.4), the GDP deflator for a given year can be defined as:

$$\text{GDP Deflator for Year t} = \left[\frac{\text{GDP measured in (current) year t prices}}{\text{GDP measured in (constant) base year prices}} \right] \quad (10.10)$$

Q_{NP} Measured as the Sum of Real Final Expenditures

When Q_{NP} is measured as the sum of all real final expenditures, the relationship between π and Q_{NP} is inverse. As π increases the size of Q_{NP} decreases; that is, the sum of all real final expenditures decreases. A more rigorous reasoning for this inverse relationship is given in the context of aggregate demand analyses in more advanced texts. An intuitive and simplistic reasoning for this is as follows. When the price of a single good rises, consumers actively search for substitutes. For example, if the price of coffee goes up, then consumers may switch to tea or some other beverage. But when π increases, we have a situation where all prices are rising simultaneously. In this context, there are three possible substitutes for 'total present expenditures' on items that make up NP. These are:

- Holding financial assets, including money
- Postponing expenditures for the future
- Making expenditures in other countries – especially neighbouring ones if they are easily accessible.

So if prices as a whole keep rising, people would rather hold their income in various financial assets rather than spend them now. They may even go overseas to make purchases, if goods are cheaper there. For example, Singaporeans often go across the border into Southern Malaysia when inflation arises in Singapore and prices are lower in Malaysia. Hence, in Figure 10.2, the relationship between Q_{NP} as the *sum of all real final expenditures* and π is inverse. This is labelled the

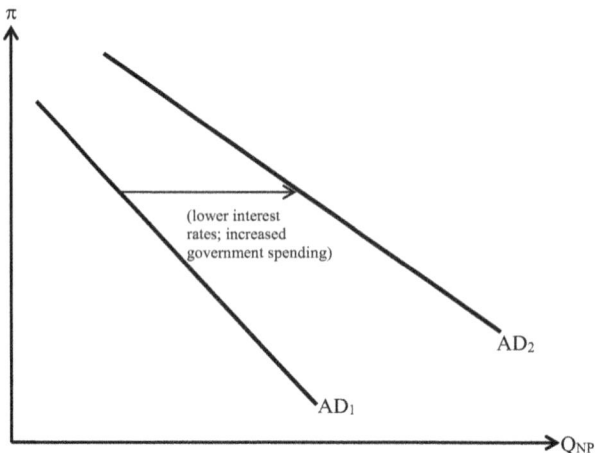

Figure 10.2 Aggregate Demand

aggregate demand (AD) curve. There is yet another reason for the AD curve to be downward-sloping. This is the response behaviour of Central Monetary Authorities (CMA) such as the Federal Reserve in the United States, The People's Bank of China, Riksbanken in Sweden and the Reserve Bank in Australia. When inflation increases, the CMAs generally raise the interest rate to curb spending (e.g. by making savings relatively more attractive, and buying on credit less attractive). Such intervention usually happens when inflation exceeds the target range. Macroeconomists regard a moderate level of inflation – namely the target rate – as desirable because it can prompt producers to hire more resources and thereby expand production.

We can further illustrate the effects of specific macroeconomic policy variables on the components of AD. For example, if income taxes are lowered, consumers have more disposable income and therefore C can increase. Such increases can prompt the AD curve to shift from AD_1 to AD_2, as illustrated in Figure 10.2. Note that a shift of the AD schedule to the right can represent a rightward movement along the horizontal axis – that is, an increase in NP, resulting in economic growth. Similarly, a lowering of interest rates can stimulate an increase in I, while a lowering of exchange rates can encourage an increase in X and a lowering of M. Thus, if policymakers wish to stimulate economic growth, they can endeavour to do so by stimulating AD within limits.

Q_{NP} Measured as the Sum of Real Incomes

When Q_{NP} is measured as the sum of all real incomes, the relationship between π and Q_{NP} is positive; that is, as π increases the size of Q_{NP}, measured as the sum of all real final incomes, increases. However, although such a relationship is generally true, this is not the case at all times. The general relationship is believed to hold true in between two extremes, at both of which NP has no bearing on π. One

Some Important Concepts in Macroeconomics 167

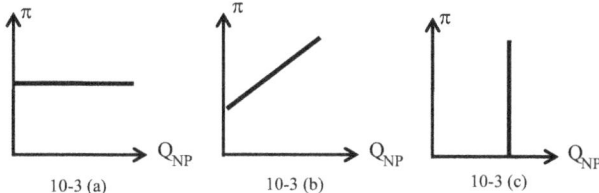

Figure 10.3 Aggregate Supply

extreme (Figure 10.3a) is where the relationship between Q_{NP} and π is a horizontal straight line, and this happens if there is excess capacity in the economy in terms of labour and resources – that is, if producers are willing to offer as many goods and services as possible for sale without any reference to prices. Such a situation normally occurs when the economy is in a severely depressed state. It is also possible that, in the short run, producers will try to sell as much as they can at the prevailing set of prices. So if π_0 represents the rate of inflation at the start of the production period, then the horizontal line at π_0 describes the short-run aggregate supply (AS) curve.

At the other extreme, the relationship is a vertical line, as in Figure 10.3c. Here the economy cannot produce beyond a certain amount of national product because of its resource constraints. This limit is called the *productive capacity* of an economy. In other words, regardless of the price level, the economy cannot generate more than a fixed amount of NP. The productive capacity of an economy usually represents the complete utilisation of L and KM; that is, the full employment of the factors of production. The full employment of L is usually approximated to the utilisation of the labour force, namely the sum of those employed as well as those looking work. However, as exposited in Frank and Bernanke (2009), the concept of unemployment is far more complex than falling short of utilising the full labour force. For example, frictional unemployment appears when workers' skills need upgrading to meet the changing needs of the economy, or when the requisite labour is located in places other than where it is needed and movement is not easily possible (e.g. because it would mean relocating families, thus increasing the opportunity costs to the workers of taking on the work that is available, in principle, to them). So during periods of accelerated expansion, the possibility of recruiting less skilled workers exists and the economy may surpass its capacity with reference to L. For our purposes, however, we will confine full employment of L to the utilisation of the labour force. As indicated above, full employment of KM represents the context of fully utilising all items KM without any one item being idle. The growth of KM stock is driven primarily by the net accumulation of investments (I). Whilst some items of KM may become idle during periods of recession, such idleness may also eventuate due to overcapitalisation from accelerated expansion. The presence of surplus holiday homes in Spain and Greece around 2009 is an example of such over-capitalization. As we will observe in the next chapter, the productive capacity of an economy must also consider the limitations imposed by KN. Recall our

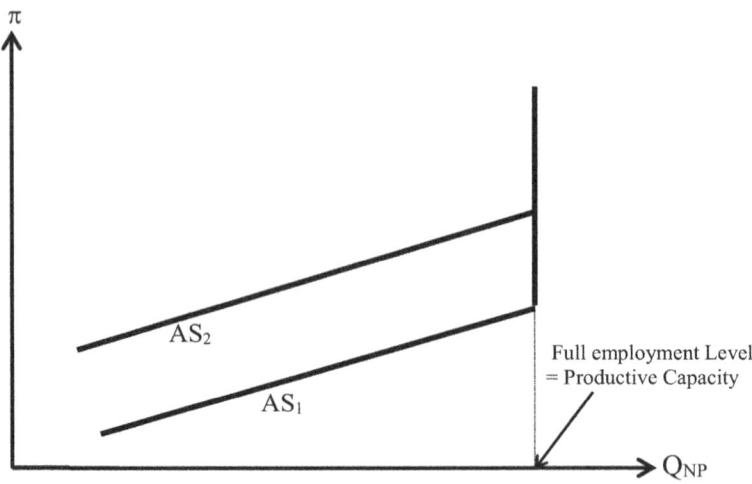

Figure 10.4 Aggregate Supply and Productive Capacity

narrative in Chapter 9, where we argued that keeping KN stocks in their preserved state is tantamount to the full employment of KN.

Figure 10.3b describes the general situation. Here, when π rises, producers employ more resources to increase output and hence the size of Q_{NP} as the sum of resource incomes increases. As indicated earlier, macroeconomists prefer a moderate level of inflation so that the recruitment of factors to expand NP can proceed. Excessive rates of inflation can prove counterproductive owing to cost blowouts. It is customary to describe the AS curve as in Figure 10.4, where the features of Figures 10.3b and 10.3c have been combined. However, the features of Figure 10.3a can be useful if the intention is to analyse a short-run issue or an economy that is severely depressed. Because Keynes (1936) argued in the context of the 1930s that the AS curve is horizontal, it is also referred to as a *Keynesian AS curve*. The vertical AS curve is also referred to as the *classical AS curve*.

In just the same way as taxes and interest rates can prompt shifts in the AD curve, corporate taxes and wage claims can cause the AS curve to shift. If workers make large wage claims or if corporate (business) taxes are raised, then the AS curve can shift from AS_1 to AS_2.

Note that in Figure 10.4, even if the AS curve has shifted, the productive capacity remains unchanged. As we will illustrate in subsequent chapters, the productive capacity of an economy could be reached sooner – in terms of Y – than later when KN is recognised.

AD–AS Framework

In the same way as in market equilibrium, the point of intersection of the AD and AS curves as shown in Figure 10.5 determines macroeconomic equilibrium.

Some Important Concepts in Macroeconomics 169

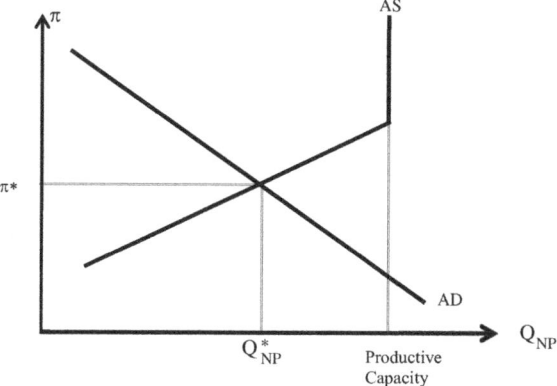

Figure 10.5 Macroeconomic Equilibrium

At this point, the equilibrium quantities of NP and inflation are defined. Note that the equilibrium quantity of NP meets the following condition:

$$Q^*_{NP} = \begin{pmatrix} \text{Size of NP as the sum} \\ \text{of all real final expenditures} \end{pmatrix} = \begin{pmatrix} \text{Size of NP as the sum} \\ \text{of all real factor incomes} \end{pmatrix} \quad (10.11)$$

Recall that the main purpose of macroeconomic policy is to stabilise the economy by preventing expansions (booms) and recessions (busts) as well as keeping inflation within a target range. We will now use the AD–AS framework to illustrate the pursuit of these goals in the context of some events that have unfolded in the past (1973–2010). The illustration of these events and the response to them is made with reference to Figure 10.6. We subsequently consider more recent events.

- 1973 marked the formation of a cartel namely Organization of Petroleum Exporting Countries (OPEC)[1] in response to the Arab–Israeli war of 1967. The cartel quadrupled the price of oil through embargos and supply restrictions. This resulted in an overall increase of all business and production costs and, consequently, the AS curve in most countries affected by the embargos shifted to the left. Such a shift caused a reduction of NP and an increase in π; that is, a recession and inflation, as illustrated by the leftward and upward pointing arrows in Figure 10.6a. The leftward shift of AS continued throughout most of the1970s and early 1980s, owing to other disturbances such as the Iran–Iraq war. During this period, many oil-importing countries embarked on a quest for alternatives to oil such as biofuels and solar energy.
- The price war in 1987, between OPEC and the then new entrants into the oil market – Britain and Norway – caused a sharp decline in the price of oil. This resulted in the AS curve shifting to the right sharply, resulting in a fall of π and an expansion of NP. This shift is illustrated by the arrow labelled (1) in Figure 10.6b. Buoyed by this expansion, spending increased rapidly, resulting

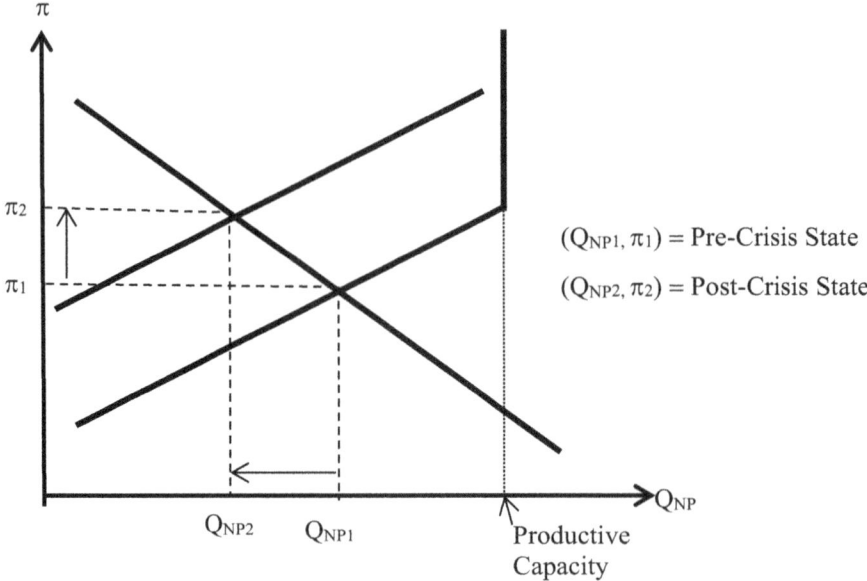

Figure 10.6 (a) OPEC Oil Crisis (1973–74)

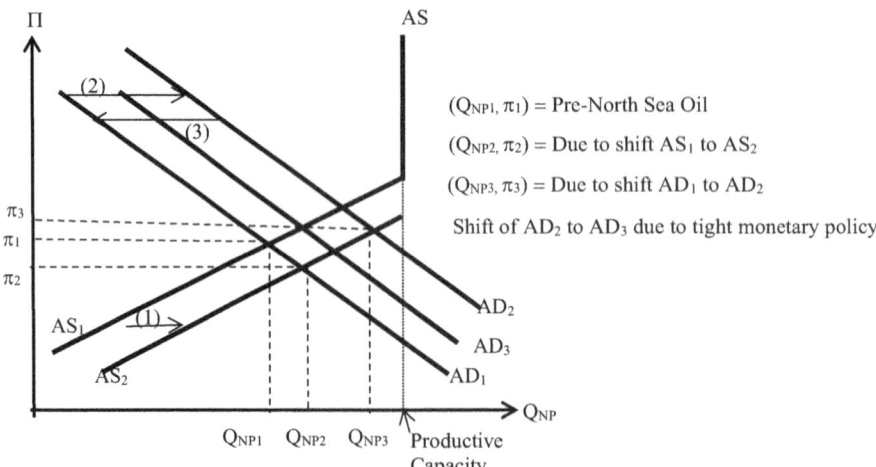

Figure 10.6 (b) Entry of North Sea Oil (1987–1992)

in a shift of the AD curve to the right (arrow labelled (2)), almost near productive capacity. This raised the fear of inflation among policymakers, who responded by raising interest rates and thereby inducing a (spending) recession; that is, the arrow labelled (3). In Australia, the then prime minister – Paul Keating – referred to this event as *the recession we had to have*. Further, the quest for alternative energy sources witnessed a significant slowdown.

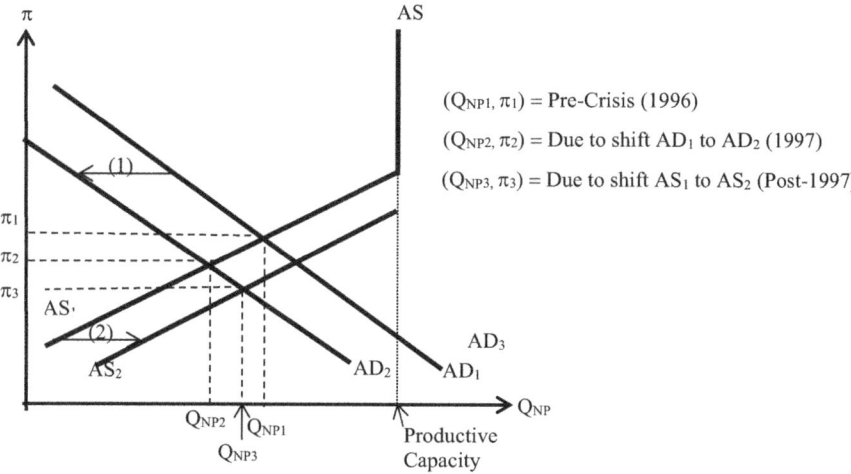

Figure 10.6 (c) The Asian Financial Crisis

- Prior to 1997, the Asian economies of Thailand, Malaysia, Indonesia, Singapore and South Korea were booming due to the massive inflow of investments. In fact, the World Bank referred to these countries as the Asian Tigers. However, around the middle of 1997, the lack of investor confidence in these countries emerged primarily owing to the evidence of underutilisation of KM. Observations of this type resulted in the AD shifting to the left, and this resulted in the lowering of π alongside a recession. This is illustrated in Figure 10.6c. The leftward shift of AD (arrow labelled (1)) was primarily due to the withdrawal of investment funds out of capital markets in Asia – primarily Thailand, Malaysia, Indonesia and South Korea. These economies responded mainly by reducing wages and lowering prices. This resulted in the rightward shift of the AS curve (arrow labelled (2)), bringing these economies back near to the levels of NP they had previously.
- In July 2000, the federal government of Australia introduced a broad-based goods and services tax (GST). The opponents of this tax argued that this tax would have both recessionary and inflationary effects due to the rise of production costs; that is, a shift of the AS schedule to the left (the arrow labelled (1) in Figure 10.6d). However, the government was able to partially counter this by granting some income tax relief and increasing expenditure on the housing sector. These measures enabled the AD schedule to shift to the right (arrow labelled (2) in Figure 10.6d), resulting in a modest amount of growth and with inflation.
- During the period leading up to the global financial crisis of 2008–2009, the AD schedule in many countries shifted to the right sharply (arrow labelled (1) in Figure 10.6e) – mainly due to the proliferation of financial products within unregulated financial markets. The realisation of the lack of tangibility in many of these goods resulted in a sharp reversal of the AD schedule (arrow labelled (2) in Figure 10.6e). This was primarily due to the withdrawal of funds out of financial institutions and the inability of the financial institutions themselves to honour the demands of their customers. There was also an upward/leftward

172 *Macroeconomics and KN*

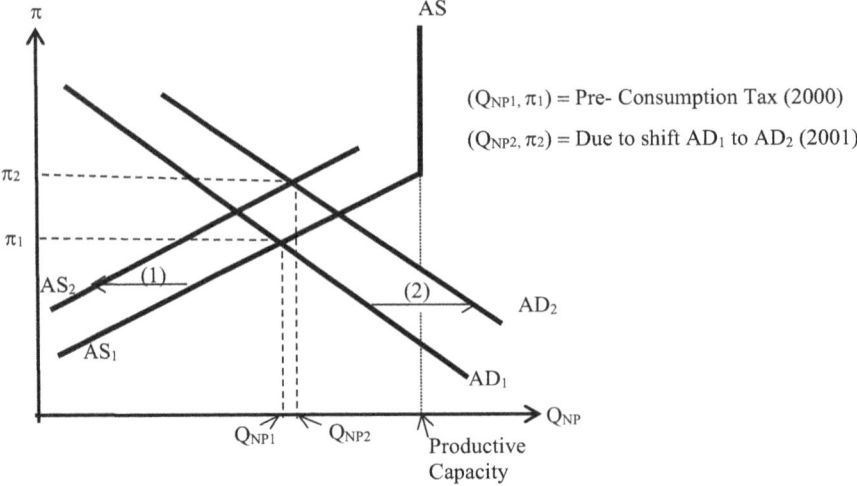

Figure 10.6 (d) The Consumption Tax in Australia

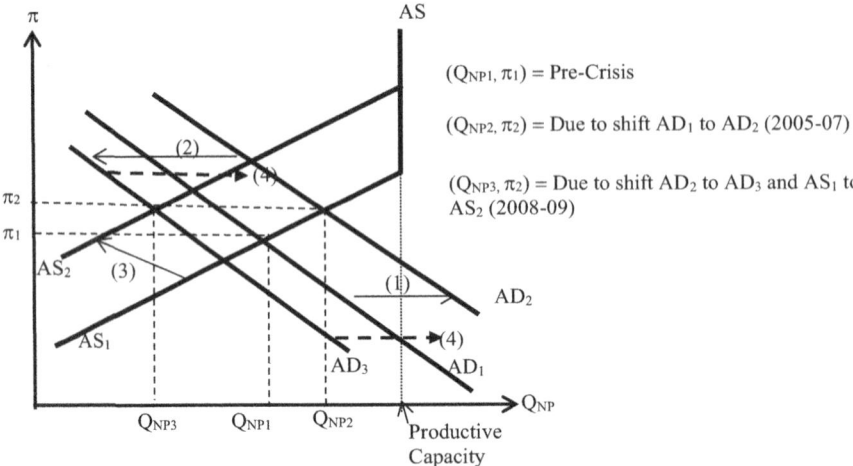

Figure 10.6 (e) The Global Financial Crisis

shift of AS (arrow labelled (3) in Figure 10.6e) because of business collapses and the like. The net result of these was, of course, a deepening of recession and unemployment. Most governments responded with stimulus measures in the hope of reversing the trends of both AD and AS. The dotted arrows labelled (4) represent these efforts. However, the stimulus did not take effect in many countries. Further, the lack of response to the stimulus had given rise to a debt crisis in some countries because incurring budget deficits financed the stimulus. In Chapter 13, we will illustrate with reference to a simple AD–AS framework that the stimulus could have been directed towards negating the depreciation of KN.

As indicated, the forgoing narrative in a series of bullet points spanned the period 1970–2010/2011. Since then, the world has been hardly free of turmoil. The same way as inadequate governance of financial markets in the United States triggered the global financial crisis, poor regulatory measures with reference to animal husbandry markets in China, triggered the global Covid-19 pandemic in 2019; Thampapillai et al. (2021). The pandemic resulted in border closures, contraction in trade and transport, and in general inward shifts of AD and AS. For their part, governments provided significant subsidies and income reliefs in a bid to prevent both AD and AS sliding inwards. The CMAs of most countries reduced interest rates to record low levels to stimulate and maintain investment activity. However, the global recovery since 2021 was primarily the result of adopting public health and safety measures. It was only in May 2023 that the World Health Organization declared that the pandemic was over.

During the global pandemic, inflation was not foreseen as a challenge. Yet in February 2022, following the Russian invasion of Ukraine, inflation surfaced as a major problem. Russia has been a traditional supplier of energy – mainly to the European Union, and Ukraine has been the major supplier of food to many parts of the world – especially Africa. The war has caused disturbances in the supply logistics of food and energy. These disturbances in turn have resulted in the shift of AS to the left in most countries. Such leftward shifts of AS are likely to prompt a rightward shift of AD and thereby further increase inflation. We attempt to illustrate this in Figure 10.7. Suppose that prior to the Russian invasion of Ukraine, the AD-AS equilibrium that prevailed in a specific country was (Y^*, π^*). The Russian invasion led to the leftward shift of AS resulting in AS^1, causing inflation to rise from π^* to π^1. If consumers now attempt to negate the shortage created, namely $(Y^* - Y^1)$, by increasing their spending, AD could shift to the right as indicated by AD^1, resulting in further increase in inflation. So, the CMAs of many affected countries raised interest rates with the intention of controlling inflation through limiting spending behaviour. Such intervention illustrates the obsession of CMAs with maintaining inflation at target rates usually in the vicinity of 2–3 per cent. As indicated earlier, the adherence to target inflation rests on the premise that economic growth must continue. And moderate rates of inflation can facilitate producers to recruit resources to increase production and thereby maintaining continued economic growth. Should inflation be excessive, then costs of production could become prohibitive dampening the prospect of further economic growth. Such is the rationale for CMAs' behaviour. However, there are other adverse effects that need to be considered – especially the heavier mortgage burdens leading in some cases to homelessness. Note though here the fundamental driver of inflation is the Russian invasion and the continuation of the war. Hence the resolution to the issue of inflation in this instance could very well lie with Mahbubani's (2022) call for peace makers and not with monetary policy. As of October 2023, yet another conflict – that between Israel and Palestine has emerged. This conflict

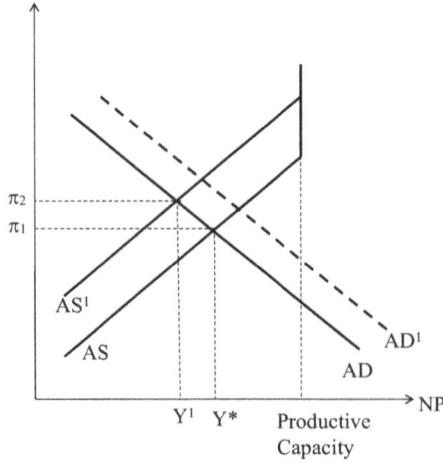

Figure 10.7 Illustrating the Inflation Effects of the War in Ukraine

too has significant inflation implications not only in the region of the conflict but also beyond. Again, the need for peace makers is explicit. Expanded war efforts mean investments in armaments such as tanks, drones and ammunition. Since these tend to be used up quickly in war efforts, their production fuels the economy in ways that neither creates direct consumer benefit nor investment in long-lasting infrastructures, as for example an investment in the health or education sector of the economy would create. As a result, prices rise without a concomitant increase in the capacity to produce economic welfare.

Besides human misery, war has devastating effects on KN, and such effects further exacerbate the misery. Bergman (2023) writes:

> If all the world's military forces were a country, they would pump out enough greenhouse gases to be the fourth-highest emitter.

Concluding Remarks

We have examined some basic concepts in macroeconomics and related them to past and contemporary events. It is now possible to illustrate how the conceptualisation of KN will influence formal macroeconomic analyses. The notion that KN degrades or depreciates can be incorporated into both AD and AS schedules. At the same time, the argument that we can either scientifically or otherwise restore KN endowments that were previously not functioning raises the possibility of introducing the concept of 'investments in KN'. Further, recognising that nature is capital can also influence the definition of productive capacity. In other words, the productive capacity of an economy will also depend (besides KM and L) on the availability of KN. We shall consider these issues in the chapters that follow.

Review Questions

1. Explain why the relationship between price level and the size of national product (Q_{NP}) is downward-sloping when Q_{NP} is measured as the sum of expenditures, and upward-sloping when Q_{NP} is measured as the sum of incomes.
2. List the factors that influence shifts in AD and AS and give at least three examples of how governments have used these factors to manage the economy.
3. Explain the effects of Covid-19 on selected economies using the AD–AS frameworks described in this chapter.

Note

1. This cartel was initially named OAPEC (Organization of Arab Petroleum Exporting countries) – and the adjective Arab was dropped following the recruitment of other oil-producing countries.

References

Bergman, J., 'Editorial', *The Conversation – Australian Edition*, 15 November 2023.
Frank, R. H. and Bernanke, B., *Principles of Economics*, McGraw-Hill, New York, 2009.
Mankiw, N. G., *Principles of Macroeconomics*, 3rd edn, Thomson South-Western, Mason, OH, 2004.
Jackson, D., *The Australian Economy*, MacMillan, Melbourne, 1989.
Mahbubani, K., 'Where are the Peacemakers?', *The Straits Times*, 19 March 2022.
Thampapillai, D. J., Bali, A., Lodi I. A. and Thampapillai, D. J., *Covid 19 and Its Contagion*, https://scholarbank.nus.edu.sg/handle/10635/201009 (posted 30 September 2021).

11 KN

Investment and Depreciation

The standard measures of economic performance introduced in the previous chapter – GDP, GNP and NNP – all exclude the direct role of KN in the economy. Hence the basis for revising macroeconomic analyses begins with redefining economic performance. The revised definition affords explicit recognition of the services provided by KN in any economy. As illustrated, such service provision falls within the purview of the laws of thermodynamics and ecological resilience. However, one should note that the term 'economic performance' here (and in most texts) is considered within a narrow scope pertaining to the size of the national product (NP). No consideration is given to the various factors that influence the quality of life in a nation. Although modifying NP for services provided by KN could go a long way towards addressing some quality-of-life issues, many others will remain unresolved. Since NNP is deemed better than the rest of the standard measures, the modified definitions below are made with reference to NNP.

KN and the Economy

In order to modify the definition of NNP, we conceptualise KN as an aggregate capital asset. That is, in their totality all natural endowments – biological and physical – will constitute KN. This line of reasoning was introduced in Chapter 8 at the firm level. We will now extend it to the broader macroeconomic level so that we can include KN in national income accounting in the same way as diverse manufactured capital assets, such as buildings and machinery are included. We have labelled, as indicated earlier, the collection of manufactured assets as KM in order to render distinct from the items that constitute KN. In the same way as national income statisticians compute stock estimates for KM, one needs to envisage the existence of a collective KN stock. That is, for example, KN of a country could represent the aggregation of its natural endowments such as forests, lakes rivers, and wildlife. Then, as illustrated below, additions to this stock will represent investment, while its diminution will represent depreciation.

As indicated in Chapter 1, the concept of KN is not necessarily recent. Recall the references in that chapter to writers like Marshall (1891), and Fisher (1904) who explicitly recognised KN in their analyses. Marshall's (1891) text, which is perhaps the first concise book on modern neoclassical economics, describes *nature* and *man* as the principal agents of production. Fisher's (1904) seminal paper is a cornerstone for studies on capital markets and further reinforces Marshall's (1891) view on KN. In his review of the definitions for capital, Fisher draws on KN assets such as lakes and rivers as analogies to explain the distinction between stocks and flows. Thus, in laying out a conceptual framework for capital, Fisher took it for granted that nature is capital and hence his recognition of KN was explicit.

Capital goods are usually distinguished from others in terms of three related features:

- Durability (they last over a long period of time)
- Provision of services that flow over time
- Deterioration (reduction in the ability to provide services over time)

These generalised characteristics of capital did in fact originate from Fisher's (1904) treatise on KN. That is, KN endowments are durable, and generate flows of services over time. They also degrade with use. As indicated in Chapter 2, the durability of KN is manifested in the three types of interrelated services that it provides: the provision of raw materials, ranging from the air we breathe to the minerals we extract; providing a receptacle for the wide array of waste emissions and discharges; and the provision of aesthetic amenities such as landscape and scenery. One may argue that the durability of KN could be prolonged over a very long time horizon if human interference in terms of rearranging nature were kept to a minimum. Such minimisation of interference is central to the main conclusions in Chapter 9, namely that 'small is beautiful' and in many instances 'less' will very well be preferable to 'more'.

Despite lessons from the history of economic thought and the conceptual validity as argued above, many contemporary economists (for example Mankiw 2014) and even some environmental economists like Stavins (2017) and Cropper and Oates (1992), fail to recognise the relevance for KN in macroeconomics. The reasons for such non-recognition rest on the microeconomic foundations of macroeconomics as exposited in most standard texts in economics. As we illustrated in Chapter 10, the sum of all real (constant price) transactions in markets for goods and services provides the basis for Aggregate Demand (AD) or NNP, whilst the sum of such transactions in factor markets provides the basis for Aggregate Supply (AS) or income (Y). The status of prices in both types markets (goods and services as well as factors of production) provides the basis for Price Level and Inflation. The overall benchmark for stability in an economy is one of 'General Equilibrium' which portrays a macroeconomic equilibrium (NNP ≡ Y) that rests on a myriad of perfectly competitive markets in equilibrium acting as supporting pillars alongside all leakages and injections being perfectly balanced. The argument made by many economists is that when issues pertaining to KN distort markets, they can be corrected by internalising externalities and/or institutional measures such as regulation. This argument is invalid conceptually and, from a practical perspective, somewhat nonsensical. It is invalid because externalities are never fully internalised and residual externalities remain and moreover do accumulate. As illustrated

Table 11.1 KN Investment Compared with Depreciation of KN

KN Investment	KN Depreciation
Activities that are designed to restore the flow of services from endowments that have **ceased to provide services**	*Activities that are designed to maintain the flow of services from endowments that* **currently provide services**
Analogy: replacement in the context of an obsolete investment	Analogy: expenditures to offset wear and tear
Examples:	Examples:
• Creation of wetlands for de-nitrification • Reforestation of areas that have been cleared for years of open-cut mining • Restoring rivers rendered unusable due to algal blooms • Detoxifying unusable soils for urban or rural development	• Pollution control activities by the public sector – municipal waste treatment • Maintaining the flow of surface- and groundwater systems • Maintaining the stocks of biological resources – forests, fish stocks

in Chindarkar and Thampapillai (2018), examples of such residual externalities that accumulate include, the permanent and growing brown cloud in Asia, fragmentation of polar icebergs; and the extinction of species. Many of the accumulated residual externalities could, in turn, trigger several other adverse externalities; for example, extreme weather events like catastrophic fires and devastating floods that keep recurring with higher degrees of intensity and frequency. Therefore, as illustrated below in this chapter and subsequent chapters, we consider the various avenues for articulating the role of KN in macroeconomics. From a practical perspective, the assumption that distortion of markets by KN can be resolved via the internalisation of externalities presumes that markets will help fix the problems created by them in the first place.

Distinguishing KN Investments from Depreciation of KN

In any economy, two types of activities, as illustrated in Table 11.1, can be undertaken in terms of the flow of services from KN. The first category results in an increase in the total volume of services from nature, while the second attempts to maintain the existing flow of services provided by nature. Therefore, it is pertinent to regard the former as KN investment and the latter as depreciation of KN.

KN Investments

KN investments may be explained by recourse to some examples. Consider the Baltic Sea. Nearly half of it (below a certain depth) is believed to be a dead resource. Ranges of anthropogenic factors, including pollution emissions and over-exploitation, have contributed to a lack of dissolved oxygen. This part of the Baltic Sea is at present incapable of generating any services. If, by incurring expenditures, the flow of services is restored, then these restoration expenditures represent a new investment. Similarly, when a badly polluted beach (say, one to which no person dares to go) is cleaned up, then that beach becomes capable of generating a flow of services. An analogy would be building a new factory when the old one has become obsolete and unproductive. In traditional national income accounting, new investments are included as gross investments. Likewise, investment-type expenditures for restoring lost (obsolete) endowments can be included in the accounting framework as a new gross investment.

The controversy in the literature concerns whether the restoration expenditures should be included in NNP, and centres around questions such as: 'Are we really adding to the stock of KN? The restored endowments were already there in the first place'. The counterargument is as follows. If, at the start of a given time period, a river is dead due to the presence of a variety of pathogens, then that river cannot be counted in the stock of KN, because it will not contribute to national output. Removing the pathogens and restoring the river is tantamount to adding to the stock of capital that contributes to national output. One cannot, of course, ignore the fact that what is restored cannot be authentic in terms of the original item of KN that became obsolete. That is, the river we restore would most probably be different from the one that existed in its pristine state. Nevertheless, one needs to acknowledge that the restoration of the river results in an increase in the total volume of services from KN, regardless of the issue of authenticity. Hence, the stock of KN at any given time will be made up of:

- Naturally occurring endowments
- Lost endowments that have been restored by human effort

When the latter are additions to the prevailing stock of KN, they can be regarded as investments. Further, if all lost (obsolete) endowments of an economy have been restored, then there is no scope for KN investments, and accounting procedures will be confined to the depreciation of the KN stock.

Depreciation of KN

The second category of activities in Table 11.1 deals with offsetting the wear and tear on KN (or its outright destruction). Therefore, the expenditures due to them should be deducted from NNP. Sometimes the distinction between KN investments and the allowances for the depreciation of KN can be confusing. The following examples could help clarify the confusion:

1 Consider a river that currently provides services while it also gets contaminated. The ongoing decontamination of the river, including the treatment of residual externalities mentioned above, is aimed at maintaining the river's role in contributing to NP. Such decontamination expenditure is similar to capital consumption because the contamination is tantamount to depletion of the river. This is different from restoring a river that was not providing any service at all and was regarded as obsolete. That is, as KN assets continue to provide services to

generate NP, their diminution in size is inevitable. Another example is the harvesting of forests for the timber industry. Despite careful rotational harvesting strategies to maintain the size of the forest stand, changes in transient canopy density could inevitably entail reduction in KN services, especially hydrologic services.
2 In the context of mineral resources, new discoveries are usually considered as additions to the stock of mineral wealth. Accordingly, the market value of the newly discovered mineral is usually regarded as an investment. Following the general theory of the mine (Hartwick and Olewiler 1986), the depreciation of a mineral resource can be explained by depletion and a reduction in the value of the mine. This reduction, as explained in Chapter 6, has usually been explained in terms of the concept of user costs; namely the loss of consumption benefits by future generations. Further, when a mineral is extracted and sold, the sale, whilst yielding consumption benefits, is also the value of the mine's depreciation. It is in such a context that the Hartwick-Rule, as explained below, becomes important towards articulating the depreciation of mineral wealth.

We can also consider the role of minerals from a different perspective. Many minerals are constituents of a complex geophysical system. Many economists think that minerals have no value in their non-extracted state beneath the earth's surface. However, a strong counterargument to this is that the minerals could in their natural (non-extracted) state offer a broader set of benefits, such as earth surface stability, and, where they are taken up by plants for example, as contributing to ecosystem productivity.

Sustainable Income

Let net national product, KN investments and the depreciation of KN be, respectively, denoted for time period t as: NNP(t), $I_{KN}(t)$ and $D_{KN}(t)$. Then, sustainable income would be defined as $[NNP(t) + I_{KN}(t) - D_{KN}(t)]$. If we suppose that current estimates of NNP(t) include the expenditures on KN investments, then the definition of sustainable income is simply $[NNP(t) - D_{KN}(t)]$. In fact, most infrastructural expenditures pertaining to I_{KN} are included in the estimates of NNP, while some are not – especially those pertaining to biodiversity. However, not all components D_{KN} are included in NNP, as explained in the next section.

We could adapt Hotelling's (1925) and Keynes' (1936) expositions of permanent income from a capital good to the case of KN. That is, the adjusted value of NP, namely $[NNP(t) + I_{KN}(t) - D_{KN}(t)]$ can be sustainable if at least two conditions are satisfied. These are:

- There is no diminution in the size of KN stock.
- The size of D_{KN} is less than the rent generated by the stock of KN.

A steady state can be defined as one in which:

- All lost endowments have been restored by way of investments.
- Positive rents net of depreciation are being maintained.

$[NNP(t) + I_{KN}(t) - D_{KN}(t)]$ can be regarded as the economic return from a nation's KN stock. Sustainable income is therefore the difference between the economic return, as measured by [NNP(t)], and the allowance made for the depreciation of KN net of the investments in KN. Therefore, the state where $[D_{KN}(t) > (NNP(t) + I_{KN}(t))]$ represents a non-viable economic system.

Recall from Chapter 10 the definition of NNP: $[C + I + G + X - M - R - K_c]$. An examination of the national income accounts of most countries is likely to reveal that some costs of D_{KN} appear as positive items in NNP; that is, they are mostly included in C, I or G. For example, costs of waste management are normally found in G. Another example pertains to expenditures by firms and households in terms of water and air filters. These expenditures are usually found in C or I. However, such expenditures are incurred because of the deterioration of KN amenities of air and water. At the same time, other non-market items such as the loss of topsoil and biodiversity are often not accounted for. Hence, in the estimation of $D_{KN}(t)$, one needs to separate out items that are currently included in NNP as well as estimate and include values for items that have been omitted. The following classification could prove useful in identifying the components of $D_{KN}(t)$.

1 *Costs involving production*: These are expenditures intended to maintain the services of KN in the context of production activities. They go beyond the costs of complying with pollution control regulations because such costs are likely to be included in market transactions. Hence here we refer to the costs of dealing with the residuals that persist beyond the ambit of regulations. In general, these are costs of treating and disposing of the wastes that are generated from production. Some of these expenses are likely to be already included in NNP (as positive items) under G. Note that in the context of weak regulations, such costs will be high. In Figure 11.1 we reproduce evidence presented by Chindarakar and Thampapillai (2018) for China and India for the period 1995 to 2015. Here, KN was assumed to be the air shed of the relevant economy and D_{KN} was measured as costs of air pollution abatement with reference to GHG emissions. In this analysis, NP was measured as GDP. As can be seen, the paths (GDP $\equiv$ Y) and (GDP $\equiv$ Y $- D_{KN}$) diverge over time implying that policy intervention for managing KN becomes a priority area.
2 *Costs involving current consumption*: These refer to expenditures that attempt to enhance the safety of consuming the services of KN; for example, water filters on taps and air filters on ventilation shafts. These could also include medical expenses due to illnesses induced by polluted environments. These expenses, too, are likely to be already included in NNP as positive items under C or I.
3 *Costs involving future consumption*: These costs arise from the depletion of the stocks of renewable and non-renewable resources and the imposition of costs on future generations by the non-availability of these resources. The loss of biodiversity belongs to this category. Most of these costs are largely

182 Macroeconomics and KN

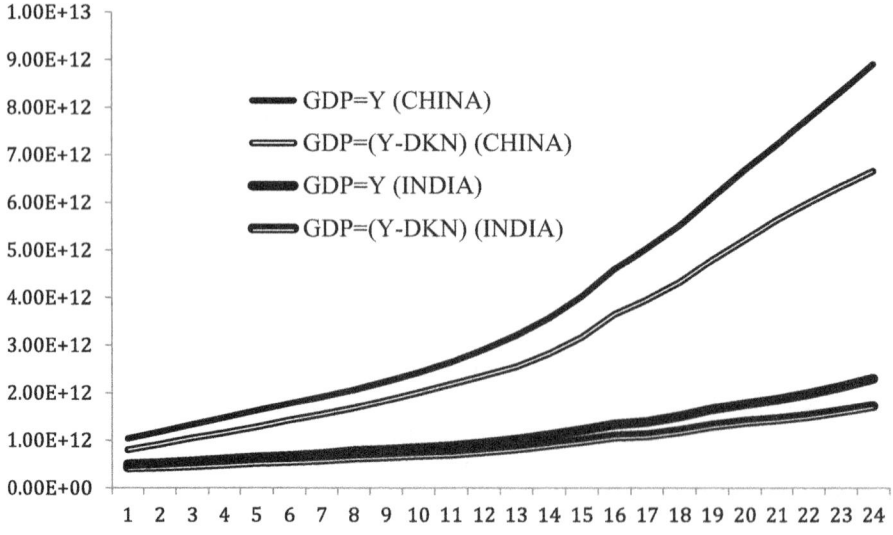

Figure 11.1 Time Paths of Y and (Y-D_{KN}) for China and India (1992–2015)
Source: Chindarkar and Thampapillai (2018)

ignored. With reference to the depletion of resource stocks, it has been customary to measure depreciation by recourse to the Hartwick-Rule already briefly mentioned above (Hartwick 1977, 1978, 1990). This rule is based on the premise that at least a portion of the revenue stream from the sale of exhaustible resources will be invested in the creation of tangible assets that have the potential to generate a new income stream. In such a context, D_{KN} would be the difference between the market value of the resource stock and the present value of the new potential income stream. Figure 11.2 illustrates this concept. The initial stock of the resource is valued at $V. Each year a constant proportion (θ) of the stock is extracted and sold. Hence annual return from mining is $($\theta$V). In the absence of any reinvestment of the annual return $($\theta$V), as shown in Figure 11.2a, the entire stock would get depleted by time period T. The paradox here is that whilst $($\theta$V) is often regarded as a component a country's NP, it is also that country's annual mineral depreciation. The Hartwick Rule calls for a proportion (ρ) to be invested in the creation of a new tangible asset stock. Suppose that ρ represents the net investment after depreciation of the new stock is also accounted for. Then, as shown in Figure 11.2b, by the end of year T, a new asset of minimum size $\{(V\theta\rho)*T\}$ could emerge. The depreciation of the original resource stock is given by $\{V[1 - (\theta\rho T)]\}$. If θ and T are both fixed, then the ability to reduce the size of mine depreciation will depend mainly on proportion of income that is saved, namely ρ. As shown in Box 11.1, some developing countries that are resource rich may not have sufficient capacity save.

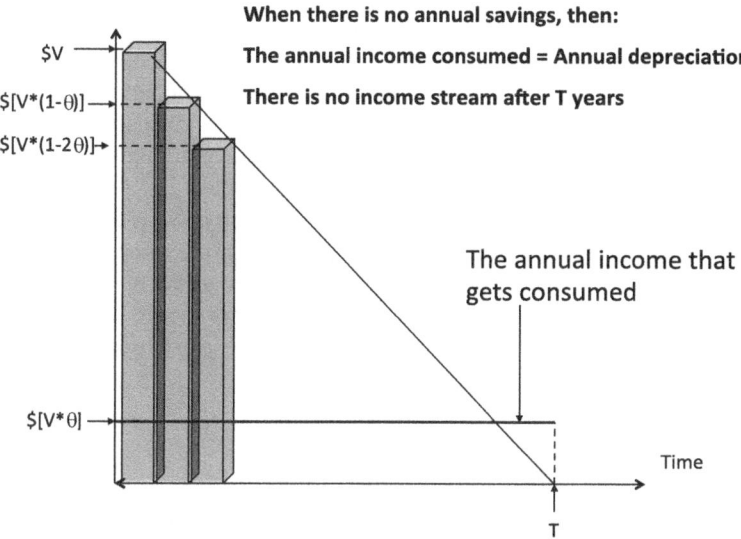

Figure 11.2A The Depreciation of the Mine without Re-Investment of Savings

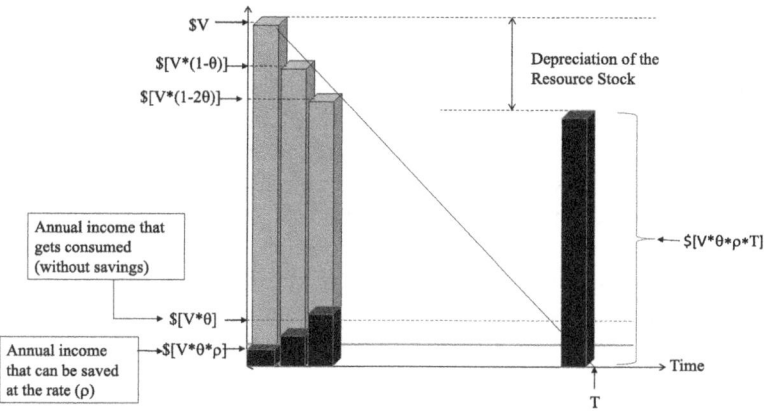

Figure 11.2B The Depreciation of the Mine with Re-Investment of Savings

Box 11.1   Mine Depreciation in Guinea (West Africa)

The West African country of Guinea is resource-rich. Nearly 50 per cent of the world's stock of bauxite is located in Guinea. Bauxite is the essential raw material for the processing of aluminum. Bauxite accounts for almost 70 per cent of Guinea's total exports. Yet bauxite mining is incapable of helping Guinea with savings that are sufficient offset the depreciation of the mine and facilitate the build-up of new stocks of tangible capital.

Recall from the discussion above that the capacity to save from mining income is an important factor in reducing the size of the mine depreciation. In turn, the capacity to save is high if resource prices are high. Unfortunately for Guinea, the price of bauxite is a mere US$28 per metric ton, whilst the costs of mining account for nearly 90 per cent of the gross revenue.

A Conceptual Framework for $D_{KN}(t)$

Consider the relationship between $D_{KN}(t)$ and NP denoted as $(Y(t))$. We now extend our firm-level reasoning in Chapter 8 to the economy overall. As per this line of reasoning, increasing $Y(t)$ would result in the increase of $D_{KN}(t)$. Recall from our earlier narrative, $D_{KN}(t)$ represents the depletion in the stock size of KN. Hence raising $Y(t)$ – which is tantamount to economic growth – would diminish the size of KN. Such diminution can render KN fragile similar to the illustration offered in Figure 8.3 in Chapter 8. Accordingly, we propose that raising $Y(t)$ is feasible only up to a threshold level of KN depletion. Any attempt to increase $Y(t)$ beyond this threshold could drive $D_{KN}(t)$ towards infinity. That is, the KN system could collapse, signified by the complete loss of assimilative and resilience capabilities. As indicated in Chapters 5–8, such a conceptualisation for $D_{KN}(t)$ places KN in the category of non-renewable resources; for example, see McInerney (1981). Besides, as argued in Chapter 8, the treatment of KN as a non-renewable item is valid, given that it is a complex system of resources and that the ability to recoup the assimilative and resilience capabilities from such a system is inevitably finite.

Several factors influence the relationship between $Y(t)$ and $D_{KN}(t)$. We propose that among these three inter-related factors are particularly important:

- Presence or absence of residual externalities that we considered in Chapters 5–8
- The rate at which D_{KN} increases in response to an expansion of Y
- The rate at which KN loses its resilience

Consider Figure 11.3, which consists of two panels. In the upper panel, we illustrate the relationship between $D_{KN}(t)$ and $Y(t)$. The lower panel illustrates how a proposed measure of the resilience of KN, denoted by (φ), reduces with increases in both $Y(t)$ and $D_{KN}(t)$.

KN: Investment and Depreciation 185

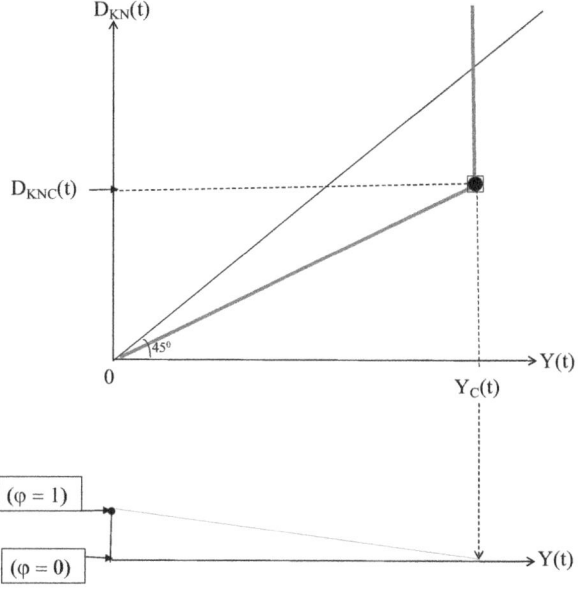

Figure 11.3 KN-Depreciation Function – Linear and Discontinuous (Residual Externalities Absent)

Consider first the upper panel of Figure 11.3. Here, we propose that $\{D_{KN}(t) = g[Y(t)]\}$ takes the form of a linear-discontinuous function as characterised by the following two expressions:

$$D_{KN}(t) = \eta[Y(t)] \text{ for } \left(0 < Y(t) < Y_C(t)\right), (1 > \eta > 0) \tag{11.1}$$
$$D_{KN}(t) \to \infty \text{ for } \left(Y(t) > Y_C(t)\right) \tag{11.2}$$

At least three observations are pertinent in terms of our hypothesised relationship for $\{D_{KN}(t) = g[Y(t)]\}$.

First, in Equation (11.1), $(Y(t) = 0)$ corresponds with $(D_{KN}(t) = 0)$. That is, if $(Y = 0)$, which implies absence of economic growth, there will be no diminution in the size of KN. This supposition also implies the absence of residual externalities.

Second, we consider a fixed domain, namely $(0 < Y(t) < Y_C(t))$. Over this domain, the gradient of the function in Equation (11.1), signifying the rate of increase of $D_{KN}(t)$ in response to raising $Y(t)$, namely η, is less than 1. This observation means that $D_{KN}(t)$ will be less than $Y(t)$ over this specific domain. Note that if η were greater than 1, then any attempt to raise $Y(t)$ would prove futile, because $D_{KN}(t)$ would exceed $Y(t)$. Hence, the distance $(0 < Y(t) < Y_C(t))$ on the horizontal axis of Figure 11.3, can be labelled the domain of 'feasible income generation'. Further, η, can be also regarded as the marginal rate of KN degradation, as shown below:

$$\eta = d(D_{KN}(t)) / d(Y(t)) \tag{11.3}$$

Third, as indicated, the status of $[D_{KN}(t) < Y(t)]$ can prevail only until the raising of $Y(t)$ reaches $Y_c(t)$. During any accounting period t, $Y_c(t)$ is the maximum target towards which Y can be raised, in terms of the assimilative capacity and resilience of KN. Any attempt to achieve an increase Y beyond $Y_c(t)$ results in irreversible KN damage, and therefore $D_{KN}(t)$ tends to infinity as presented in Equation (11.2). Hence the relationship $\{D_{KN}(t) = g[\Delta Y(t)]\}$ is discontinuous.

Consider now the lower panel of Figure 11.3. Here, we introduce a resilience coefficient (φ) for KN, the size of which ranges between 0 and 1. ($\varphi = 1$) represents perfect resilience of KN and corresponds with $[D_{KN}(t)] = 0$; that is, the absence of any diminution in the size of KN. Alternatively, complete non-resilience corresponds with ($\varphi = 0$), and happens when $[(D_{KN}(t)) \to \infty]$. A convenient definition for φ would be:

$$\varphi(t) = \{D_{KNC}(t) - D_{KN}(t)\} / [D_{KNC}(t)] \tag{11.4}$$

Hence, ($\varphi = 1$) when $[D_{KN}(t) = 0]$, and ($\varphi = 0$) when $[D_{KN}(t)) = D_{KNC}(t)]$. Note that in Equation (11.4), $D_{KNC}(t)$ corresponds with the upper limit pertaining to the domain of feasible income generation, namely $Y_c(t)$, and hence borders on the state where $[D_{KN}(t) \to \infty]$.

The presence of residual externalities implies that the accounting period commences with an outstanding volume of KN depletion that has to be dealt with. Figure 11.4 provides an illustration of this context. Here, the residual externality,

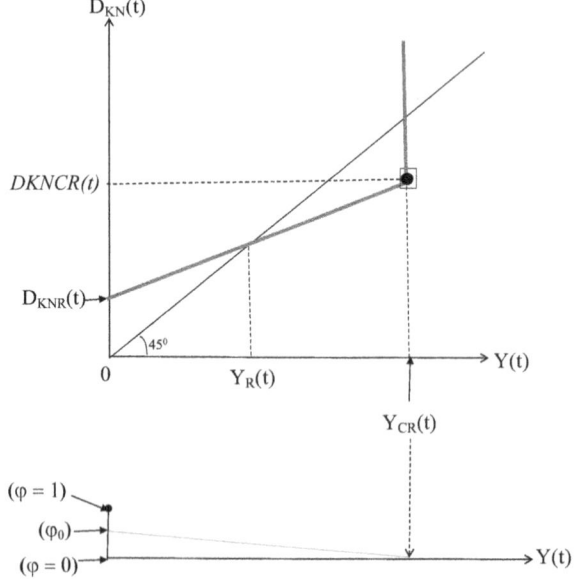

Figure 11.4 KN-Depreciation Function – Linear and Discontinuous (Residual Externalities Present)

the aggregate value of outstanding KN depletion is represented by $D_{KNR}(t)$. Note that in the upper panel of Figure 11.4, when $Y(t) = 0$, $D_{KN}(t)$ is no longer zero, but a positive value representing that of the residual externality. Hence, the linear-discontinuous function describing $\{D_{KN}(t) = g[Y(t)]\}$ now reads as:

$$D_{KN}(t) = D_{KNR}(t) + \eta_R (Y(t)) \text{ for } (0 < Y(t) < Y_{CR}(t)) \tag{11.5}$$

$$D_{KN}(t) \rightarrow \infty \text{ for } Y(t) > Y_{CR}(t) \tag{11.6}$$

A comparison of the two situations – presence and absence of residual externalities – reveals the following:

1 In the presence of residual externalities, the domain of feasible income generation, $(Y_R(t) < Y(t) < Y_{CR}(t))$, is narrower than that prevails in the absence of residual externalities;
2 The gradient of the function describing $\{D_{KN}(t) = g[\Delta Y(t)]\}$ – marginal rate of KN degradation – is possibly greater when the residual externalities are present than when they are absent; that is (η_R in Equation 11.5) > (η in Equation 11.3).
3 As $Y(t)$ is raised, the upper limit of the domain of feasible income generation, could appear sooner in the presence of residual externalities than in their absence; that is, ($Y_{CR}(t)$ in Equation 11.5) < ($Y_C(t)$ in Equation 11.3)
4 As can be observed from the lower panel of Figure 11.4, the resilience coefficient no longer begins with 1, when $Y(t)$ is zero. This is due to the presence of the residual externalities. Instead, it starts with a value less than 1. Based on the definition of the resilience coefficient offered in Equation (11.4), the value of (φ) when $[Y(t) = 0]$, namely (φ_0) is given by:

$$(\varphi_0) = [(D_{KNCR}(t) - D_{KNR}(t)) / (D_{KNCR}(t))] < 1 \tag{11.7}$$

The status of KN as portrayed in Figure 11.4 is likely to be closer to reality than that presented in Figure 11.3. This is because the persistent pursuit of economic growth and the accumulation of several residual externalities have left KN in a fragile state in many economies. Please see Box 11.2, where we illustrate how the pursuit of economic growth in Australia has inevitably led to the depletion of KN. Such examples are of course widespread across several economies, notwithstanding the fact that several changes to KN were often inflicted over many centuries through various processes such as war, colonisation, and trade

---

Box 11.2 Some Selected Examples of KN Depletion That Supported Economic Growth in Australia

Australia was once heralded as the only OECD country to have displayed nearly 30 years (1988–2018) of continued economic growth. When this period of growth ended, Martin (2020) surveyed a panel of some

24 Australian economists. It is noteworthy that neither the questions nor the responses included any reference to the potential role of KN in the 30-year growth and the subsequent decline.

Deforestation – Forests product industry supports several economic activities such as housing and commerce and is a major source of deforestation. Forest logging licences are easily given because economic growth takes priority. We illustrate this with wood-chipping in Eden – an idyllic coastal township in Southern New South Wales of Australia and a serious casualty of 2019–2020 summer forest fires. Since 1977, Harris-Daishowa, a Japanese wood-chip exporter has been logging old-growth forests on the escarpment around Eden. This activity – spanning more than 80,000 hectares of virgin (old growth) forest – has been carried out under leaseholds that were formally approved by the Federal as well as the State governments. Periodic opposition by conservation groups to this activity could not overcome the premise that wood chipping in Eden contributes to Australia's GDP as well as local and regional employment. But the long-term damages caused by forest logging – especially of old-growth forests far exceed the monetary benefits of wood chipping. Recent scientific research from the University of Michigan reveals how forest canopy changes become important drivers of drought and reductions in soil moisture through changes to the hydrologic cycle. Had such research evidence been overwhelmingly available in the 1970s and 1980s, then it is possible that the logging leases might have been denied not only around Eden but also in several other areas – especially old growth forests. The chip mill in Eden and significant areas of surrounding forests have now been destroyed in the fires of 2019–2020 and would unlikely be a source of income and employment for a considerable length of time.

Water Abstraction from River Systems – Australia's hydrologic cycle is also disturbed by excessive depletion of water from river systems right across the Australian continent. The abstractions of water are primarily made for agriculture and mining. Over a period exceeding 200 years, the Murray-Darling Basin (MDB) has been transformed from a robust to a fragile KN asset. The main reasons for this transformation are, amongst others, both regulated as well unregulated withdrawals of water. The MDB Authority reports show that in just the four-year period from 2015 to 2018, some 68 billion litres of water had been abstracted, under licences, and that too, from just a small section of the basin in Queensland's Barwon-Darling tributaries. Annual withdrawals from the entire MDB, for irrigation could add up to a mammoth volume of 3.8 trillion litres. If one is searching for the drivers of the Australian drought and climate change – they are also right here in Australia.

Mining – Another activity that causes disturbances to both the surface as well as sub-surface ecosystems is mining. Besides, mining is water-intensive. The extraction of one ton of coal requires 250 litres of *fresh* water. So, for example, the Adani mine with an intended to extraction volume some

60 million tons of coal per year from the Galilee Basin in Central Queensland, would withdraw 15 billion litres of water per year from the Carmichael River. Such an abstraction volume will undoubtedly and adversely affect the Great Artesian Basin – largest groundwater basin in the world. Some mistakenly argue that the withdrawal of water is from the river and not the groundwater basin. The reality is that the river is part of nature's recharge mechanism for the groundwater resource system. Australia does have efficient technologies for recouping water from mining. However, the adoption these technologies are not as widespread as they should be. Further, the clearing of vegetation for mining and the scarring of the landscape would inevitably alter the hydrologic cycle for the reasons given above.

Sand mining in coastal areas and from the seabed is an important contributor to economic growth through housing and urban development. Sand volumes, more than 70 million tons, have been removed from the sand dunes in Kurnell Peninsula to support mainly building activity in Sydney. North Stradbroke Island in Queensland, which generates roughly 500,000 tonnes of sand annually, has been the source of controversy due to the severe ecological impacts. Important among such impacts is the loss of mangroves from coastal and river systems where sand mining occurs. Mangroves are nature's filters for air and water and important players in the hydrologic system.

Practically every economic activity utilises KN. It is for this reason, as explained in Chapter 1, that Alfred Marshall (1891) deemed nature to be the ultimate capital. That is, if one were to take any item and disaggregate it into its components until one can disaggregate no more, the ultimate components come from nature. It is therefore essential that decisions pertaining to economic growth must not be confined to GDP alone – but must equally consider changes to KN. It is from this perspective that Australia's so-called 30-year record performance of economic growth must be reviewed. Stewardship towards KN must become an important economic activity and must enter the portfolio for employment creation – especially in the transition towards a post-COVID world.

(The above narrative was taken from Thampapillai (2020)

We return to the domain of feasible income generation in Figure 11.4, that is $(Y_R(t) < Y(t) < Y_{CR}(t))$. Preliminarily, it would appear that an economy should be capable of raising $Y(t)$ above $Y_R(t)$ in order to reap positive income benefits. This is because for all values of $Y(t)$ falling below $Y_R(t)$, the economy would incur a loss; that is, $D_{KN}(t)$ would exceed $Y(t)$ over the range $[0 < Y_R(t)]$. Hence, it might appear prudent to choose an income target that falls within the domain of feasible income generation, namely $[Y_R(t) \leftrightarrow Y_{CR}(t)]$. However, this is a comparative static (snap-shot) observation. From a dynamic perspective, raising $Y(t)$ would inevitably prompt increases in $D_{KN}(t)$, and this could not exclude raising the function $\{D_{KN}(t) = g[Y(t)]\}$ above the 45^0 line. Should such a context unfold, then the

190 Macroeconomics and KN

economy would have lost its productive capacity. Hence, a prudent policy measure would be to include KN investments such as reforestation and restoration of mangroves to push $\{D_{KN}(t) = g[\Delta Y(t)]\}$ below the 45° line. Further, the upper limit of this domain of feasible income, $Y_{CR}(t)$, denotes the target income that maximises $Y(t)$ in excess of $D_{KN}(t)$. However, from Equation (11.6), we can see that $Y_{CR}(t)$ is also the income level that brings the economy to the brink of an KN disaster. This difficulty is due to the linearity and discontinuity assumptions that have been made for $\{D_{KN}(t) = g[\Delta Y(t)]\}$ in Equations (11.5) and (11.6). A non-linear relationship such as an exponential function is perhaps more appropriate. Yet the linear function can prove useful, especially for joint consideration in linear macroeconomic models and when we are able to assume that $Y_{CR}(t)$ is sufficiently large to be ignored.

Consider now an exponential cost function of the following form:

$$D_{KN}(t) = D_{KNR}(t) e^{\eta E(Y(t))} \tag{11.8}$$

In Equation (11.8), ηE represents the rate of increase of $D_{KN}(t)$ with reference to $Y(t)$, and $(D_{KNR}(t))$ represents the specific value of $(D_{KN}(t))$ when $(Y(t) \to 0)$. That is, as indicated above, $D_{KNR}(t)$ is the aggregate value of outstanding KN depletion, namely the residual externality. From Equation (11.8), it is possible to define η_E as:

$$\eta_E = \left[\frac{\ln(D_{KN}(t)) - \ln(D_{KNR}(t))}{Y(t)} \right] \tag{11.9}$$

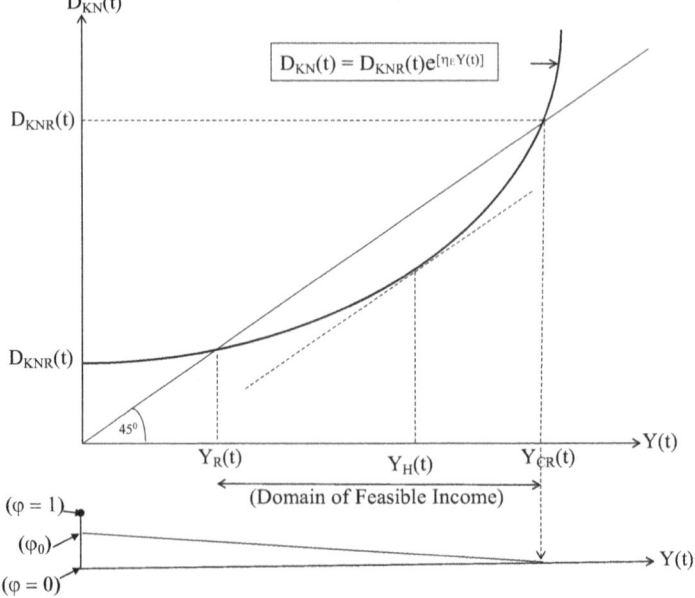

Figure 11.5 KN-Depreciation Function – Exponential

The upper panel of Figure 11.5 presents $\{D_{KN}(t) = g[Y(t)]\}$ as an exponential function. Here, the domain of feasible income generation is: $\{Y_R(t) < Y(t) < Y_{CR}(t)\}$. Further, note that for this domain to be visible, the function $\{D_{KN}(t) = g[\Delta Y(t)]\}$ needs to intersect the 45° line. Such an intersection, in turn, can occur only when the gradient of the function is < 1. Alternatively, when the gradient is > 1, the function $\{D_{KN}(t) = g[\Delta Y(t)]\}$ lies above the 45° line, indicating that KN is heavily degraded, and the economy is without productive capacity. As indicated above, one needs to distinguish the inferences from the context of comparative statics from those of dynamics. That is, from a dynamics perspective, pursuing income targets that are apparently feasible could still lead to lifting the function $\{D_{KN}(t) = g[\Delta Y(t)]\}$ upwards. Hence, KN investments need to be present in the activities that constitute $Y(t)$ in order to retain productive capacity.

As indicated above, the gradient of $D_{KN}(t)$ is also the marginal rate of KN degradation and can be defined with reference to the exponential function as:

$$\frac{d(D_{KN}(t))}{d(Y(t))} = D_{KNR}(t)\eta_E e^{\eta_E Y(t)} \qquad (11.10)$$

The maximising value of feasible national income, namely $Y_H(t)$ in Figure 11.5, can be derived by equating Equation (11.10) to 1. That is:

$$[Y_H(t)] = -(\ln D_{KNR}(t) + \ln \eta_E)/\eta_E \qquad (11.11)$$

As can be observed from the lower panel of Figure 11.5, the maximising value of $Y(t)$ may compromise the extent of the resilience of KN.

Some Empirical Evidence

The relationship $\{D_{KN}(t) = g[Y(t)]\}$ was tested as a time series comparison of national income in the United States with KN depreciation data from Daly and Cobb (1989). The following result was obtained for the exponential function, when $D_{KN}(t)$ was equated to the sum of costs associated with water, air and noise pollution and the loss of wetlands and farmlands:

$$\ln D_{KN} = 4.94 + 0.00045\,Y \qquad (11.12)$$
$$\phantom{\ln D_{KN} =\ }(183.8)\ \ (35.13)\quad r^2 = 0.972$$

The Daly and Cobb data can be briefly described as follows. Costs of water pollution were based on lost recreation benefits due to point source discharges and the costs of dredging operations against siltation. Air pollution costs included damages to agricultural vegetation, materials damage and the effects of acid rain. These were estimated using hedonic prices and wage differentials. The costs of noise pollution were taken from estimates provided by the World Health Organization. The loss of wetlands was valued on the basis of land values and KN quality scores for attributes such as habitats and aesthetics. The loss of farmland due to soil degradation was valued in terms of lost agricultural output, and the loss of non-renewable resources was estimated as the forgone value of mineral output. Long-term KN damage was assumed to be caused by non-renewable energy consumption and was valued in terms of an energy consumption tax. Y in Daly and Cobb data was NNP.

A similar test with Australian data (Thampapillai and Uhlin 1994), where $D_{KN}(t)$ was assumed to be made up of the expenditures of firms dealing with waste management, revealed the following relationship, which also qualifies the exponential form:

$$\ln D_{KN} = \underset{(30)}{4.0} + \underset{(7.7)}{0.005} Y \quad r^2 = 0.72 \tag{11.13}$$

In both cases, t-values (in parentheses) and r^2 are satisfactory, and Y in the Thampapillai and Uhlin study was GDP. And in both cases, an increase in Y corresponded to an increase in D_{KN}.

Concluding Remarks

To summarise, we have considered in this chapter the basis for modifying national product in order to derive an expression for sustainable income. This basis rests on the conceptualisation of KN as a stock that is analogous to the estimates of manufactured capital stock (KM) that are published in the national accounts. This basis then enables us to distinguish between KN investments and the depreciation of KN. A major challenge is the measurement of the depreciation allowance for KN. Empirical evidence seems to suggest that the relationship between this allowance $D_{KN}(t)$ and Y is an exponential function. Until systems of environmental accounting are properly set up, the estimation of $D_{KN}(t)$ – and similarly $I_{KN}(t)$ – has to use proxies. In the next chapter, we will consider the internalisation of the $D_{KN}(t)$ into selected macroeconomic models.

Review Questions

1 Critically review the following statement: 'All expenditures incurred with respect to KN represent depreciation allowances and it is not possible to justify any environmental expenditure as an investment'.
2 Explain how $D_{KN}(t)$ could become a determinant of productive capacity.

References

Chindarkar, N. and Thampapillai, D. J., 'Rethinking teaching of basic principles of economics from a sustainability perspective', *Sustainability*, 10(5): 1486, 2018.
Cropper, M.L. and Oates, W.E., 'Environmental economics: A survey', *Journal of Economic Literature*, 30(2): 675–740, 1992.
Daly, H. E. and Cobb, J. B., *For the Common Good: Redirecting the Economy towards Community, the Environment, and a Sustainable Future*, Beacon Press, Boston, MA, 1989.

Fisher, I., 'Precedents for defining capital', *Quarterly Journal of Economics*, 18(3): 386–408, 1904.

Hartwick, J., 'Intergenerational equity and the investing of rents from exhaustible resources', *American Economic Review*, 66: 972–974, 1977.

Hartwick, J., 'Investing returns from depleting renewable resource stocks and intergenerational equity', *Economic Letters*, 1(1): 85–88, 1978.

Hartwick, J., 'Natural resources, national accounting and economic depreciation', *Journal of Public Economics*, 43(3): 291–304, 1990.

Hotelling, H., 'A general mathematical theory of depreciation', *Journal of the American Statistical Association*, 20(151): 340–353, 1925.

Keynes, J., *The General Theory of Employment, Interest and Money*, Macmillan, London, 1936.

Mankiw, N.G. *Principles of Economics*, 7th ed., Cengage Learning, Stamford, CT, 2014.

Marshall, A., *Principles of Economics*, Macmillan, London, 1891.

Martin, P., '2020 survey: No lift in wage growth, no lift in economic growth and no progress on unemployment in year of low expectations', *The Conversation*, 28 January 2020.

McInerney, J., 'The simple analytics of natural resource economics', in J. A. Butlin (ed.), *Economics and Resources Policy*, Longman, London, 1981.

Stavins, R., "The evolution of environmental economics: A view from the inside", *Singapore Economic Review*, 62, 251–274, 2017.

Thampapillai, D. J., *GDP Growth Versus Sustainability*, Henry Halloran Trust, University of Sydney, 2020.

Thampapillai, D. J. and Uhlin, H.-E., *On the Measurement of Environmental Depreciation in Macroeconomic Analysis,* Working Paper 12, Department of Economics, Swedish University of Agricultural Sciences, Uppsala, 1994.

12 Environmental Macroeconomics
Short-Run Analysis-I

As indicated in Chapter 10, the goals of macroeconomics are differentiated in terms of the time period considered. In the short run, the goal is to stabilise economies against wayward oscillations of the national product (NP), prices or inflation (denoted by π) and employment (represented usually by the amount of labour (L) and manufactured capital (KM) utilised). Such stabilisation is expected to yield a smooth transition across time periods in such a way that the long run is represented by a steady upward trajectory of NP, L and KM with π kept under control. Recall our discussion in Chapter 10, and the narrative surrounding Figure 10.1, on the subject of steady state. We have reiterated these ideas in Figure 12.1. The steady upward trajectory of NP, as illustrated in Figure 12.1a, is referred to as the *steady state* by most macroeconomists. The recognition of sustainability prompts a serious departure from the standard conceptualisation of the steady state. This is because the finiteness of KN dictates that NP cannot increase indefinitely. As indicated previously, increasing NP would involve the depletion of KN. Suppose now that the term 'employment' needs to include KN besides L and KM. In this context, employment of KN would mean preserving KN stocks in their natural state. Hence, as illustrated in Figure 12.1b, the steady state would involve maintaining NP at some given level, say Y^*, instead of pursuing an upward trajectory that could deplete KN. Loss of resilience is an inevitable outcome of KN-depletion.

In this chapter and the next, we will illustrate the departure from standard analysis with reference to short-run stabilisation. The changes to long-run analysis form the content of Chapter 14. Most macroeconomists claim that errors made in the short run could be corrected in the long run. For macroeconomists, errors are the deviations from a dedicated upward trajectory of NP. However, note that the distinction between 'short run' and 'long run' can be elusive. This is because the long run is in fact a collection of several short runs. Hence the synchrony between short run and long run is indeed important, especially with reference to maintaining KN stocks. The drivers of so-called errors can invariably be rooted in KN-related factors such as prolonged droughts, catastrophic forest fires and devastating floods.

In this chapter, we introduce a basic Keynesian framework of income determination. It is useful to start with this basic Keynesian framework because it is, perhaps, the cornerstone of the various macroeconomic models that have evolved over time. The analytic basis underlying this framework is the equilibrium between aggregate

Environmental Macroeconomics: Short-Run Analysis-I 195

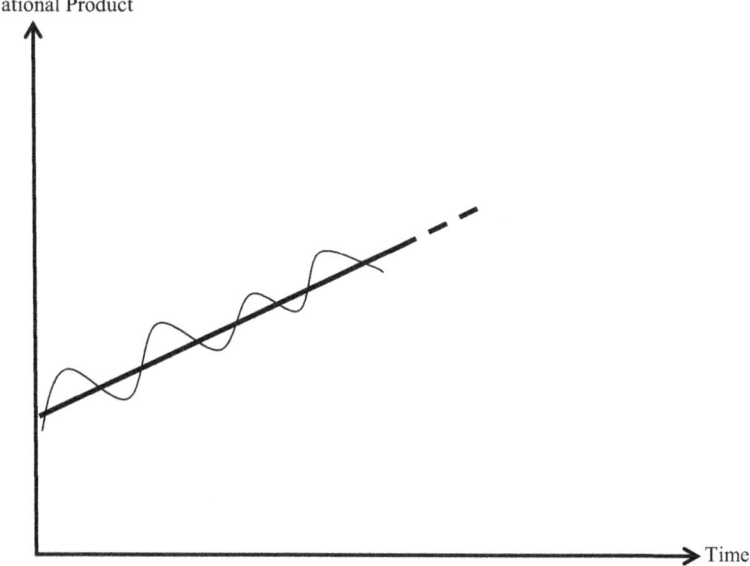

Figure 12.1a Steady-State in Standard Macroeconomics

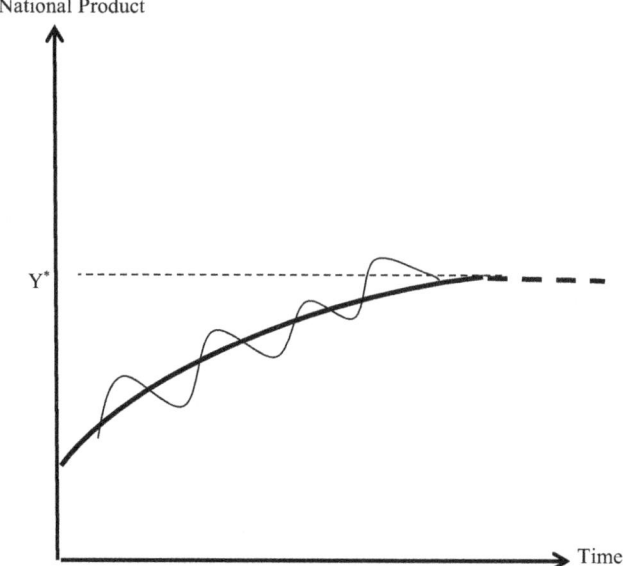

Figure 12.1b Steady-State in Environmental Macroeconomics

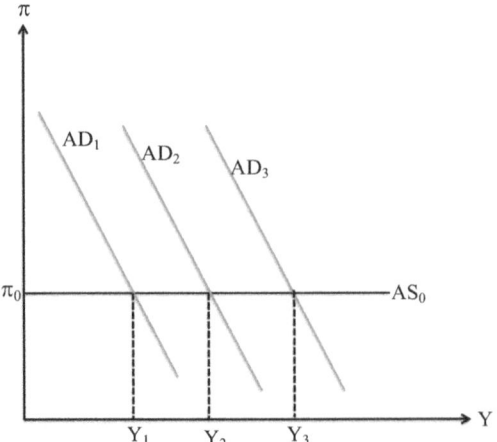

Figure 12.2 Basis for Income (Y) being determined by Expenditure (AD)

demand (AD) and aggregate supply (AS), namely AS ≡ AD. Recall from Chapter 10 that AS is the sum of real factor incomes constituting national income (Y). Also recall from Chapter 10, that AD represents the size of NP measured as the sum of real final expenditures. Such expenditures may be measured as either gross domestic product (GDP), or gross national product (GNP), or net national product (NNP). As indicated in Chapter 11, NNP is the better of these three measures, and our intention therein was to extend the measurement to include the important role played by KN in any economy. To recapitulate, such a role includes the potential to make KN investments and the need to explicitly recognise the depreciation of KN (D_{KN}).

Two main assumptions in the Keynesian framework are:

- Price level or inflation is constant.
- Y is determined by AD.

The first assumption implies that the aggregate supply (AS) curve is horizontal. The horizontal AS curve was explained in Chapter 10 (Figure 10.3a). We have reproduced this context in Figure 12.2 and included the presence of AD to illustrate the significance of the second assumption. That is, given AD_1 – equilibrium income will be determined as Y_1, and for AD_2 – income is Y_2, and so on. The first assumption also removes the need to consider inflation targeting as a specific goal. Both assumptions together imply that an economy is capable of producing any amount of Y at a given level of inflation. We remove these two assumptions in the next chapter. But we introduce in this chapter the notion of productive capacity that constrains the possibility of limitless expansion of Y.

Further, the size of expenditure as specified in AD becomes the basis for policy and planning around Y. For example, if AD were defined as GDP, then the equilibrium between AS and AD will be Y ≡ GDP, implying that policy formulation would proceed on the basis that the size of Y is GDP. Alternatively, had AD been defined as NNP, then the relevant equilibrium would be Y ≡ NNP, and would in turn, imply that the size of Y for policy formulation is NNP. The size of Y for policy and planning based on NNP would be smaller compared to the context based on GDP. The size of Y for policymaking would be even smaller when AD is defined in terms of sustainable national product and the equilibrium is Y ≡ NNP − D_{KN}.

The framework for determining Y is illustrated in three steps. In the first, we will consider the determination of Y without any reference whatsoever to KN. In the second step, we will introduce KN and assume that the function describing the depreciation of KN, namely D_{KN}, is linear. However, as indicated in Chapter 11, the linear framework may not properly manifest the reality of the constraints surrounding KN and so, in the third step, D_{KN} will be explained in terms of a non-linear exponential function along the lines suggested in Chapter 11. With both the linear and nonlinear frameworks, we consider the implications of residual externalities being present.

As illustrated below, the main policy tool here is aggregate demand (AD). Recall again from Chapter 10 that AD is the sum of all real final expenditures. Based on the underlying premise that expenditure determines Y, AD is managed towards either expanding or contracting the economy as required. For example, in the presence of unemployment, policymakers guided by the Keynesian dogma, would be inclined to stimulate expenditure. Such stimulus could prompt producers to increase employment of resources and thereby reduce unemployment, and in the meantime raise Y. In the event the economy has expanded beyond its productive capacity, then policymakers would do the reverse; that is, adopt measures to contract expenditure.

Equilibrium Income without D_{KN}

In a given time period, t, the equilibrium between the aggregate demand (NNP(t)) and income Y(t) will be:

$$Y(t) \equiv NNP(t) \tag{12.1}$$

For reasons of simplicity, assume that all components of NNP except consumption (C) are constant. We denote this constant by Φ(t) and following the definitions given in Equations (10.7), (10.8) and (10.9) of Chapter 10, {Φ(t) = I + G + X − M − R − K_c}. In most macroeconomic texts, consumption in a given year C(t) is usually defined as the following linear function:

$$C(t) = \alpha(t) + \beta(t)\big[Y(t) - T(t)\big] \tag{12.2a}$$

In Equation (12.2a), α(t) is *autonomous consumption,* namely the consumption that is needed regardless of income and includes items that constitute basic needs.

198 Macroeconomics and KN

β(t), more commonly termed the *marginal propensity to consume,* is the amount by which consumption will increase when disposable income [Y(t) − T(t)] increases by one unit. When the average taxation rate is denoted by τ(t), then taxes collected is defined as [T(t) = τ(t)(Y(t))], and hence Equation (12.2a) can be stated as:

$$C(t) = \alpha(t) + \beta(t)[Y(t) - \tau Y(t)] = \alpha(t) + \beta(t)Y(t)[1 - \tau(t)], \qquad (12.2b)$$

We will assume, for convenience and simplicity, that α(t) is contained in G of Φ(t). By definition, the estimate for β is {ΔC/[Δ(Y − T)]}. However as in some of the studies considered below (in Section 12.4), a point-estimate for β for a given time period, say β(t), could be elicited from the national accounts as {C(t)/[Y(t) − T(t)]}. The justification for such an approach rests on the premise that the values given in the national accounts are 'value added' estimates.

NNP(t) can be now written as:

$$NNP(t) = \Phi(t) + \beta(t)Y(t)[1 - \tau(t)] \qquad (12.3)$$

The value of Y that satisfies the identity [Y(t) ≡ NNP(t)] will be denoted as Y(t)*. This value can be found by substituting Equation (12.3) into (12.1) and solving for Y. That is, {Y(t) ≡ Φ(t) + β(t)Y(t)[1 − τ(t)]} could be rearranged in the form of an equation as {Y(t)[1 − β(t)(1 − τ(t))] = Φ(t)}. When this equation is solved for Y, we get:

$$Y(t)^* = \frac{\Phi(t)}{[1 - \beta(t)(1 - \tau(t))]} \qquad (12.4)$$

In Figure 12.3, this equilibrium income is determined by the point of intersection of the AD schedule describing {NNP(t) = Φ(t) + β(t)Y(t)[1 − τ(t)]} with the 45° line. Note that the 45° line describes all points where {Y(t) = NNP(t)}.

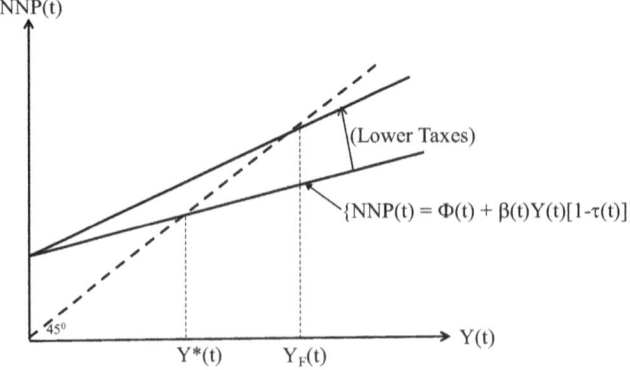

Figure 12.3 Equilibrium Income within an Elementary Linear Keynesian Model

Suppose now that Y has to be raised from Y(t)* to $Y_F(t)$ because policymakers believe that they can achieve full employment at an income level of $Y_F(t)$. The income differential $[Y_F(t) - Y^*(t)]$ as illustrated in Figure 12.2 is a *recessionary gap*, because $[Y_F(t) > Y^*(t)]$. Should a situation arise where $[Y_F(t) < Y^*(t)]$, then we would be presented with an *expansionary gap*. As discussed in Chapter 10, an expansionary gap could emerge when an economy receives a rapid inflow of investments. The aim of stabilisation policy is to close these gaps by either, stimulating increases Y in the context of a recessionary gap, or adopting to reverse Y for an expansionary gap.

As indicated, the Keynesian framework assumes that Y will be determined by expenditure, namely AD. So, the policy tools for closing the gaps indicated above would be those that bear influence on AD. These include taxes, interest rates and government spending. For example, a lowering of taxes, to increase (Y − T) and hence C, could shift the aggregate demand schedule, as shown in Figure 12.2. However, the manipulation of tools may not always yield expected outcomes. For example, raising interest rates with the intention of curbing spending behaviour may be offset by the inflow of investments in the expectation of higher dividends. Similarly, the lowering of income taxes could also result in increased borrowing activities and hence exert upward pressure on interest rates. Such increases could, in turn, counter the anticipated stimulatory effects of tax relief. Thus, policymakers need to study the potential outcomes of manipulating shifts in aggregate demand in order to achieve specific income targets. Such manipulation is referred to as 'pump-priming'. As we will see below, pump priming takes on an added dimension when we internalise KN into aggregate demand. It is also important to note that the quantity $\{1/[1 - \beta(t)(1 - \tau(t))]\}$ in Equation (12.4) is referred to as the 'multiplier' in standard macroeconomic theory. That is, if AD is raised (lowered) by one unit, then Y will increase (decrease) by an amount equal to the multiplier. As we will show below, incorporating D_{KN} into the model above leads to a decrease in the size of the multiplier.

Equilibrium Income with D_{KN}: Linear Framework

We start our analysis with the equilibrium between the aggregate demand and sustainable Y. For a given time period t, this will be defined, following the definition given in Chapter 11, as:

$$Y(t) \equiv NNP(t) + I_{KN}(t) - D_{KN}(t) \qquad (12.5a)$$

As indicated in Chapter 11, some components of $I_{KN}(t)$, such as reforestation and expenses pertaining to infrastructural intervention – say for establishing wetlands and protecting hydrologic features of a landscape – are included in G or I. We assume here that $I_{KN}(t)$ is already accounted for within NNP, and hence the equilibrium is simplified as:

$$Y(t) \equiv NNP(t) - D_{KN}(t) \qquad (12.5b)$$

200 *Macroeconomics and KN*

We have already defined NNP(t) as $\{\Phi(t) + \beta(t)Y(t)[1 - \tau(t)]\}$ in Equation (12.3). This has to be redefined in terms of (12.5b) as: $\{\Phi(t) + \beta'(t)Y(t)[1 - \tau'(t)]\}$. In this redefinition, the revised marginal propensity to consume, namely β' will be the ratio $[C/(Y - T - D_{KN})]$ and the revised taxation rate, namely τ' will be $[T/(Y - D_{KN})]$. In defining D_{KN}, we will make two simplifying assumptions in relation to the definitions offered in Chapter 11. First, we will assume that the upper limit for Y that renders ($D_{KN} \to \infty$), namely ($Y_c(t)$) as in Equation (11.1) of Chapter 11, is large enough (relative to current Y) to be ignored. Second, we will initially assume that residual externalities are absent. This context was described in Equation (11.1) and Figure 11.3. Given the assumptions made here, and the synonymity between Y and NNP, $D_{KN}(t)$ can be defined as a linear proportion, $\eta(t)$, of NNP(t). Note that $\eta(t)$ being the ratio $[D_{KN}(t)]/[NNP(t)]$ is synonymous with the gradient of the D_{KN} function described in Equation (11.1). Hence:

$$D_{KN}(t) = \eta(t)\big[NNP(t)\big] = \eta(t)\Phi(t) + \eta(t)\beta'(t)Y(t)[1-\tau'(t)] \qquad (12.6)$$

The equilibrium between AS and AD can now be stated as:

$$Y(t) \equiv \Phi(t) + \beta'(t)Y(t)[1-\tau'(t)] - \eta(t)\{\Phi(t) + \beta'(t)Y(t)[1-\tau'(t)]\} \quad (12.7)$$

With some algebra, the value of sustainable equilibrium Y, namely ($Y(t)^{**}$) will be defined as:

$$Y(t)^{**} = \frac{\Phi(t)\big[1-\eta(t)\big]}{\{1-\beta'(t)[(1-\eta(t))(1-\tau'(t))]\}} \qquad (12.8)$$

Consider now the case where residual externalities are present. This context was identified in Equation (11.2) and Figure 11.4, and $D_{KN}(t)$ can be defined in terms of NNP as:

$$\begin{aligned}D_{KN}(t) &= D_{KNR}(t) + \eta_R(t)\big[NNP(t)\big] = \\ &D_{KNR}(t) + \{\eta_R(t)\Phi(t) + \eta_R(t)\beta'(t)Y(t)[1-\tau'(t)]\}\end{aligned} \qquad (12.9)$$

Recall that $D_{KNR}(t)$ represents the size of the residual externalities and η_R is the revised gradient of the D_{KN} function. The equilibrium between AS and AD can be now revised as:

$$\begin{aligned}Y(t) \equiv \Phi(t) &+ \beta'(t)Y(t)[1-\tau'(t)] - \{D_{KNR}(t) + [\eta_R(t)\Phi(t) \\ &+ \eta_R(t)\beta'(t)Y(t)[1-\tau'(t)]]\}\end{aligned} \qquad (12.10)$$

We denote equilibrium Y in the context of residual externalities as $Y(t)_R^{**}$, which can be defined as follows:

$$Y(t)_R^{**} = \frac{\Phi(t)[1-\eta_R(t)] - D_{KNR}(t)}{\{1 - \beta'(t)[(1-\eta_R(t))(1-\tau'(t))]\}} \quad (12.11)$$

The determination of $Y(t)^{**}$ and $Y(t)_R^{**}$ are illustrated in the upper panel of Figure 12.4 by the point of intersection of the relevant AD function with the 45° degree line. As one would expect, the magnitudes of $Y(t)^{**}$ and $Y(t)_R^{**}$ are directly proportional to $\Phi(t)$ and $\beta'(t)$, and inversely proportional to $\eta(t)$, $\eta_R(t)$ and $\tau'(t)$. Hence, the policymaker would adopt measures that would influence these coefficients to either expand or contract the size of equilibrium Y.

In an economy portraying the characteristics of Figure 12.4, if $D_{KN}(t)$ is ignored, then income determination will be based on NNP alone. As illustrated in Figure 12.4, the AD schedule based on [NNP(t)] results in income being determined as $Y(t)^*$. Hence, the difference $[Y(t)^* - Y(t)^{**}]$ or $[Y(t)^* - Y(t)_R^{**}]$ may be regarded as the current consumption of income at the expense of maintaining a near sustainable flow of income. We use the phrase 'near sustainable' because sustainability is a relative concept. Setting aside the allowance for the depreciation of KN, namely incurring $D_{KN}(t)$, extends the longevity of KN and thus results in a smaller but steadier flow of income. The recognition of $D_{KN}(t)$ also reduces the size of the multiplier. That is, for example when KN is ignored, the size of the multiplier is $\{1/[1 - \beta(t)(1 - \tau(t))]\}$, and this reduces to $\{1/[1 - \beta'(t)[(1 - \eta(t))(1 - \tau'(t))]]\}$, in the absence of residual externalities. In the presence of residual externalities there will be a further reduction to $\{1/[1 - \beta'(t)[(1 - \eta_R(t))(1 - \tau'(t))]]\}$ because it is likely that ($\eta_R > \eta$) and the sizes of beta and tau are also likely to be larger in the presence of residual externalities than in their absence.

The definition of full employment income, $Y_F(t)$, will depend on the type of production function that is recognised. Recall from Chapter 8, where we illustrated the distinction between $\{Y = f(L, KM)\}$ and $\{Y = g(L, KM, KN)\}$. We have now introduced a third function to account for the presence of residual externalities, namely $\{Y = g_1(L, KM, KN_R)\}$. The two-factor function involving only KM and L overstates both performance and capacity of an economy compared to the two functions that include the role of KN. The three functions are illustrated in the lower panel of Figure 12.4.

For reasons of illustrative convenience, we will assume that KM is fully employed, and unemployment pertains to only L. Following the rationale introduced in Chapter 8, we could argue that the utilisation of a given level of L would mistakenly indicate a higher level of Y in $\{Y = f(L, KM)\}$ than in $\{Y = g(L, KM, KN)\}$ or $\{Y = g_1(L, KM, KN_R)\}$. It is possible to demonstrate (as we formally do in the next Chapter) that the distance along the domain of Y, between $\{Y = f(L, KM)\}$ and $\{Y = g(L, KM, KN)\}$ or $\{Y = g_1(L, KM, KN_R)\}$ represents a measure of D_{KN}.

202 Macroeconomics and KN

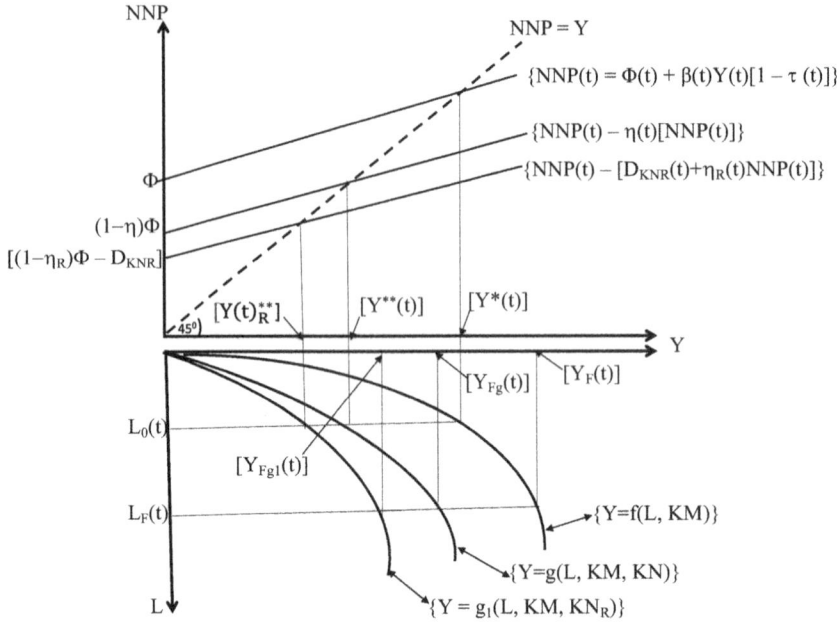

Figure 12.4 Sustainable Income within an Elementary Linear Keynesian Model

Now consider the following sets of comparisons, which are illustrated in the lower panel of Figure 12.4 with reference to L and Y. Suppose that $L_F(t)$ and $L_0(t)$ along the domain of L represent respectively magnitudes of full employment and current prevailing employment.

- To achieve full employment, $L_F(t)$, the two-factor production function will dictate that the income required is $Y_F(t)$. However, with reference to {Y = g(L, KM, KN)} - the magnitude of Y required is $[Y_{Fg}(t)]$, and with reference to {Y = g_1(L, KM, KN_R)} - the amount of Y required is $[Y_{Fg1}(t)]$.
- For reasons of illustrative convenience, we have assumed the equilibrium values of Y correspond with the current level of employment. That is, the employment of L0(t) persons while corresponding with the equilibrium of income of $Y^*(t)$ in the two-factor function, would correspond with $Y^{**}(t)$ or $Y(t)_R^{**}$ in the relevant 3-factor function involving KN.

Consider now the potential for distortion in policy formulation as illustrated in Figure 12.4. The policymakers who ignore KN, think that the economy is situated at $Y^*(t)$. Hence, they would deem it appropriate to pump-prime the economy towards $Y_F(t)$ to achieve full employment. However, when one does acknowledge the role of KN, the following observations unfold:

- The reality is that the economy is situated at either $[Y^{**}]$ or, $\left[Y(t)_R^{**}\right]$ and not at Y^*; and

- If the economy is to be pump-primed towards full employment ($L_F(t)$), then the target level of income to aspire to should be either [$Y_{Fg}(t)$] or [$Y_{Fgl}(t)$] and not $Y_F(t)$.

Hence policymakers who fail to acknowledge the role of KN could be mistakenly pushing the economy beyond its productive capacity. And as we argue in the next chapter, such mistaken efforts could be a source of inflation.

Further, note that the upward shift of AD through pump priming may also be achieved by lowering $\eta(t)$ or $\eta_R(t)$; that is, by reducing $D_{KN}(t)$. There are basically two broad approaches for reducing $D_{KN}(t)$. The first is a technology-based solution: to find, if one might coin a phrase, *KN-saving* technologies. These could include a very wide range of options involving cost-effective methods of utilising KN for economic activities such as advances in molecular biology to remove algal blooms; improved methods of waste treatment and recycling involving the closed-loop production systems discussed in Chapter 8; and the use of solar energy for domestic and commercial purposes. The second approach is a lifestyle option in terms of managing with less Y rather than with more. Being on the three-factor production function could also mean that economic agents, whether they are producers or consumers, adopt conscious decisions to moderate their behaviour and thereby inflict less harm on KN. Such behavioural changes include the use of public transport instead of private transport, less energy-intensive facilities and in general a curbing of extravagant consumption. To illustrate: in the Northern Hemisphere some people generally lounge around in sleeveless shirts in midwinter with the thermostat set at around 25°C. Other extravagances include the proliferation of devices that substitute for human effort, such as electric toothbrushes and carving knives. Further, when relationships in factor utilisation are properly articulated in terms of L, KM and KN, then returns to L, namely wages would be lower compared to when these relationships are confined to L and KM alone. We will consider this in the next chapter.

A difficulty with the linear framework is that it overlooks the inherent irreversibility of changes in KN. For this reason, a non-linear framework is considered below.

Equilibrium Income with D_{KN}: Non-Linear Framework

When irreversible environmental changes occur, $D_{KN}(t)$ will tend to increase at a rate greater than the fixed rate implied by a linear function. Therefore, we shall use the exponential function that was proposed in Equation (11.8) of Chapter 11 noting the synonymity between Y and NNP:

$$D_{KN}(t) = D_{KNR}(t) e^{\eta_E(t)\{\Phi(t) + [\beta' Y(t)(1-\tau'(t))]\}} \qquad (12.12)$$

Recall from Equation (11.9) that $\eta_E(t)$ is the proportion $\{[\ln D_{KN}(t) - \ln D_{KNR}(t)]/[NNP(t)]\}$. Thus the relevant expressions to define the economy are Equations (12.1), (12.3) and (12.12), and sustainable Y can be defined as:

$$Y(t) = [\Phi(t) + \beta'(t) Y(t)(1-\tau'(t))] - D_{KNR}(t) e^{\eta_E(t)\{\Phi(t) + [\beta'(t) Y(t)(1-\tau'(t))]\}}$$

$$(12.13)$$

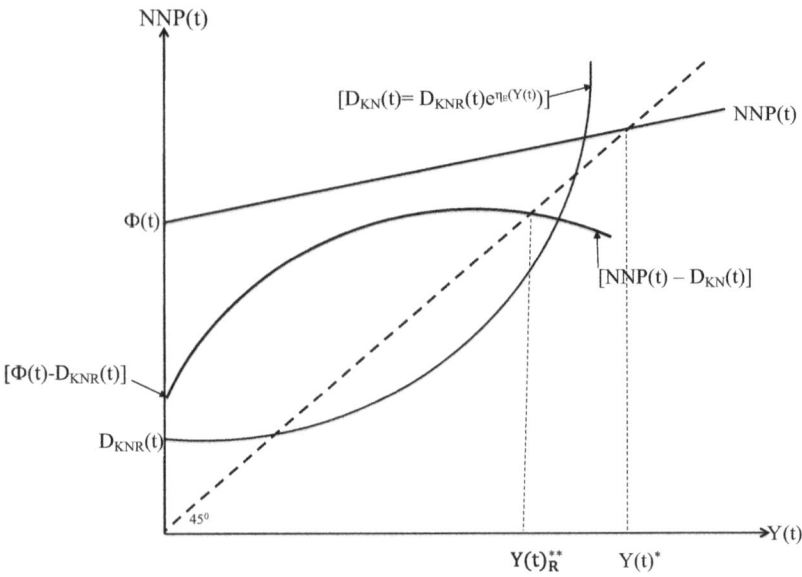

Figure 12.5 Non-linear System and Sustainable Income within a Keynesian Model

For a given value of η_E, a positive equilibrium is feasible, so far as values of NNP exist such that the gradient of D_{KN} (as in Equation (11.10)), namely $[D_{KNR}(t)\eta e^{\eta_E(0)[NNP(t)]} < 1]$. As illustrated in Figure 12.5, the equilibrium value of Y, namely $\left[Y(t)_R^{**}\right]$ results from the intersection of the AD function labelled $[NNP(t) - D_{KN}(t)]$ with the 45°° line. The analytic determination of $\left[Y(t)_R^{**}\right]$ in Figure 12.5 could involve a computational approach; for example, iteratively changing the value of Y(t) until the left-hand side of Equation (12.13) equals its right-hand side.

The main difference between the linear and non-linear frameworks concerns the extent to which pump-priming can continue by prompting upward shifts of the components of AD. With the linear framework, pump priming can continue indefinitely, so long as the gradient of $(NNP - D_{KN})$ does not exceed 1. With the non-linear framework, prompting the upward shift, for example by raising $\Phi(t)$, could at some point shift the D_{KN} schedule above the 45° line (that is, render the gradient to exceed 1) and thereby render a positive equilibrium infeasible.

Income Determination and Policy Analyses

The linear and non-linear formulations of income determination frameworks can be empirically applied for policy analyses when D_{KN} can be valued. The review of the literature on environmental accounting reveals that several proxies have been used for D_{KN}. These include:

1 A range of proxies for quality of life considerations (Daly and Cobb 1989)

2 Cost of fertiliser application for soil erosion (Repetto 1986; Tan and Thampapillai 2011)
3 Stumpage value of forests for deforestation (Repetto and Gillis 1988; Repetto and Magrath 1989; Cruz and Repetto 1992)
4 Market value of fish for over-fishing (Cruz and Repetto 1992)
5 Loss in agricultural output for soil degradation (Young 1992)
6 Total costs of mitigating greenhouse gas emissions (Thampapillai and Uhlin 1997; Thampapillai et al. 2010; Tan and Thampapillai 2011)

We illustrate below selected applications of the frameworks discussed above. These applications, however, did not include the presence of residual externalities. In each of these applications, the values of Y^* and Y^{**} were estimated over specific time domains primarily to ascertain the extent of divergence between the time paths of these two values. The greater the divergence, the greater is the lag in adopting policies that prompt sustainability. Further, for the purposes of demonstrating projected values of Y^* and Y^{**}, with reference to the time periods when the studies were undertaken, these applications relied on trend equations for each of the pertinent coefficients, namely $\Phi(t)$, $\beta(t)$, $\tau(t)$, and $\eta(t)$. Such equations, derived from time series data, for example, take the form $\{\Phi(t) = f[\Phi(t-1)]\}$. We illustrate the applications with reference to the countries that were analysed.

United States

The work by Thampapillai and Uhlin (1997) is one of the earliest applications involving the internalisation of D_{KN} in a Keynesian framework. This study was with reference to the United States for the period 1980–1991. Here, D_{KN} was measured by recourse to data from Daly and Cobb (1989) on diverse sources of KN losses such as: costs of pollution pertaining to air and soil, noise pollution, loss of wetlands, farmlands and non-renewable resources, and long-term environmental damage. The time paths of Y^* and Y^{**} are presented in Figure 12.5. And as illustrated here, the time path of Y^{**} derived from the non-linear framework shows possible convergence with the path of Y^*, whilst the potential for such convergence is absent in the comparison of the path of Y^* with that of Y^{**} derived from the linear framework. If one deems that the non-linear framework is more relevant than the linear one, the potential convergence may be interpreted as improvements in efficiency with reference to how KN had been utilised in the United States during the period in question. In fact, this study indicates that full employment would have been feasible in the United States had the value of η in the nonlinear framework reduced by about twenty per cent.

However, some caution needs to be exercised with the interpretation of the above results for two reasons. First, the results rest essentially on KN data, which were inevitably estimated by indirect methods of valuation. Second, the equilibrium incomes determined here rely on the assumption of a fixed price level. Hence, the possible convergence between the paths of Y^* and Y^{**} need not necessarily imply improvements in KN efficiency.

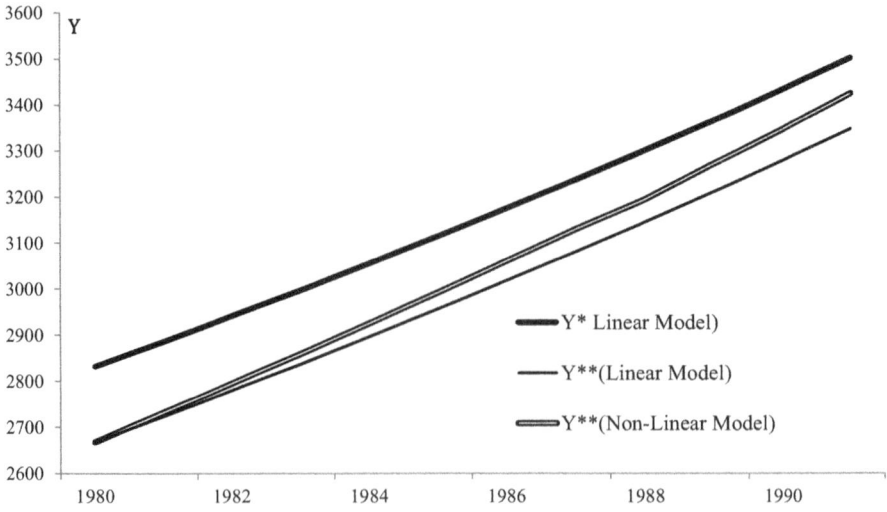

Figure 12.6 Time Paths of Y* and Y** for United States (1980–1991)

All values in constant 1982 US Dollars and based on Thampapillai and Uhlin (1997)

China Indonesia and Vietnam

The illustrations presented here draw on Thampapillai, Quah and Thangavelu (2005), and Tan and Thampapillai (2011, 2014) for China; and Thampapillai, Quah and Thangavelu (2003) for Indonesia; and Thampapillai, Wu and Tan (2010) for Vietnam. Each of these studies employed a linear Keynesian framework and confined KN to the air shed. The study on Vietnam included coastal waters as an added element of KN besides the air shed. D_{KN} with reference to air shed was measured by estimating the cost of mitigating GHG emissions. Further, the study of Vietnam included the possibility of reducing D_{KN} by constructing coastal sewerage plants. The main argument here based on scientific studies is that reducing the volume of sewage contamination in the oceans would reduce the adverse impacts on phytoplankton, which in turn facilitate the oceans to act as a carbon sink.

Annual data on GHG emissions (in CO2 equivalents) was taken from World Development Indicators provided by the World Bank and the mitigation cost was estimated at US$40 per ton and the appropriate exchange rate and GDP deflator. In each study, a marginal tax (ranging between 2 and 4 per cent) was introduced, and the tax earnings were then transferred as government expenditure towards the restoration of KN; that is an identity of $(\Delta T \equiv \Delta G)$ was established. The main restoration activity through ΔG for Indonesia and China was reforestation. For Vietnam, the construction of sewerage treatment plants was also included as an additional restoration activity. Hence the extent of the restoration activity would depend on the size of the marginal tax and the duration of the tax. These studies assumed that reforestation could reduce CO_2 loads in the atmosphere at the rate of 0.016 billion tons per million hectares, whilst one sewerage treatment plant could reduce CO_2 loads by 14,500 tons per year.

Across these studies, reforestation of up to 14 million hectares was possible if taxation persisted over a period of at least six years.

As indicated, the pertinent coefficients (Φ, β, τ, η) were forecasted from an initial reference year into the future and hence used to simulate the paths of Y^* and Y^{**}. The effects of KN restoration activity get captured in the estimation of η. For example, consider the time paths of Y^*, Y^{**} and $Y^{**}(RF)$ for Indonesia in Figure 12.7. $Y^{**}(RF)$

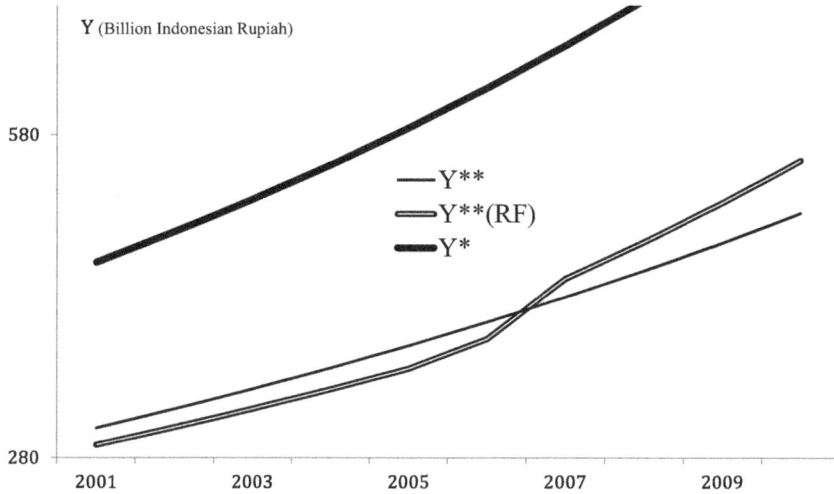

Figure 12.7 Effects of KN Restoration in Indonesia

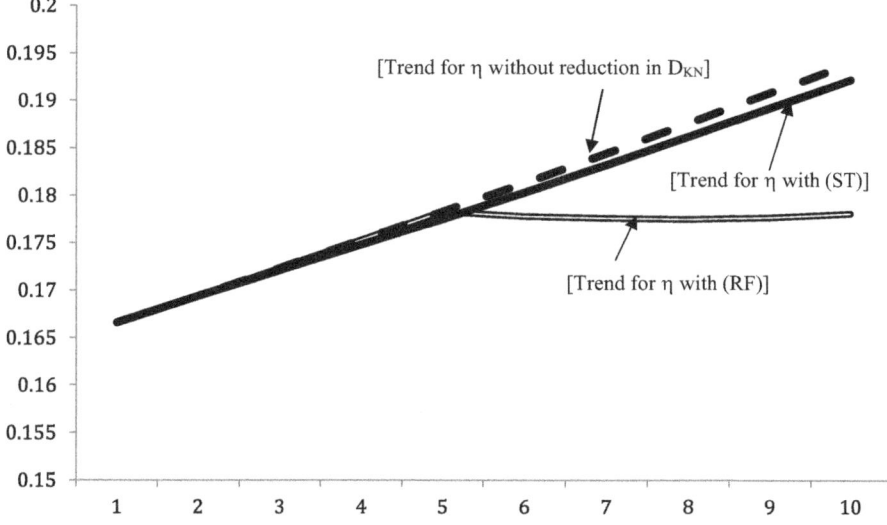

Figure 12.8 Impact of Reforestation (RF) and Sewerage Treatment (ST) on η in Vietnam

is the equilibrium income that results when CO_2 reduction due to reforestation is recognised. Although the paths of Y* and Y**(RF) display divergence, it is clear that taxation and reinvesting that taxation on KN restoration has beneficial impacts. Figure 12.8 illustrates the effect of reforestation and sewerage treatment on η in Vietnam. Although the effect of sewerage treatment on CO_2 reduction is minimal, other health benefits, which are not captured here, are indeed significant.

Mongolia

Mongolia is a resource-rich nation where mining accounts for nearly eighty per cent of export revenue. Thampapillai, Hansen and Bolat (2014) illustrate the effect of mine depreciation on the Mongolian economy with reference to just one of Mongolia's largest coal deposits, within the Tavan-Tolgoi coalfield in the Southern Gobi desert. In this study, simulated paths of Y* and Y** were presented for the period 2010–2017 using a linear Keynesian framework and D_{KN} was estimated by recourse to the Hartwick Rule that was discussed in Chapter 11. Recall from Chapter 11 that for a non-renewable resource, mining revenue will equal its depreciation, unless a portion of the revenue is reinvested in the creation of a new form of capital that could become an alternate source of Y. Depreciation, then, is the difference in value between the size original mineral stock and the size of the new capital stock that is created. Therefore, the extent of mine depreciation depends on the extent of savings that can be generated from mining revenues for mobilisation towards new investments away from mining. This study identifies the Resource Rent Tax (RRT) as a potential source of savings that could lead to a set of diversified investments; see box 12.1 for an explanation of the RRT.

Box 12.1   The Resource Rent Tax

The RRT is the economic rent that is owed to the rightful owner of a natural resource following its extraction.

The following premises underlie the definition of the RRT.

- The rightful owner of a natural resource is the society that is endowed with the resource.
- A mining firm, if assigned the task extracting the resource, is entitled to a fee. This fee is deemed normal profit, which is usually estimated in the range between 10 and 20 per cent (or even more) of all capital outlays. Such profit includes returns to management and the dividends owed to shareholders and investors of the firm.
- RRT is the surplus that remains, after all costs including normal profit owed to the mining firms have been accounted for.

Following the analysis of costs in Chapter 8, RRT = {(PQ) – MC} where PQ represents total revenue and MC is marginal costs. As per the description

Environmental Macroeconomics: Short-Run Analysis-I 209

above, MC would also include the normal profit that should accrue to the mining form.

In the Mongolian study considered in this Chapter, the RRT was estimated as nearly $578 Million per year with the price of coal being assumed to be $90 per tonne and the annual extraction as 15 million tonnes.

As in the other illustrations, nominating an initial reference year, and then projecting the values of Φ, β, τ and η, into the future by using trend equations enabled the simulation of income paths for Y^* and Y^{**}. The results of the simulation are displayed in Figure 12.9. Note that Y^{**} was differentiated into three categories, based on how the RRT and other revenues were reinvested:

- No reinvestment of RRT and related government revenue $[Y^{**}(0)]$
- Reinvestment of RRT only $[Y^{**}(1)]$
- Reinvestment of RRT and related government revenue $[Y^{**}(2)]$

As shown, the path that is based on the reinvestment of both the RRT and related government revenue, namely $Y^{**}(2)$ reveals a distinct path of ascension. Such ascension is modest with the path of $Y^{**}(1)$ which is based on the reinvestment of RRT alone. The path that ignores any reinvestment, namely $Y^{**}(0)$, bears the complete value of the depreciation and shows a decline over time. However, when D_{KN} is ignored, the path of Y^* displays a false image that the economy is performing well.

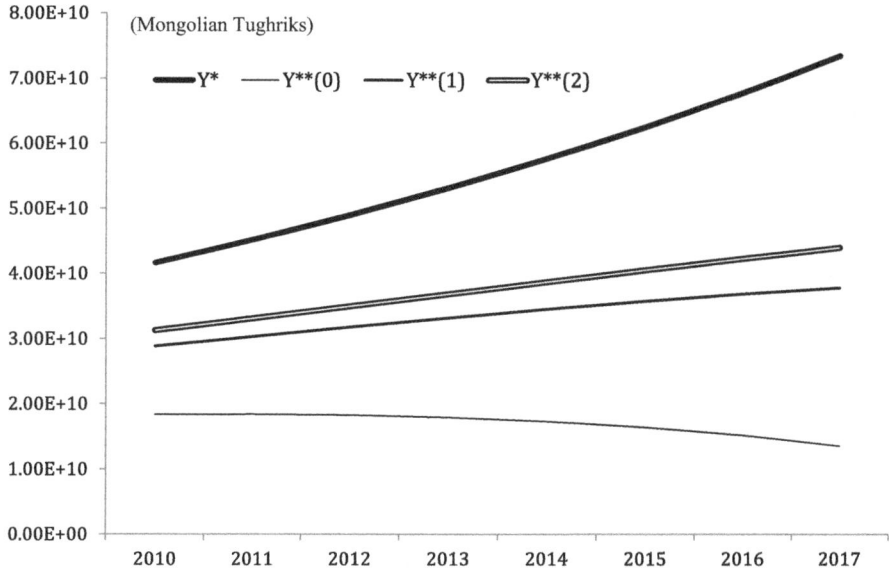

Figure 12.9 Simulated Time-Paths of Y* and Y** for Mongolia

Further Thoughts on D_{KN} and KN Investments

Although the national accounts do not provide sufficient disaggregation, it is possible to envisage a special class of investments termed 'KN-investments' as illustrated in this chapter. These can be important shifters of the income schedule and, as discussed in Chapter 11, they can be defined as the restoration of lost endowments – analogous to replacement investments in the context of sunk costs. To recapitulate an example: if, at the start of a given period, a river is declared dead due to a variety of pathogens, then it cannot be counted in the stock of KN because it will not contribute towards national output. That is, its status will be similar to that of a sunk cost. Removing the pathogens and restoring the river is tantamount to adding to the stock of KN. Hence, activities such as reforestation of mined-out areas, the creation of wetlands in water courses destroyed by nitrogen loading, and the detoxification of contaminated land to enable new development are examples of KN-investments. Since these investments are reproducible capital, they can be regarded as special cases of the 'Hartwick Rule' in practice, making reinvestments in reproducible forms of capital to maintain a permanent flow of income (Hartwick 1977, 1978; Solow 1986).

Concluding Remarks

Although we introduced a simple framework in this chapter, it was possible to illustrate some important policy applications. The assumption that AS is horizontal – that is, 'whatever demanded is produced' – could seem to be restrictive. Nevertheless, this assumption can be applied within limits in economies that have a fair degree of unemployment. The important message from this chapter is that, during times of unemployment and a general economic recession, investment in KN can also be a feasible avenue for stimulating the economy. However, policymakers have traditionally opted for lowering taxes or interest rates or spending on items such as roads and other infrastructure, even when such items are in abundance. Alternatively, there are several options for reducing D_{KN} and, as a result, prompting increases in Y. For example, reforestation provides bio-fuel, reduces energy imports, traps carbon from the atmosphere and as result makes the workforce healthier. A healthier workforce is more efficient and hence raises income.

Review Questions

1 Suppose that the following data are available for an economy in a given year:

$F = 1622.5$; $b = 0.715$; $b' = 0.842$; $t = 0.228$; $t' = 0.258$; $g = 0.116$; $h_E = 7.4*(10^{-12})$

12.25

Estimate Y^* and Y^{**} using the linear framework

Use a computational method and estimate the size of Y^{**} by applying the equality in Equation (12.13) above.

2 Discuss whether the non-linear framework illustrated in this chapter is likely to provide a policy outcome that is more sustainable compared to the linear framework.

References

Daly, H. E. and Cobb, J. B., *For the Common Good: Redirecting the Economy towards Community, the Environment, and a Sustainable Future*, Beacon Press, Boston, MA, 1989.

Cruz, W. and Repetto, R., *Environmental Effects of Stabilization and Structural Adjustment Programs: The Philippines Case*, World Resources Institute, Washington, DC, 1992.

Hartwick, J., 'Intergenerational equity and the investing of rents from exhaustible resources', *American Economic Review*, 66: 972–974, 1977.

Hartwick, J., 'Investing returns from depleting renewable resource stocks and intergenerational equity', *Economic Letters*, 1(1): 85–88, 1978.

Repetto, R., 'Soil loss and population pressure on Java', *Ambio*, 15(1): 14–18, 1986.

Repetto, R. and Gillis, M. (eds), *Public Policy and the Misuse of Forest Resources*, Cambridge University Press, Cambridge, 1988.

Repetto, R. and Magrath, W. B., *Wasting Assets: Natural Resources in National Income Accounts*, World Resources Institute, Washington, DC, 1989.

Solow, R. M., 'On the intergenerational allocation of natural resources', *Scandinavian Journal of Economics*, 88(1): 141–149, 1986.

Tan, S. L. and Thampapillai, D.J., 'Assessment of fiscal intervention measures in China: Perspectives from environmental macroeconomics', in H. Ashiobar, L. Kreiser and J. Sirisom (eds), *Environmental Taxation in China and Asia Pacific: Achieving Environmental Sustainability through Fiscal Policy*, Edward Elgar, London, 2011.

Tan, S. L. and Thampapillai, D. J., "Environmental taxation for a sustainable future: Perspectives from environmental macroeconomics', in Krieser, L. et al. (eds), *Environmental Taxation and Green Fiscal Reform,* Edward Elgar, Cheltenham, UK, 2014

Thampapillai, D. J., Hansen, J. and Bolat, A., 'Resource rent taxes and sustainable development', *Energy Policy*, August 2014.

Thampapillai, D. J., Wu, X. and Tan, S. L., 'Fiscal balance: Environmental taxes and investments', *Journal of Natural Resources Policy Research*, 2(2): 137–147, 2010.

Thampapillai, D. J., Quah, E. and Thangavelu, S., 'An introductory macroeconomic framework for China: Implications for west china development', in Ding, L. and Nielson, W. A. (eds), *China's West Region Development: Domestic Strategies and Global Implications*, World Scientific, 2005.

Thampapillai, D. J., Quah, E. and Thangavelu, S., The Internalization of Environmental Capital into a Simple Macroeconomic Model: Illustrations with reference to Indonesia, Paper Presented to the 32nd Conference of Economists, October 2003, Canberra ACT.

Thampapillai, D. J. and Uhlin, H.-E., 'Environmental capital and sustainable income: Basic concepts and empirical tests', *Cambridge Journal of Economics*, 21(3): 379–394, 1997.

Young, R., 'Evaluating long-lived projects: The issue of intergenerational equity', *Australian Journal of Agricultural Economics*, 36(3): 207–232, 1992.

13 Environmental Macroeconomics
Short-Run Analysis II

The main assumption underlying the Keynesian framework illustrated in the previous chapter is that Aggregate Demand (AD) can determine Aggregate Supply (AS) without any reference to the rate of inflation. The direct implication of this assumption is that the incurrence of expenditure is rewarded by producers recruiting resources to increase employment, output and income (Y). Note that in most macroeconomic texts, the terms 'output' and 'income' are interchangeably used, and both are denoted by Y. In his treatise on *Employment Money and Interest*, Keynes (1936) provides an extreme example of expenditure stimulus to revive an economy out of severe depression. That is, to bury old bottles filled with treasury notes, and then provide an opportunity for private enterprise to discover them. Keynes' argued that private enterprise would then use the discovered wealth towards generating employment and Y without any state intervention. Apart from obvious time lags between expenditure and output response, the main limitations of the Keynesian framework are the nonrecognition of inflation (π) control, and AS management. Inflation control and AS management are both important macroeconomic goals besides the management of AD. Hence, in this chapter, we make π and AS explicit components of the policy analytic framework alongside AD. As in the previous chapter, we first introduce the standard macroeconomic framework and then proceed to revise it to acknowledge the presence of KN.

We demonstrate the various components of our proposed framework by recourse to the *analytics of point estimates and assumed functional forms*. That is, we assume that specific functional forms can portray the components of the macroeconomic framework. We then use the data of a specific point in time, to estimate coefficients pertinent to the functional forms, and thereby display the framework for policy analysis. In this chapter, we replace GDP with gross value added (GVA) as the basis for the underlying equilibrium; that is, (Y $\equiv$ GVA). Given that (GVA = GDP – T), we subtract T (net taxes) from both the expenditure and income accounts. In the absence of T, as illustrated below, Y is distributed solely between the factors of production.

The Standard Framework

A simple version of the standard macroeconomic framework comprises the joint display of frameworks for: AD, AS and production (factor-utilisation).

DOI: 10.4324/9781003408574-16

Each of these components can be illustrated with the use of point estimate data from the national accounts. In order to facilitate such an illustration, we make use of the macroeconomic indicators for Australia shown in Table 13.1 for the five-year period from 2005 to 2009. Although this time period might appear somewhat dated, we choose this period for a specific reason. It is during this period that global financial markets displayed several irregularities that culminated in the Global Financial Crisis (GFC). The GFC presented a unique opportunity for several governments to restructure and stabilise their respective economies by including KN in the policy mix. Australia's response was typical of those attempted by several other countries, namely the adoption of pump-priming efforts to revive the economy out of perceived recession. Such revival efforts were instrumental in inflicting further damage to KN, and thereby accelerating its destabilisation. As we illustrate below, the Australian economy (like most others) was operating beyond its productive capacity when KN is recognised.

Table 13.1 is structured in six sections. The first four sections provide the pertinent data from the national accounts. The first section provides data on money stock (M1) and price levels (P). M1 indicates the volume of money that is readily accessible and is freely circulating in the economy. P is measured by the GDP deflator and as we show below is the basis for measuring inflation (π). The sections on expenditure and income accounts illustrate how estimates of GDP can be arrived at through the two standard methods, namely aggregation of real final expenditures and factor incomes. Note that the section on resources is confined to labour (L) and manufactured capital (KM), as per the standard framework. The final two sections describe information that is pertinent to the standard macroeconomic framework that we will develop below. We explain the estimation of this pertinent information as each component of the framework is developed.

Production Function (Factor Utilisation)

Most macroeconomic texts describe factor utilisation in terms of a Cobb–Douglas (CD) function of constant returns to scale involving two factors, namely manufactured capital stock (KM) and labour (L):

$$Y_t = \alpha_t KM_t^{\theta_t} L_t^{\lambda_t} \tag{13.1}$$

where θ_t and λ_t represent the factor shares of national income (Y_t) in time t accruing respectively to KM and L, and assuming constant returns to scale ($\theta_t + \lambda_t = 1$). The assumption of constant returns to scale enables factor shares of income to be elicited, as point estimates, directly from the income accounts where the following identity prevails for ($Y \equiv GVA$):

$$Y_t \equiv \{\text{Compensation to employees } (CE_t)\} \\ + \{\text{Gross Operating surplus}(GOS_t)\} \tag{13.2}$$

Table 13.1 Selected Macroeconomic Indicators and Estimates for Australia (2005–2009)

	2005	2006	2007	2008	2009
1 Money and prices					
GDP deflator	1	1.05	1.09	1.16	1.17
M1 ($ million)	182,761	206,777	231,359	242,579	251,663
2 Expenditure Accounts ($ million)					
Consumption (C)	565,338	581,856	601,181	595,427	618650
Investment (I)	279,164	293,046	320,393	306,202	292947
Government (G)	171,641	177,105	185,150	185,873	195931
Exports (X)	195,676	205,219	214,309	245,434	219764
Imports (M)	211,030	217,820	236,859	240,862	213947
Discrepancy	0	0	0	−11749	−25322
GDP Expenditure Account	1,000,789	1,039,406	1,084,174	1,080,325	1,088,023
3 Income accounts ($ million)					
Compensation of Employees (CE)	487,157	508,320	527,898	523,716	521,242
Gross Operating Surplus (GOS)	406,958	422,464	443,667	445,868	440,010
Net Taxes (T)	106,672	108,575	112,609	104,106	104,594
Discrepancy	0	0	0	6586	22,177
GDP Income Account	1,000,787	1,039,359	1,084,174	1,080,277	1,088,023
Y = GVA = (CE + GOS)	894,115	930,784	971,565	969,584	961,252
4 Resources					
Employment (L) '000 Persons	10,049	10,306	10,591	10,830	10,866
Full Employment (L_F) '000 Persons	10,580	10,822	11,073	11,307	11,505
Capital Stock (KM) $ Million	4,128,355	4,283,789	4,461,389	4618877	4,757,862
5 Estimates: Factor Utilisation					
θ = (GOS/Y)	0.455	0.454	0.457	0.460	0.458
λ = (CE/Y)	0.545	0.546	0.543	0.540	0.542
α	248.35	254.02	246.67	230.25	230.69
6 Estimates: AS – AD					
π (rate)	1.042	1.050	1.038	1.064	1.009
V	4.895	4.73	4.577	4.636	4.47
Φ = (X − M)	−15354	−12601	−22,550	4572	5817
β = C/(Y − T)	0.718	0.708	0.700	0.688	0.722
τ = (T/Y)	0.119	0.117	0.116	0.107	0.109
Y_F	919,562	955,947	995,330	992,416	991,489
$Y_F − Y$	25,447	25,163	23,765	22,832	30,237
π_F	1.014	1.023	1.013	1.040	0.975

With the exception of M1, all monetary estimates are in constant 2005 Australian dollars.

Source: OECD iLibrary. (https://www.oecd-ilibrary.org/economics/data/main-economic-indicators_mei-data-en)

Note that an increase in Y is attributed to expansion of both KM and L. Because CE and GOS represent respectively payments accruing to L and KM, point estimates for the factor shares of Y can be elicited directly from income accounts as follows:

$$\left[\lambda_t = \frac{CE_t}{Y_t}\right]; \left[\theta_t = \frac{GOS_t}{Y_t}\right] \qquad (13.3)$$

The coefficients θ and λ in Table 13.1 were derived through the application of definitions given in Equation (13.3). The estimates of these coefficients, together with the point estimate data on KM and L, enable the estimation of the total factor productivity measure, α, through a rearrangement of Equation (13.1):

$$\alpha_t = \frac{Y_t}{KM_t^{\theta_t} L_t^{\lambda_t}}.$$

It is now possible to illustrate the CD function for each year. For example, from the data and estimates in Table 13.1, the factor-utilisation function of 2009 would read as:

$\{Y = 230.69 KM^{0.46} L^{0.54}\}$. The CD functions so estimated can then enable the identification of capacity (full employment) income (Y_{Ft}) by substituting the value of L_{Ft} for L in the CD function. It is then possible to estimate the output gap ($Y_{Ft} - Y_t$). These estimates are illustrated in Table 13.1 and, as shown below, will assist in the display of AS. Please note that in this illustration, we have assumed KM stocks to be fully utilised. Hence in our illustration, the term unemployment pertains to L alone.

Aggregate Supply (AS)

The display of AS in most texts usually proceeds on the basis of associating expectations concerning inflation with decisions pertaining to Y. As discussed in Chapter 10, policymakers prefer moderate levels of π that would promote the expansion of Y. As a result, macroeconomists envisage a positive (upward-sloping) relationship between π and Y spanning a target range of π. However, the empirical display of AS is invariably surrounded by difficulties pertaining to modelling of expectations with reference to π and Y. Therefore, we simplify the exposition of AS here by differentiating capacity supply (AS_F) from a short-run response (AS_{SR}). We nominate the utilisation of the labour force (L_F) as full employment and hence the basis for capacity.

AS_{SR} is assumed to be Keynesian. That is, producers expect the prevailing level of π to persist in the short run and hence will strive to produce as much as possible ($Y \to \infty$) at this level. The following conditional statement can describe this context:

$$AS_{SRt} : \{(\pi = \pi_t) | (Y \to \infty)\} \qquad (13.4)$$

The interpretation of Equation (13.4) is that, given the inflation rate of π_t, producers will attempt to produce as much as possible at the prices dictated by π_t. Consider the inflation rates provided in Table 13.1. We adopt the following definition for π_t: $\{[(P_t - P_{(t-1)})/(P_{(t-1)})] + 1\}$. This definition would enable P_t to be in turn defined as $[P_{(t-1)} \pi_t]$. As a result, ($\pi = 1$), will represent the stationary (zero) level of inflation. On the basis of such information, the AS_{SR} for 2009 would be defined as:

$$AS_{SR\ 2009} : \{(\pi = 1.009) \mid (Y \to \infty)\}.$$

The definition of AS_F is based on the premise that capacity income (Y_{Ft}) in a given time period t cannot be exceeded. That is, AS_{Ft} assumes the classical shape described in Chapter 10 and is defined as:

$$AS_{Ft} : \{(Y = Y_{Ft}) \mid (\pi \to \infty)\} \qquad (13.5)$$

As in Chapter 12, we premise the utilisation of L_F as the basis for capacity and assume that KM is fully utilised. Hence, Y_{Ft} can be estimated substituting the value of L_F for L in Equation (13.1). As illustrated in Table 13.1, AS_F for 2009 would be defined as:

$$AS_{F\ 2009} : \{(Y = 991{,}489) \mid (\pi \to \infty)\}.$$

Aggregate Demand (AD)

In Chapter 12, AD was deemed the only determinant of income. Therefore, it was sufficient to describe AD in terms of an expenditure statement, such as that given in Equation (12.3). However, when the influence of π has to be also considered, we need to acknowledge that the relationship between π and Y (measured as the sum of real final expenditures) is inverse. The basis for such a relationship was discussed in Chapter 10. That is, for example, when prices increase consumers will either postpone or relocate or simply cut back their expenditures. With the onset of π, central monetary authorities will also raise the interest rate to maintain the real interest rate at a steady level. This reinforces the inverse relationship.

A convenient way to capture this inverse relationship would be to premise (following Mankiw 2004; Flath 2005) AD on the quantity equation. As in most standard texts, the quantity equation is $[P_t Y_t = M_t V_t]$, where (P_t, Y_t, M_t, V_t) represent respectively the price level; real national income, money stock and velocity of circulation, each in time period t. Although M_t is differentiated into several types (in terms of accessibility) such as M1, M2 and so forth, here we use M1, namely the volume of money in circulation that can be easily accessed. Velocity of circulation is a measure of the number of times a unit of currency gets exchanged in the economy. In order to get an expression that conforms to $\{\pi = f(Y)\}$, we adopt the definition for π_t developed above, namely: $\{\pi_t = [(P_t - P_{(t-1)})/(P_{(t-1)})] + 1\}$. This definition would enable P_t to be in turn defined as $[P_{(t-1)} \pi_t]$. By substituting $[P_{(t-1)} \pi_t]$ in place of P_t in the quantity equation, the following definition of AD is possible.

$$\pi_t = \left[\frac{M_t V_t}{Y_t P_{t-1}} \right] \tag{13.6}$$

The display of AD and the elicitation of likely changes in AD due to possible methods of policy intervention can be further aided by the following set of simplifying assumptions:

1. Given π_t and Y_t in time t, a short-run equilibrium, namely $\{AS_{SRt} = AD_t\}$, does exist for (Y_t, π_t). A second set of coordinates for the AD function, namely (Y_{Ft}, π_{Ft}), can be resolved by estimating the value of π_{Ft} for $(Y = Y_{Ft})$ from Equation (13.6). These values are also shown in Table 13.1. Although the AD function in Equation (13.6) is strictly non-linear, the two sets of coordinates $\{(Y_t, \pi_t); (Y_{Ft}, \pi_{Ft})\}$ permit a linear approximation over a limited domain, namely $(Y_t \leftrightarrow Y_{Ft})$.
2. As indicated, the definition of money stock is confined to narrow money (M1). The changes in M1 in response to changes in the interest rate (Δr) can be given by (dM/dr) which, as indicated below, is based on time trends of M and r.
3. Expected changes in Y in a given time t, (ΔY_t), are drawn from changes in tax rates ($\Delta \tau$) influencing consumption (C); government spending (ΔG); and interest rates (Δr) influencing investment stock (I).
4. Velocity of money (V_t) during a given time period remains fixed and can be ascertained as point estimate from data on π, M and Y as shown in Table 13.1.

Given the above assumptions, the following definitions, namely Equations (13.7), (13.8), (13.9) and (13.10) can be made with reference to likely changes in AD, when interventions in terms of fiscal and monetary policies ($\Delta \tau$, ΔG and Δr) are enforced.

$$M_t = \left[M_{t-1} * \left(\left(\frac{dM}{dt} \right) + \left[\left(\frac{dM}{dr} \right) * (\Delta r_{t-1}) \right] \right) \right] \tag{13.7}$$

In Equation (13.7), ($\Delta r_{t-1} = r_{t-1} - r_{t-2}$) is based on the appropriate point estimates for the interest rates and $\left(\frac{dM}{dr} \right) = \left(\frac{dM/dt}{dr/dt} \right)$.

$$GVA_t = \Phi_t + \beta_t Y_t (1 - \Delta \tau_t) + G_t - T_t + I_t \tag{13.8}$$

In Equation (13.8), Φ_t and β_t are respectively a constant comprising net exports, and marginal propensity to consume.

$$\Delta I_t = \left[I_{t-1} * \left(\left(\frac{dI}{dt} \right) + \left[\left(\frac{dI}{dr} \right) * (\Delta r_{t-1}) \right] \right) \right] \tag{13.9}$$

218 *Macroeconomics and KN*

In an Australian study (Thampapillai 2012) spanning the period 2001–2010, the values for (dM/dt) and (dI/dt) were identified respectively as 1.042 and 1.0114. The values for (dr/dt) were 1.052 for the periods of rise in r, and 0.922 for the periods of decline in r.

The assumption of a short-run equilibrium implies that, for a given π_t, (GVA$_t$ = Y_t) and hence from the foregoing, an expression for AD in time t could be provided as follows:

$$\pi_t = \frac{M_t V_t}{\left[\Phi_t + \beta_t Y_t (1 - \tau_t) + G_t - T_t + I_t\right] P_{t-1}} \tag{13.10}$$

The expression in Equation (13.10) enables an assessment of the impacts on AD when fiscal and monetary policy variables are altered. These impacts are confined not only to changes in Y, as with the standard Keynesian framework, but also variations involving π.

Display of the Standard Framework

It is now possible to assemble the three components considered above and display the standard macroeconomic framework. In Figure 13.1, this framework is illustrated as a sketch for 2009 by recourse to the point estimate information presented in Table 13.1. This display rests on the assumption that a short-run equilibrium (AS$_{SR}$ = AD) does exist and that a linear approximation of AD over the domain of the output gap is feasible. This output gap is clearly recessionary because (Y_t < Y_{Ft}). The closure of this gap could involve, as shown in standard texts, one or a mixture of the following measures:

- The choice of appropriate fiscal and monetary measures to stimulate a shift of the AD function to the right and/or
- The lowering of wages and prices to facilitate the AS$_{SR}$ function to shift downwards to a level complying with (π_{F2009} = 0.975). That is, had prices and wages been reduced towards lowering inflation by about 3.4 per cent [($\pi_t - \pi_{Ft}$) = (1.009 − 0.975) = 0.034], then the attainment of full employment might have been feasible.

As indicated in an Australian study (Thampapillai 2012), a similar effect on the AS$_{SR}$ could have also been achieved by reducing the inequality of incomes through lowering wages at the higher end of the income spectrum. However, Australia did not vigorously pursue such policy options. For that matter, neither did most other countries battling the economic downturn experienced during the early 2000s.

Almost all countries attempted to move AD to the right by lowering interest rates and tax rates. The impacts of such interventions on AD as defined in Equation (13.10) can be assessed by recourse to Equations (13.7), (13.8) and (13.9). Further changes in the gradient of AD in response to an intervention measure, say

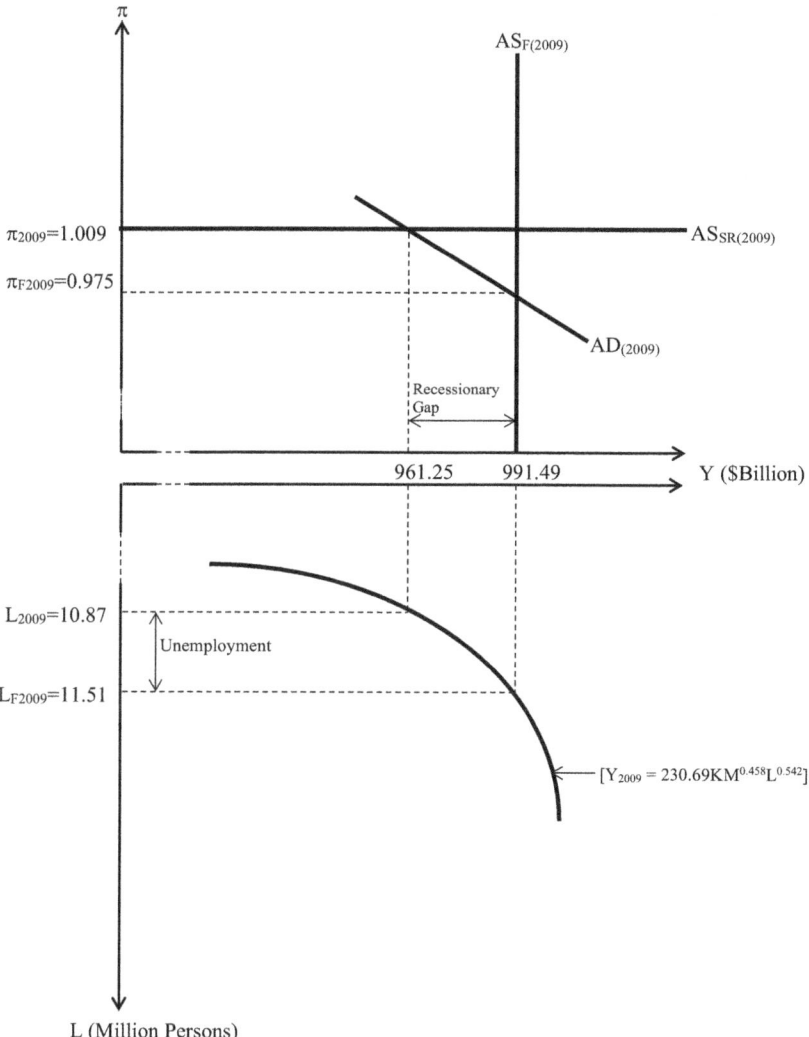

Figure 13.1 The Australian Economy (2009) – Standard Macroeconomic Framework

a tax change, such as that defined below could also help assess the responsiveness of AD.

$$\frac{d\pi}{d\tau} = \frac{M_t V_t \beta_t Y_t}{\left[\Phi_t + \beta_t Y_t (1-\tau_t) + G_t - T_t + I_t\right]^2 P_{t-1}} \quad (13.11)$$

At least two aspects capture the important distinction between the standard framework and the environmental macroeconomic framework. The first is the policy domain. This is the income domain within which the policymaker will try to resolve for inflation, employment and Y. As shown in Figure 13.1, this domain ($Y_t \leftrightarrow Y_{Ft}$)

for 2009 in Australia was (961.25 ↔ 991.49) when expressed in dollars billion. The second aspect is the responsiveness to intervention, such as those suggested above in Equation (13.11).

We illustrate below that on both counts the policymaker would be in error. That is, when KN is ignored, for example, the income domain ($Y_t \leftrightarrow Y_{Ft}$) in Figure 13.1 can be labelled a 'mistaken domain' and Equation (13.11) overstates the responsiveness of the intervention measure.

The Environmental Macroeconomic Framework

We now proceed to revise the framework developed above to recognise KN. As in Chapter 12, we will assume that KN investments are already included in GVA and that it is KN depreciation, namely D_{KN} that has to be accounted for. Hence the basic equilibrium for developing and illustrating this framework would be the identity ($Y \equiv GVA - D_{KN}$). This identity stipulates that the magnitude one perceives as that of Y is not GVA as with the standard framework, but instead, it is GVA less the amount that we need to set aside for dealing with the depreciation of KN. We now use the information in Table 13.2 for the illustration. Some of the information from Table 13.1 has been reproduced here for comparison purposes. As in Chapter 12, we confine KN to the air shed of the economy, and therefore the values of D_{KN} provided here are clearly underestimates. We equate D_{KN} to the costs of abating greenhouse gas (GHG) emissions measured in CO_2 equivalents. These costs, as per the Stern (2007) report, have been set at US$100 per ton. As illustrated below, the various coefficients for the macroeconomic framework estimated in the previous section, require revision due to the acknowledgement of KN and D_{KN}.

Reformulated Production Function (Factor Utilisation)

The main argument here is that Y cannot be explained by the interaction between KM and L alone. This is because KN plays a central role in the formation of Y. For example, one cannot envisage the formation of Y without clean air or water and services from a range of natural endowments. And, as argued in Chapter 8, we need to consider the distinction between the accumulation and the depletion of KN. Recall the argument in terms of the laws of thermodynamics and ecosystem resilience providing the basis for diminishing marginal returns. That is, progressive attempts to increase Y require progressively greater depletion of KN as well as loss of resilience of KN. Therefore, in the standard factor-utilisation function as given in Equation (13.1), the contribution of KM and L is overstated. The basis for this overstatement was illustrated in Figure 8.4 of Chapter 8. Recall from the left-hand panel of this figure that the efforts for achieving successive increases of Y will involve not only the raising of KM and L, but also increases in the depletion of KN by amounts such as $\Delta k_1,...,\Delta k_3$. The right-hand panel of Figure 8.4 results in two production functions – one that excludes the role of KN and the other that includes the contribution of KN. As we proposed in Chapter 8, at the microeconomic level,

Table 13.2 Environmental Macroeconomic Indicators and Estimates for Australia (2005–2009)

	2005	2006	2007	2008	2009
1 From Table 13.1					
Money and Prices					
GDP Deflator	1	1.05	1.09	1.16	1.17
M1 ($ million)	182,761	206,777	231,359	242,579	251,663
Y (National Income = CE + GOS)	894,115	930783.8	971565.1	969584.5	961252.1
Employment (L) '000 Persons	10048.65	10305.63	10591.03	10830.43	10866.16
Full Employment (LF) '000 Persons	10579.81	10821.51	11073.12	11307	11504.61
Capital Stock (KM) ($ million)	4,128,355	4,283,789	4,461,389	4,618,877	4,757,862
$\theta = (CE/Y)$	0.455	0.454	0.457	0.460	0.458
$\lambda = (GOS/Y)$	0.545	0.546	0.543	0.540	0.542
α	248.35	254.02	246.67	230.25	230.69
2 Environmental accounts					
CO_2 Kilo Tons	363,967	371,471	373,739	379,902	386,066
$D_{KN} = CO_2$ abatement cost ($ million)	40,070	40,896	41,146	41,824	42,503
$Y - D_{KN}$ ($ Million)	854,045	889,888	930,420	927,760	918,750
3 Estimates: Factor-Utilisation					
η	0.045	0.044	0.042	0.043	0.044
KN ($ Million):	9,391,617	10,004,699	11,334,911	11,108,299	11,692,168
4 Estimates: AD – AS					
Y_S	854,045	889,888	930,420	927,760	918,750
Y_{FS}	878,353	913,946	953,197	949,593	947,639
$Y_{FS} - Y_S$	24,308	24,058	22,778	21,833	28,889
$\bar{\pi}$	1.091	1.098	1.084	1.112	1.055
$\bar{\pi}_F$	1.060	1.069	1.058	1.087	1.023

With the exception of M1, all monetary estimates are in constant 2005 Australian dollars.

Source: OECD iLibrary.

KM and KN are both measurable on the same scale resulting in a composite measure (K) for capital encompassing both KM and KN.

We now extend this conceptualisation to the macroeconomic level and also integrate the concept of resilience that was developed in Chapter 11 into the definition of KN depletion. Further, suppose that each (Δk) as indicated above represents net depletion with reference to gross utilisation such that $[\Delta k = (1 - \varphi) KN]$. And hence the composite measure can be defined as follows: $[K = KM + (1 - \varphi) KN]$. In this definition of K, φ explains the resilience of KN and ranges between 0 and 1. Perfect resilience of KN is represented by ($\varphi = 1$), whilst ($\varphi = 0$) signifies complete

222 Macroeconomics and KN

non-resilience. So, when KN displays perfect resilience, that is ($\varphi = 1$), then the depletion of KN non-existent, and [K = KM]. At the other extreme, where KN has lost its resilience completely, that is ($\varphi = 0$), then the complete complement of KN that is required gets depleted and [K = KM + KN].

Consider now Figure 13.2. This is in fact a replication of the right-hand panel of Figure 8.4 – extended to the context of the macroeconomy. As per the function labelled {Y = f(KM, L)}, income level Y_t would be attributed to the role of KM_t and L_t. In this display, if we assume that L_t is held constant whilst KM is variable, then, the formation of Y_t would be solely attributed to KM_t. Such attribution overstates the contribution of KM_t, because KN also contributes to the formation of Y_t. It is possible to quantify the exacerbation of factor-contribution by differentiating the equilibrium that recognises the role of KN, namely (Y ≡ GVA − D_{KN}) from that which does not, namely (Y ≡ GVA). Note that we used this differentiation in Chapter 12 to illustrate the determination of Y in terms of aggregate demand. We will now use the same principle for factor-utilisation that enables the explanation of aggregate supply.

The domain for K is displayed beneath the horizontal axis in Figure 13.2. The vital role of KN in determining Y_t can then be explained by the fact that the amount of K needed to explain Y_t, namely K_t is indeed greater than KM_t. Hence, for every point on the locus describing {Y = f(KM, L)}, there exists a corresponding quantity of K that is in excess of KM. The locus of these points, namely the coordinates of (K, Y), constitutes the function involving KN besides KM and L; that is {Y = g(KM, L, KN)}. When we acknowledge the relevance of KN, then the three-factor function {Y = g(KM, L, KN)} is the valid descriptor of Y.

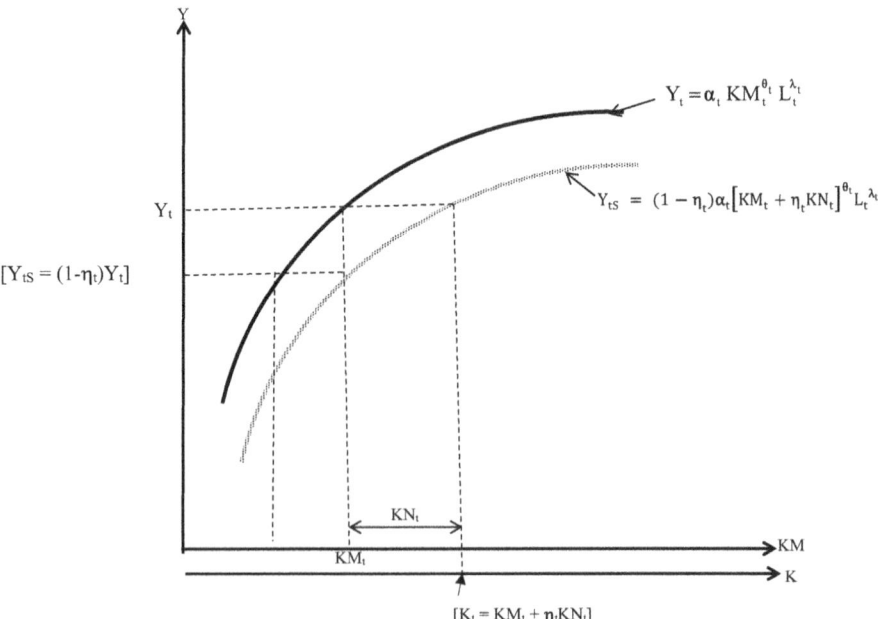

Figure 13.2 The Conceptual Basis for the Estimation of KN

Therefore, the production function {Y = g(KM, L, KN)} must depict at least two properties. First, the value of Y that gets resolved must equate to $(Y - D_{KN})$; and second, this function must also make explicit the contribution of KN that gets depleted.

Consider now how the Cobb–Douglas function in Equation (13.1) could be transformed to the contexts of $(Y \equiv GVA - D_{KN})$ and {Y = g(KM, L, KN)}. Following our narratives in Chapters 11 and 12, suppose that D_{KN} is a constant proportion η of Y. That is, $\eta = [D_{KN}/Y]$. Then multiplying Equation (13.1) by $(1 - \eta)$ would be tantamount to resolving for $(Y - D_{KN})$. To capture the contribution of KN, we replace KM with K, and assume that the composite capital (K) will receive the same factor share of income as KM in Equation (13.1). We denote the estimates of Y originating from {Y = g(KM, L, KN)} during a given time period t as Y_{tS} so that they are distinct from the estimates derived from the standard two-factor function, namely Y_t.

Given our definition of K above, the modified Cobb–Douglas function will read as:

$$Y_{tS} = (1-\eta_t)\alpha_t \left[KM_t + (1-\varphi_t)KN_t\right]^{\theta_t} L_t^{\lambda_t} \tag{13.12}$$

Further, the following observations permit the establishment of a relationship between (η) and (φ) that can be introduced into Equation (13.12).

1 Perfect resilience, namely (φ = 1), would coincide with D_{KN} being absent, that is (η = 0);
2 And complete non-resilience, namely (φ = 0), could occur when D_{KN} is as high as GVA or GDP, that is, (η = 1).

Hence, it follows that (φ = 1 − η), and substituting this definition of φ into Equation (13.12), we get:

$$Y_{tS} = (1-\eta_t)\alpha_t \left[KM_t + \eta_t KN_t\right]^{\theta_t} L_t^{\lambda_t} \tag{13.13}$$

Note that, in the formulations offered in Equations (13.12) and (13.13), should ($\eta_t \to 0$), then ($KN_t \to 0$) and the definition of factor-utilisation would tend towards that stipulated in Equation (13.1).

The basis for identifying KN is provided in Figure 13.2. As illustrated here, the role of KN can be identified for explaining a given level of income, say Y_t. Note that in Figure 13.2, Y_t is attributed to KM_t in {Y = g(KM, L)}; and the same quantum of Y_t is also attributed to K_t in {Y = g(KM, L, KN)}. Because ($K_t = KM_t + \eta_t KN_t$), the quantity of KN_t utilised in the formation of Y_t would be $[(K_t - KM_t)/\eta_t]$.

The value of KN_t can be resolved by equating Equations (13.1) and (13.13) resulting in the following expression:

$$KN_t = KM_t \left[\left(\frac{1}{\eta_t(1-\eta_t)^{(1/\theta_t)}}\right) - \frac{1}{\eta_t}\right] \tag{13.14}$$

Note that the estimates of KN_t in Table 13.2 are more than twice the size of KM. That is, (KN:KM) ratio suggests that each unit of KM requires nearly twice the amount of KN to enable the formation of Y. Note that in our analysis, the size of D_{KN} has been confined to the cost of CO_2 abatement. Had other components of D_{KN} such as biodiversity loss, soil erosion, deforestation and water pollution been considered, then the size of KN depleted would be considerably larger.

Reformulated AD–AS Framework

When all arguments in Equation (13.13) are known, it is possible to revise the values of observed and capacity income (Y_t, and Y_{Ft}) towards values that recognise the role of KN. These are identified in Table 13.2 as Y_S and Y_{FS}. Hence capacity AS would be redefined as:

$$AS_{FtS} : \{(Y = Y_{FtS}) \,|\, (\pi \to \infty)\} \tag{13.15}$$

To illustrate for 2009:

$$AS_{F2009S} : \{(Y = 947{,}639) \,|\, (\pi \to \infty)\}.$$

A review of section 4 in Table 13.2 reveals that the values of Y_{FS} are less than the values of Y reported with reference to the standard model (section 3 in Table 13.1). As discussed below, this observation indicates that the Australian economy has been performing beyond the capacity imposed by KN.

Note that AD_{tS} represents the revised description of AD in the context of recognising KN. Following our treatment in Chapter 12, D_{KN} can be internalised into AD by redefining aggregate expenditure in Equation (13.8) as:

$$GVA_t = (1-\eta_t)\left[\Phi_t + \beta_t Y_t(1-\Delta\tau_t) + G_t - T_t + I_t\right] \tag{13.16}$$

Hence the revised definition for AD is:

$$\bar{\pi}_t = \frac{M_t V_t}{(1-\eta_t)\left[\Phi_t + \beta_t Y_t(1-\tau_t) + G_t - T_t + I_t\right] P_{t-1}} \tag{13.17}$$

The short-run supply response is also revised as:

$$AS_{SRtS} \{(\pi = \bar{\pi}_t) \,|\, (Y \to \infty)\} \tag{13.18}$$

As illustrated in Table 13.2, the size of $\bar{\pi}_t$ in Equation (13.18) is higher than π_t in Equation (13.4), owing to the internalisation of KN in Equation (13.18); that is, ($Y_{tS} < Y_t$). The coordinates of the short-run equilibrium for $\{AS_{SRt} = AD_{tS}\}$

is revised as $(Y_{tS}, \bar{\pi}_t)$. As with the standard framework, a second set of coordinates for the AD function, namely $(Y_{FtS}, \bar{\pi}_{Ft})$, can be elicited from the definition in Equation (13.18) for the context of $(Y = Y_{FS})$. As illustrated in Figure 13.3, the AD_{tS} function is marginally steeper than the AD_t function. The implications of this observation are discussed below.

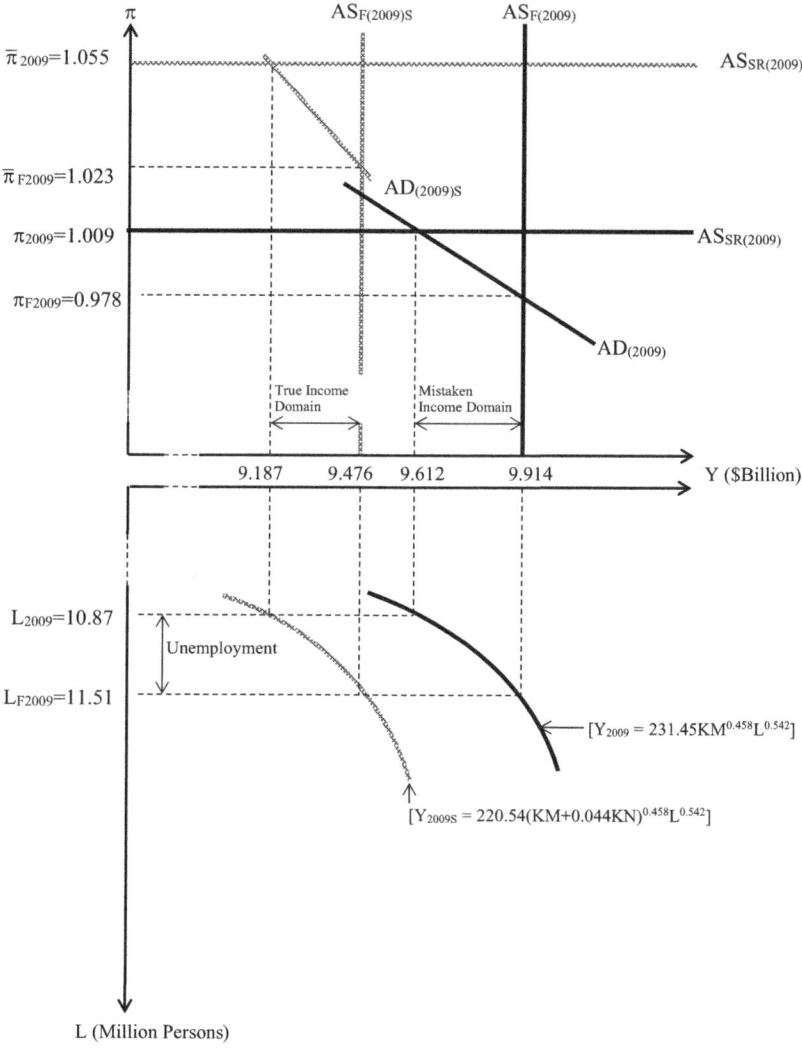

Figure 13.3 The Australian Economy (2009) – Environmental-Macroeconomic Framework Compared with the Standard Framework

Display of the Environmental Macroeconomic Framework

The display of the framework will follow the demonstration of Equations (13.12), (13.17), and (13.18) from the relevant point estimate data. This display for 2009 is illustrated in Figure 13.3. We have also included in this figure the display in terms of the standard framework (Figure 13.1) for comparison.

As with the standard framework, the expected changes in the snapshot for the subsequent period will be in part determined by the responsiveness to intervention measures. The counterpart of Equation (13.11) in the revised context would be:

$$\frac{d\bar{\pi}}{d\tau} = \frac{M_t V_t \beta_t Y_t}{\left[\Phi_t + \beta_t Y_t (1-\tau_t) + G_t - T_t + I_t\right]^2 (1-\eta_t) P_{t-1}} \tag{13.19}$$

A comparison of Equations (13.11) and (13.19) reveals that the inflationary response to changes in τ would be higher in the context of the environmental macroeconomic framework than in that of the standard framework. Recall the observation that the AD function is steeper with reference to the environmental macroeconomic framework than it is with reference to the standard framework. This implies that if prices (and wages) are regarded as a policy tool for closing the output gap ($Y_F - Y$), then prices must in fact be prompted to adjust faster than the standard framework may suggest.

Perhaps the most striking observation that emerges from the comparison presented in Figure 13.3 is that the Australian economy was performing beyond its true productive capacity in 2009. Note that the productive capacity in 2009 when KN is recognised was $947,639 million. This capacity is less than the observed level of Y in 2009, namely $961,252 million. An examination of Tables 13.1 and 13.2 reveal this to be the case for each year in the period under review. This means that the income domains for policy analyses should have been different. For example, in 2009, the policymakers had focused on the income domain of ($961,252M ↔ $991,478M). Instead, they should have focused on the domain of ($918,750M ↔ $947,639M). Hence we differentiate these domains and label ($961,252M ↔ $991,478M) as the 'mistaken domain', as opposed to the 'true domain' of ($918,750M ↔ $947,639M). Australia being on the mistaken domain was also driven by the adoption of rescue measures to stimulate the economy based on the perception that Australia too (like several other countries) was a victim of the GFC. These measures included the reduction of interest rates and increased government spending; (Thampapillai 2012). As argued in Box 13.1, the drivers of the GFC were also drivers of KN damage and climate risks. However the rescue packages themselves, became further drivers of KN and climate risks.

Box 13.1 The Global Financial Crisis and the Environmental Crisis

The financial crisis first surfaced in the United States around 2005–2006 due to excessive lending and the promotion of risky financial products. Because several investors from both within and outside the United States placed their savings in these markets and the markets failed – the crisis burgeoned into a global crisis in 2008–2009. The rescue measures were speedy, intense and even overwhelming. For example, the United States Congress approved the allocation of up to $780 Billion as a recovery fund. While some countries have made questionable recoveries, others displayed either stagnation or a slow shuffle forward, with some trapped in yet another crisis – the debt crisis. Both the financial crisis and the environmental crisis have common drivers, many in the making over the past several decades. Dominant among these factors is the continued reliance and expansion of fossil-fuel utilisation to drive economic growth. Regardless of this connection, the gravity of the environmental crisis stemming from growth without safeguards has been on the policy agenda for several decades. During both the lead-up to the global financial crisis and recovery efforts, little attention was paid to the environmental crisis already gripping the globe. Though belated, there remains a clear need to revisit the rescue packages, re-navigating them towards dealing with the environmental crisis.

Drivers behind the financial crisis have not only hindered genuine efforts to limit environmental risks, but also exacerbated these risks. In such a context, the rescue packages should have been directed towards economic activities that strive to minimise environmental risks. Prominent among these should have been the establishment of an energy infrastructure that would quickly replace fossil-fuel utilisation. Instead, many rescue efforts themselves – support for the automobile industry and expansion of road infrastructure – depended on and even expanded fossil-fuel resources. Hence, the rescue packages have turned out to be drivers that magnified the risks.

As many commentators concede, the main driver of the financial crisis was the prevalence of gross irregularities within an unregulated financial sector. Foremost of these irregularities is perhaps the excessive wages commandeered within the financial sector – driven primarily by the pervasion of asymmetric information and the formulation and delivery of risky financial products – which in turn to sustain high wages. The emergence of an elite high-income group delivered clear demand-price signals for the market entry of goods and services that are energy-intensive and environmentally damaging. Consider the manufacture of:

- Airbus A-380 and Boeing 787 Dreamliner, massive aircrafts, which despite claims of fuel efficiency dig deep into fossil-fuel dependency
- The Oasis Ocean Liner, which boasts of shopping promenades and energy-consumption regimes that could power more than 100,000 homes
- Record numbers of SUVs for meeting modest family needs

The notion that expanded use of such items would hike greenhouse gas emissions and deepen fossil-fuel dependency went unnoticed by *policymakers*.

(*Source:* https://archive-yaleglobal.yale.edu/content/economic-fixes-should-not-worsen-environmental-crisis)

Concluding Remarks

As illustrated above, relative to the environmental macroeconomic framework, the application of the standard macroeconomic framework overstates targets and performance. This was illustrated above with reference to the mistaken income policy domains. As a result, policymakers would tend to pursue mistaken income targets. As illustrated, such targets exceed true productive capacity. Hence higher-than-usual inflation outcomes would be inevitable, and this has been the case in Australia. Further, if prices and wages are used as policy tools, there is a need to adjust prices at a rate faster than that implied by the standard framework.

Also, note that intervention measures such as tax and interest rate relief may not be as effective as the policymaker would anticipate. This can be also explained by recourse to the multiplier. For example, the multiplier for consumption expenditure in the context of the standard macroeconomic framework is certainly larger than that of the environmental framework; that is,

$$\left\{\frac{1}{(1-\beta)(1-\tau)}\right\} > \left\{\frac{1}{(1-\beta)(1-\tau)(1-\eta)}\right\} \quad (13.20)$$

We began this chapter with an explanation for our choice of macroeconomic data for the period 2005–2009, namely that this was the period when the GFC erupted. The remedial (pump priming) measures that were adopted during this period – across the world – ignored environmental and climate risks. They did in fact exacerbate these risks. In our illustration, we showed that the Australian economy was performing beyond its true capacity.

Had the environmental macroeconomic framework been adopted for the recovery from the GFC, then, policymakers would have been prompted to search for policy initiatives that target the goal of sustainability. Such initiatives would have focused on minimising the extent of KN depreciation. Examples of these would have included the development of renewable and low greenhouse emission technologies instead of further exploration for fossil fuels and the promotion of innovative closed-loop production systems that reuse wastes and emissions. The subject of wage policy also would have been considered, because wage reductions can help attain sustainable income and employment targets. Such wage reductions would constitute the essence of acknowledging the relevance of the production function based on L, KM and KN.

Review Questions

1 Discuss the observation that the persistence of inflation despite the prevalence of recessionary conditions can be explained when an environmental macroeconomic framework is considered.
2 Under what conditions could AS_{SR} remain the same in both the standard and environmental macroeconomic frameworks?

3 Illustrate how the environmental macroeconomic framework used in this chapter necessitates the lowering of wages as a policy option.
4 Based on the content of this chapter, indicate the specific circumstances under which the following function would be valid: $Y_t = (1-\eta_t)\alpha_t(KM_t + KN_t)^{\theta_t} L_t^{\lambda_t}$.

References

Flath, D., *The Japanese Economy*, Oxford University Press, New York, 2005.
Mankiw, N. G., *Principles of Macroeconomics*, 3rd edn, Thomson South-Western, Mason, OH, 2004.
Stern, N., *The Economics of Climate Change—The Stern Review*, Cabinet Office, HM Treasury, London, 2007.
Thampapillai, D. J., 'Macroeconomics vs environmental macroeconomics', *Australian Journal of Agricultural and Resource Economics*, 56(3): 332–346, 2012.

14 Environmental Macroeconomics
Long-Run Analysis[1]

Introduction

As we have iterated previously, texts and reviews on environmental economics (for example, see Stavins 2017, 2007; Cropper and Oates 1992), let alone those on economics more generally, fail to include the role of KN in macroeconomic performance. We were able to illustrate, with reference to short-run stabilisation frameworks, policy errors that would emerge when KN is ignored. Recall the distinction between the 'mistaken domain' and the 'true domain' in Chapters 12 and 13. An accumulation of 'mistaken domains' would no doubt render the task of long-run stabilisation exceedingly difficult. It was in this context that Daly (1991) articulated the need for recognising environmental macroeconomics as a distinct discipline more than two decades ago. The basis for a steady state as illustrated in Figure 12.1(b) does originate from Daly's writings. Further, the practice of environmental accounting (Ahmad et al. 1989), as an extension of national income accounting in countries like Canada and Sweden, dates back to the early 1970s.

The basic premise underlying Daly's (1991) articulation and the practice of environmental accounting is that no economic activity can proceed without the supporting role of KN. As we have illustrated before in Chapter 1, this premise was central to the discourse of early neoclassical economics (Marshall 1891; Fisher 1904). Whilst Marshall's (1891) narrative deemed KN as the ultimate capital, Fisher's (1904) formulation of capital theory rested on the properties of KN. Most notably perhaps, the British philosopher and civil servant, John Stuart Mill (1848) emphasised that economic growth and development processes must not violate biophysical constraints. Recall, from Chapter 2, Georgescu-Roegen's (1971) formal validation of this premise by recourse to the laws of thermodynamics. Further validation and extensions of the premise can be found in Daly and Cobb (1989), Daly (1996), Ruth (1993, 2018), Cleveland and Ruth (1997) and Gowdy and Messener (1998). Hence as we have argued in Chapters 11–13, the explanation of national income (Y) must include KN alongside manufactured capital (KM) and labour (L); that is, Y = f(KM, L, KN). This explanation is quite distinct from that of contemporary neoclassical economics, namely Y = g(KM, L). The exclusion of KN from the analytic frameworks of Y commenced perhaps with Samuelson (1948) and Solow (1956). Such exclusion, whilst guided by analytic convenience, was also based on the mistaken premise that many KN endowments are infinite. Subsequent developments on the explanation of Y involved the introduction of some components of KN, namely extractive natural resources (Hartwick 1990; Solow 1974; Dasgupta and Heal 1974, 1979). However, such extractive resources are only a subset of the larger domain of KN. And the recognition of KN as an aggregate endowment that is analogous to KM has remained largely elusive to many economists including environmental economists.

As indicated in Chapter 10, the confinement of KN to microeconomics stems also from the explanation of macroeconomics as one that rests on microeconomic foundations. That is, the symmetry between the exhaustive list of discrete markets, each in equilibrium, and the overall macroeconomic equilibrium. The latter is defined as the identity between the sum of all real final market transactions in product and factor markets leading respectively to gross value added (GVA) and national income (Y), namely (Y ≡ GVA).[2] Hence for this overall macroeconomic equilibrium (Y ≡ GVA) to hold, all individual markets must remain in equilibrium without any distortion. Herein lies the problem. Leave alone imperfections in market organisation, externalities do emerge.

The focus in this chapter, as in the previous three chapters, pertains to externalities stemming from the exploitation of KN. It would hardly suffice to claim that market distortions are corrected when externalities are internalised. As argued in Chapter 11, externalities are *never fully internalised* neither by regulation nor by interventions through prices. Moreover, internalisation of (negative) externalities typically occurs after harm has been observed. The fact is that residual externalities exist and they accumulate. As indicated in earlier chapters, several examples validate the presence and accumulation of residual externalities. These include amongst many others: the permanent and growing brown cloud in Asia (Ramanathan et al. 2008); the fragmentation of polar icebergs (Bouhier et al. 2017); the hole in the ozone layer (Fahey and Hegglin 2011); and the accumulations of plastic and other debris in oceans (Eriksen et al. 2014). Hence, the definition of the macroeconomic equilibrium should be revised to identify the damaging effects of the residual externality. This revision is (Y ≡ GVA − D_{KN}), where D_{KN} represents the aggregate measure of the residual externality that remains. Having noted the effect of this revision with reference to short-run stabilisation in Chapters 12 and 13, the focus of this chapter is to examine selected long-run stabilisation frameworks in macroeconomics. These are: the Harrod–Domar model, the Swan-Solow steady state analysis and endogenous growth models. We illustrate the revision and application of these frameworks with reference to the South Korean economy.

The Steady State

The approaches to long-run stabilisation are founded in the literature on economic growth. The theories of economic growth, some of which are considered in this chapter, rest on the premise that the ideal growth path is a stable increase in the national product over time. Neoclassical economics defines such a growth path as a steady state, a smooth upward trajectory of real national product over time. This interpretation of a steady state is distinct from the one offered by Daly (1997), who points out that, because of KN constraints, growth cannot proceed indefinitely. When he speaks of a steady-state economy, material and energy inputs into the economy and outputs of waste products from it all fall within the ecosystem's ability to provide these inputs and assimilate these outputs in ways that do not undermine KN. In short, Daly (1997) suggests that the indefinite upward trajectory for growth is impossible when KN is recognised. The goal, then, is development within biophysical limits – a qualitative improvement in the standards of living and quality of life, for example – rather than the expansion of material wealth.

Macroeconomists often distinguish the short run from the long run and assume that a steady state would eventuate in the long run despite aberrations in the short run. That is, the short-run goal of macroeconomics is to recognise the oscillations of the key variables (output, employment, and prices) and policymakers must strive to minimise these oscillations. However, the attainment of a steady state in the long run cannot be separated from the stabilisation efforts in the short run. After all, the long run is a collection of short runs.

As indicated, this chapter deals with three theories that have contributed to the explanation of long-run stabilisation:

1 The Harrod-Domar growth model
2 The Swan-Solow steady-state model
3 The more contemporary endogenous growth models

Although the endogenous growth models are in vogue now, the previous models warrant consideration here because they were dominant during different time periods and help shed light on a particular kind of thinking that left their mark on macroeconomics. Several macroeconomics texts, for example (Dornbusch and Fischer 1994; Glahe 1977; Abel et al. 2008), outline the basic principles and steps for the development of these models. Our intention here is primarily to exposit the changes that would unfold when KN is recognised. In each theory, the underlying production function confines the explanation of Y to two factors namely (L) and (KM); that is, $\{Y = f(KM, L)\}$. We extend the function to be of the form $\{Y = g(KM, L, KN)\}$. Following our approach in Chapters 12 and 13, we modify the conceptualisation of the production function that underlies macroeconomic growth models as follows:

1 The conceptual basis for KN in revising these frameworks rests on the revised equilibrium, namely $(Y \equiv GVA - D_{KN})$.
2 We use the standard two-factor Cobb-Douglas production function as defined in equation (13.1); that is $\left\{Y_t = \alpha_t KM_t^{\theta_t} L_t^{\lambda_t}\right\}$, to describe the approach adopted in each theory.
3 We then revise each theory using the reformulated production function that explicitly recognises KN as equation (13.13) of the previous Chapter. That is, $\left\{Y_{tS} = (1-\eta_t)\alpha_t \left[KM_t + \eta_t KN_t\right]^{\theta_t} L_t^{\lambda_t}\right\}$.

Recall from our narrative in Chapter 13 that the reformulated function rests on the premise of aggregate capital (K) encompassing KM and KN – both measurable in the same metric and embodying the resilience of KN – does exist; that is (K = KM + ηKN).

The Harrod–Domar (H–D) Growth Model

The Standard Version of the H-D Growth Model

This model follows the work of Harrod (1939) and Domar (1946). It focused on savings as well as the efficiency of capital as the determinants of economic growth. To illustrate the derivation of the H–D growth model, suppose that the stock of KM and Y during a given period t and the subsequent period (t+1) are denoted respectively as: (KM_t, Y_t) and $[KM_{t+1} = (KM_t + \Delta KM_t)]$, $[Y_{t+1} = (Y_t + \Delta Y_t)]$.

Further, the capital-output ratio (denoted as κ) is assumed to remain constant over time. Then:

$$\kappa = \left[\frac{KM_t}{Y_t}\right] = \left[\frac{KM_t + \Delta KM_t}{Y_t + \Delta Y_t}\right] = \left[\frac{\Delta KM_t}{\Delta Y_t}\right] \quad (14.1)$$

Hence it follows that the additions to KM, which are in fact the investment (I) made during a period, can be defined as:

$$I_t = \Delta KM_t = \kappa \Delta Y_t \quad (14.2)$$

The H–D framework, (like most macroeconomic frameworks), assumes the existence of a savings-investment identity; $(S \equiv I)$.

If savings (S), is defined by a proportion (ρ) of national income, then given the savings-investment identity, it follows that:

$$I_t = S_t = \rho Y_t = \Delta KM_t \quad (14.3)$$

Given that from Equation (14.2) above ΔKM_t is also equal to $(\kappa \Delta Y_t)$; the proportion of income saved can be expressed as:

$$\rho Y_t = \kappa \Delta Y t \quad (14.4)$$

The basic H-D model is the result of dividing both sides of Equation (14.4) by κY_t. Hence:

$$\left[\frac{\Delta Y_t}{Y_t}\right] = \left[\frac{\rho}{\kappa}\right] \quad (14.5)$$

The left-hand side of Equation (14.5) is in fact the rate of economic growth. The inference from Equation (14.5) is that the rate of economic growth is directly proportional to the savings ratio and is inversely proportional to the capital-output ratio. That is, ρ and κ become the determinants of the growth path. The policy implication from this simple analysis is the need to encourage savings and make investments in efficient forms of KM to achieve higher rates of economic growth. Note that the lower the value of κ, the higher the efficiency of KM.

The H-D Model Revised with KN

The recognition of KN warrants the following revisions to the H–D model:

1. The determination of Y is not confined to KM alone, but the aggregate measure of capital (K) that includes KN. In Chapter 13, we defined this aggregate measure as (K = KM + ηKN). Recall that η is the proportion of D_{KN} in GVA and is also an indicator of resilience with reference to KN.
2. Further, given that the pertinent identity is (Y ≡ GVA − D_{KN}), we replace (Y_t) with (Y_{tS}). Recall from our narrative in Chapter 13 that $\{Y_{tS} = (Y_t - D_{KN}) = (1 - \eta)Y_t\}$.

We assume that the capital-output ratio and the proportion of D_{KN} in GVA, both remain constant over time. Hence, the definition of the revised capital-output ratio, $\bar{\kappa}$, is as follows:

$$\bar{\kappa} = \left[\frac{KM_t + \eta KN_t}{Y_{tS}}\right] = \left[\frac{(KM_t + \eta KN_t) + (\Delta KM_t + \eta \Delta KN_t)}{(Y_{tS} + \Delta Y_{tS})}\right] = \frac{(\Delta KM_t + \eta \Delta KN_t)}{\Delta Y_{tS}}$$

(14.6)

Given that ($Y_{tS} < Y_t$), it is probable that the magnitude of savings could be reduced. We however assume that the savings ratio, ρ remains unchanged, and hence define savings and investment as:

$$I_t = S_t = \rho Y_{tS} = (\Delta KM_t + \eta \Delta KN_t) = \bar{\kappa}\Delta Y_{tS}$$

(14.7)

It now follows, as per reasoning made above, that the basic H-D model (14.7) could be rewritten as:

$$\left[\frac{\Delta Y_{tS}}{Y_{tS}}\right] = \left[\frac{\rho}{\bar{\kappa}}\right]$$

(14.8)

Note that the rate of economic growth identified in Equation (14.8) will be certainly less than that revealed in Equation (14.5) because ($\bar{\kappa} > \kappa$). It would be even smaller if ρ has also decreased due to ($Y_{tS} < Y_t$).

Hence, reducing the reliance on KN becomes an additional avenue for raising the rate of economic growth. Such reduction is tantamount to reducing the depreciation of KN, namely D_{KN}. To recapitulate, we demonstrated in Chapter 13 that every unit of KM accumulated involves the depletion of nearly twice the amount

KN. If investors were made aware of this perverse relationship between KN depletion and KM accumulation, then the search for types of KM that would minimise damage to KN would become imperative. However, the singular focus of the H–D model and those that we review below was to merely elicit strategies for enhancing economic growth – that is increasing the size of Y.

The Swan-Solow Model

The Standard Version of the Swan-Solow Model

As exposited in most standard economics texts, the Swan-Solow model follows the concurrent work of an Australian economist – Trevor Swan and the American Nobel Laureate – Robert Solow. Coincidentally, they both published their conceptual premises for this framework in the same year 1956 [Swan (1956) and Solow (1956)]. The main feature of this model is the identification of a pattern of capital accumulation that leads to growth culminating in a steady-state equilibrium. The basic assumptions are: constant technology and an underlying production function $\{Y = f(KM, L)\}$ that displays diminishing marginal returns. Further assumptions include the absence of external shocks and the presence of smooth and stable economic performance guided by savings (S) becoming directly translated into investment (I); that is, $[S \equiv I]$. The steady-state equilibrium is in fact the context where the amount of added KM per worker is just sufficient to meet the needs of new workers and the depreciation of existing KM stock. Note that the consideration of the depreciation of KM is dependent on the definition of Y. If for example, Y is defined in terms of net national product, then the determination of the steady state equilibrium will proceed only on the basis of the rate at which new workers enter the workforce. This rate is ultimately driven by the rate of population growth. But, because Y is defined here in terms of GVA, we consider both the rate of depreciation of KM as well that of the entry of new workers.

In Figure 14.1, the curve labelled $\{y = f(k)\}$ explains how output per worker $(y = Y/L)$ increases as capital per worker $(k = KM/L)$ gets accumulated. The curve labelled $\{\rho y = \rho f(k)\}$ results when $\{y = f(k)\}$ is multiplied by the savings ratio (ρ). This ratio is the same as that considered above with the H–D growth model. The straight line from the origin conforms to the equation $\{y = z(k) = [(\delta + v)k]\}$, where δ and v represent the rates respectively of KM depreciation and workforce expansion. This straight line therefore describes the amount of KM stock that is required to replace the depreciated KM stock and the needs of new workers. The steady-state equilibrium is defined by the point of intersection between the straight line for $\{z(k) = (\delta + v)k\}$ and the curve describing savings, that is, $\{\rho y = \rho f(k)\}$.

When $\{y = f(k)\}$ is derived from Equation (13.1), that is $\{Y = \alpha KM^\theta L^{(1-\theta)}\}$, the following quantity of KM per worker describes the steady-state equilibrium (k^*):

$$k^* = \left\{ \left[\frac{\rho\alpha}{(\delta+v)} \right]^{[1/(1-\theta)]} \right\} \tag{14.9}$$

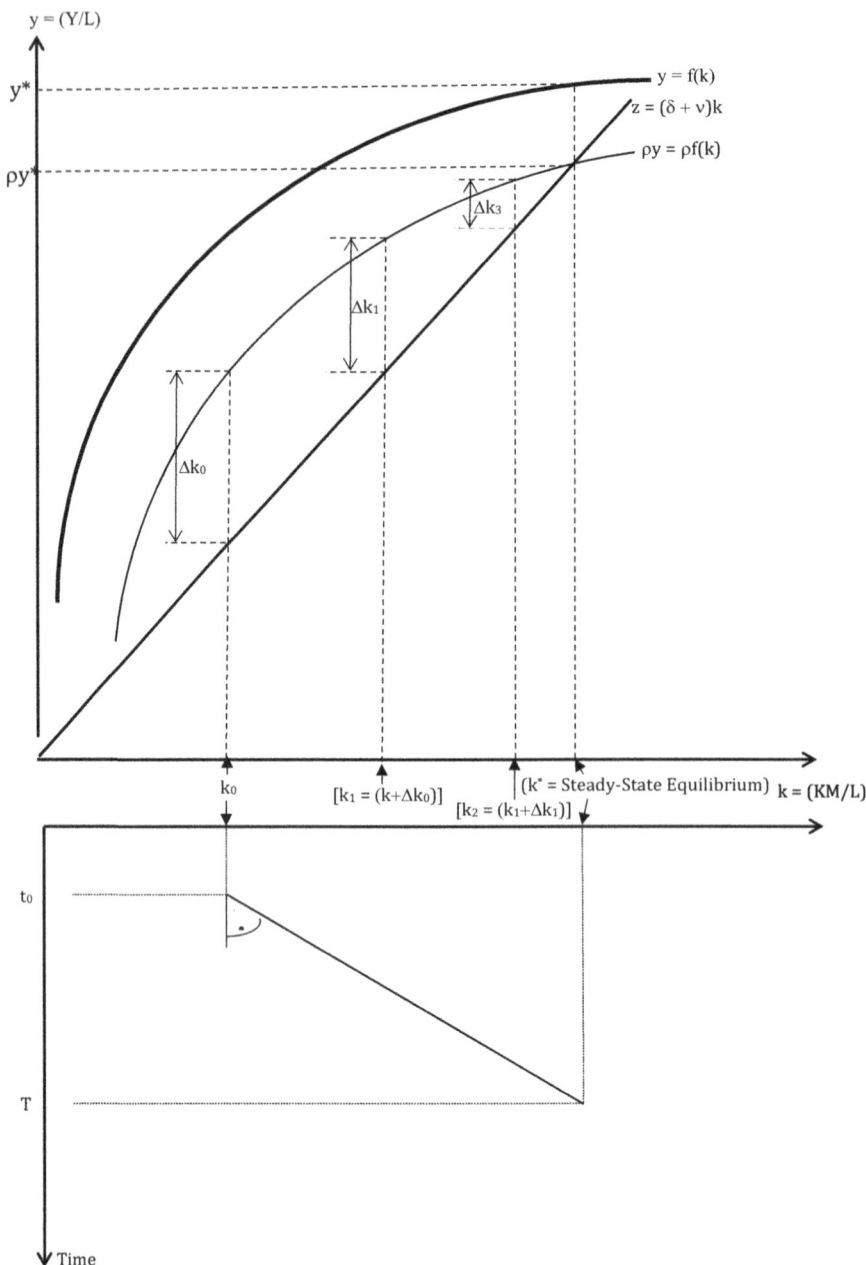

Figure 14.1 Swan-Solow Model Capital – Accumulation and the Time-Path

The derivation of Equation (14.9) is displayed in Appendix 1 of this Chapter.

In Figure 14.1, suppose that the initial stock of KM per worker is k_0. Then the amount added to this stock is the vertical distance at k_0 between the curve describing savings per worker, that is, $\{\rho y = \rho f(k)\}$, and the straight line that describes the needs of depreciation and new workers, namely $\{z(k) = (\delta + v)k\}$. This vertical distance is Δk_0 in Figure 14.1. So the added KM stock per worker becomes $(k_1 = k_0 + \Delta k_0)$. Given the context of k_1, it would then be possible to add a further quantity Δk_1, and so on until k^* is eventually reached.

The time taken to reach the steady state will depend on the rate of accumulation of k per unit of time and is defined as: $\left[\dot{k} = (dk/dt)\right]$. A linear time path, as shown in the lower panel of Figure 14.1, presumes that $(\dot{k})$ is constant. Further, the time-path shown in Figure 14.1 is based on the premise that the production function identified in t_0 remains unchanged throughout. Given these assumptions, the rate of capital accumulation becomes a determinant of the rate of economic growth as follows:

$$\frac{\dot{y}}{y} = \theta\left[\frac{\rho}{\kappa} - (\delta + v)\right] \qquad (14.10)$$

The derivation of Equation (14.10) is displayed in Appendix 2 of this Chapter. The left-hand side of Equation (14.10) represents the rate of economic growth defined as the change in output per worker per unit of time as well as a unit of output per worker itself. The determinants of the rate of economic growth are found on the right-hand side of Equation (14.10). An increase in θ and ρ, which contribute to greater capital accumulation, would accelerate growth. So would a decrease in κ, which is the capital-output ratio identified in the H–D model. But such acceleration is constrained by the rate of depreciation of KM, namely δ; and the rate of entry of new workers, namely v. Because v does ultimately depend on the rate of population growth, reducing this rate was perceived in the past to be desirable, especially when some capacity for capital accumulation and growth had to be created. This is because reducing v could lead to an increase in k^* and thereby create capacity for KM accumulation and growth. Nevertheless, such perceptions were during a time when an ageing population was not a challenge.

However, the accumulation of KM beyond the steady-state equilibrium would depict a state of over-capitalisation and inefficiency. In such a context, it would be prudent to either cut back on KM accumulation or search for improvements in factor productivity that would shift the production function upwards and extend the steady-state equilibrium further to the right. Hence, from a policy perspective, it is useful to assess, at any given time, the ratio (k^*/k). If this ratio is greater than 1 there is capacity for KM accumulation. The reverse – the ratio being less than one – portrays the case of over-capitalisation and inefficiency. Such policy diagnostics would of course change when one revises the Swan-Solow model by including KN. This is considered next.

238 *Macroeconomics and KN*

The Revised Version of the Swan-Solow Model

As indicated above, the premise for revision is that the valid descriptor of Y is $\{Y = g(KM, L, KN)\}$ and not $\{Y = f(KM,L)\}$. The primary conceptual tools in the Swan-Solow model, namely output per worker $[f(k)]$, savings per worker $[\rho f(k)]$ and the schedule of depreciation and needs of new workers $[z(k)]$, have to be now redefined in terms of $\{Y=g(KM, L, KN)\}$.

Given the assumption of a composite measure for capital – K encompassing both KM and KN, the composite capital per worker is:

$$\bar{k} = \frac{(KM + \eta KN)}{L} = \left[\left(k = \frac{KM}{L}\right) + \left(k_N = \frac{\eta KN}{L}\right)\right] \qquad (14.11)$$

Hence it follows that output per worker and savings per worker can be redefined as:

$$\frac{Y}{L} = y = \frac{g(KM, L, KN)}{L} = g(\bar{k}) \qquad (14.12)$$

$$\rho y = \rho g(\bar{k}) \qquad (14.13)$$

As indicated above, $\{Y = g(KM, L, KN)\}$ is based on the identity that recognises the depreciation of KN; that is $(Y \equiv GVA - D_{KN})$. Hence, the content of the third primary tool, namely the schedule $[z(k)]$ remains unaltered. Further, as illustrated earlier in Chapter 13, KM and KN can be measured on the same metric scale. Hence, it follows that k and $\bar{k}$ can be represented in the same domain as illustrated in Figure 14.2. Further, as with the H–D model, we assume that the savings ratio remains the same – although it is probable that this ratio might get smaller. Herein two sets of savings functions, namely $[\rho f(k)]$ and $[\rho g(\bar{k})]$ are displayed. Also note that $[z(k)]$ in Figure 14.1 and $[z(\bar{k})]$ in Figure 14.2 are the same for reasons given earlier, namely that D_{KN} has been accounted for in the production function described in Equation (14.2).

The results of the internalisation of KN are evident in Figure 14.2 and are as follows.

1 The steady-state equilibrium is now reached before the standard equilibrium $(k^*, \rho y^*)$ and is defined by $(\bar{k})^*, [(1-\eta)\rho y^{**}]$. This is determined by the intersection of the revised savings function $[\rho g(\bar{k})]$ and $[z(\bar{k})]$. When $[g(\bar{k})]$ is assumed to be a Cobb-Douglas function such as that displayed in Equation (13.13), in the previous Chapter:

$$\left(Y = (1-\eta)\alpha[KM + \eta KN]^\theta L_t^\lambda\right)$$

The steady-state value of composite capital per worker is defined as:

$$\bar{k}^* = \left[\frac{\rho(1-\eta)\alpha}{\delta + v}\right]^{\frac{1}{1-\theta}} \qquad (14.14)$$

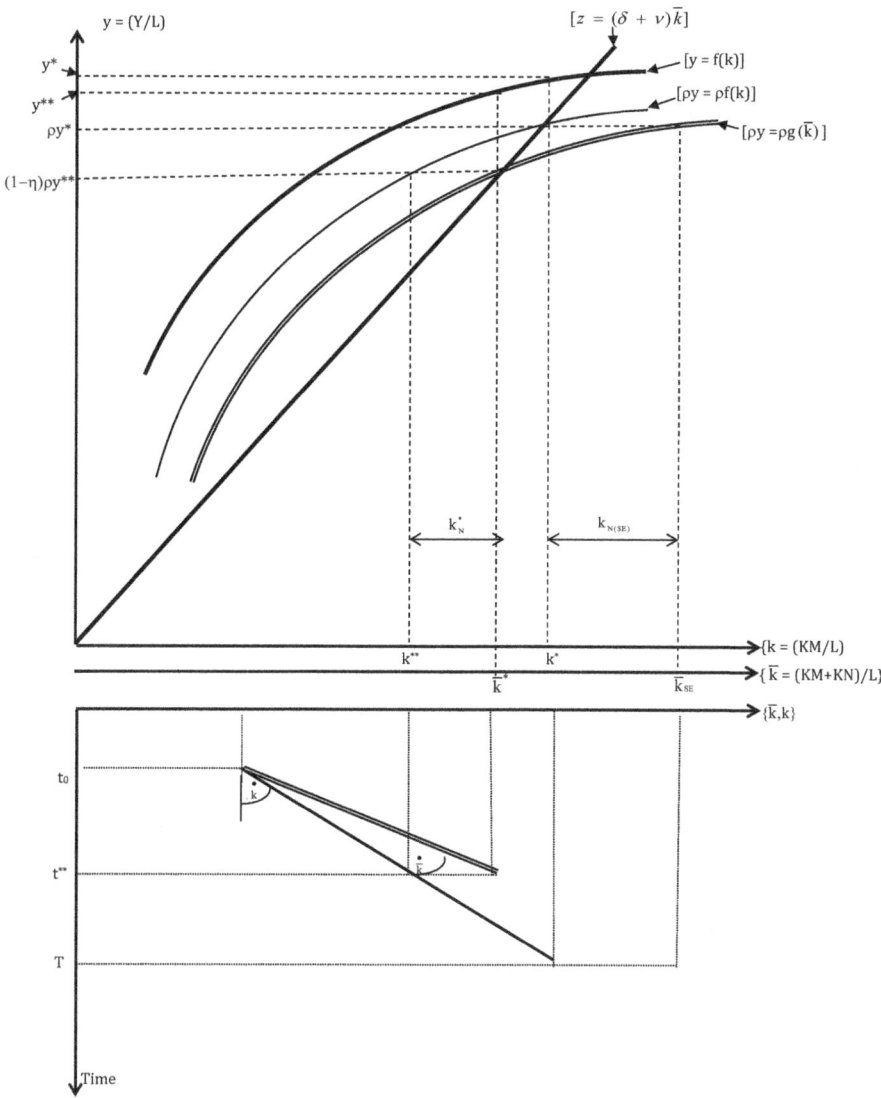

Figure 14.2 Revised Swan-Solow Model Capital – Accumulation and the Time-Path

The derivation of Equation (14.14) is displayed in Appendix 3 of this Chapter.

2 Following the rationale introduced here, the steady-state quantum of composite capital per worker, $\bar{k}^*$, consists of KM per worker amounting to k^{**} and KN per worker amounting to k_N^*.
3 The issue of over-capitalisation (or for that matter under-capitalisation) has to be now reviewed from a different perspective. In the standard Swan-Solow model,

the ratio (k*/k) proved to be a useful indicator. For example, the observation [(k*/k) < 1] portrays the case of over-capitalisation. With the revised framework, because $\bar{k}$ comprises both k and k_N, over-capitalisation of KM begins when k exceeds k**. Hence the indicator for the appropriate level of capitalisation in the revised model will be (k**/k).

4 Further, with reference to k_N, we need to introduce a concept that describes the appropriate level of utilisation (as opposed to capitalisation). Given the revised steady-state value of $\bar{k}^*$, the acceptable level of per worker KN utilisation is k_N^*, and any level of KN utilisation in excess of this amount will represent a case of over-utilisation. Note that the utilisation of k_N^*, is contingent on the per worker KM accumulation being k**.

In light of the above observations, the standard Swan-Solow steady-state equilibrium [(k*, ρy*) in Figure 14.2] represents both an over-utilisation of k_N amounting to $(k_{N(SE)} - k_N^*)$ and an over-capitalisation of k amounting to (k* − k**). When the accumulation of k and the utilisation of k_N are taken together, as shown in the lower panel of Figure 14.2, the time gradient becomes steeper; that is $\left(\dot{k} < \dot{\bar{k}} \right)$. The rate of economic growth in the revised context is:

$$\frac{\dot{y}}{y} = \theta \left[\frac{\rho}{\bar{\kappa}} - (\delta + \nu) \right] \tag{14.15}$$

Note that the only difference now is that the H-D capital-output ratio (κ) has been replaced with the revised H–D ratio involving composite capital (KM+KN), namely ($\bar{\kappa}$). Because ($\bar{\kappa} > \kappa$), the rate of growth dictated by Equation (14.15) will be smaller than that indicated by Equation (14.10) in the standard model. Hence the main outcome of internalising KN into the Swan-Solow model is moderation in economic growth targets and this in turn involves moderation in the accumulation of KM and the utilisation of KN.

Long-term trends in the ratio (k*/k**) could inform the direction in which an economy is headed. If this ratio converges towards 1, then this could imply that economic management decisions are cognisant of the role of KN.

Endogenous Growth Models

In the two models considered thus far, economic growth is explained in terms of changes in the stocks of KM and L, and in the utilisation of KN. However, factors such as advances in technology, build-up of knowledge, and improvements in organisational, labour and managerial skills remain exogenous to these models. The influence of such factors can be explained through certain variables in the models. For example, a lower value of the capital-output ratio (κ) in the H-D model and a higher value in the total factor productivity coefficient (α) in the Swan-Solow model can explain improved technical efficiency. The endogenous growth models attempt to bring these influential variables within the model – rather than leave them external to the model. Some of the earliest attempts towards this end came from Arrow (1962), Uzawa (1965) and Romer (1986). Initially we consider a simplified version of Romer's (1986) model. The treatment of this model follows the adaptation (for illustrative purposes) chosen by Todaro and Smith (2004). We use this to provide the basis for a more generalised endogenous model. In each case, the effects of awarding explicit recognition to KN is considered.

The Romer Model

The model is based on the assumption that aggregate production in the economy stems from a set of symmetrically uniform firms and industries. As in the Swan-Solow model, each firm displays constant returns to scale with reference to KM and L. But, each firm is also influenced by the presence of an economy-wide stock of KM that generates positive knowledge-based externalities. As a result, the factor-utilisation function of an individual firm or industry is as follows:

$$Y_i = \alpha KM_i^\theta L_i^{(1-\theta)} KM^\beta \tag{14.16}$$

Note that the positive externality is reflected in the coefficient β. The assumption of symmetry enables the aggregation across all industries and firms. Because $\sum_i KM_i = KM$ and $\sum_i L_i = L$, the aggregate function will be one that displays increasing returns to scale, namely

$$Y = \alpha KM^{(\theta+\beta)} L^{(1-\theta)} \tag{14.17}$$

If $\beta = 0$, then Equation (14.17) reduces to the Swan-Solow framework considered above. Further, Romer and others suggest that the differences in performance across countries could be explained by differences in β, (besides the accumulation of KM and L). This is captured in the definition of the rate of economic growth in per worker terms:

$$\frac{\dot{y}}{y} = (\theta+\beta)\left[\frac{\rho}{\kappa} - (\delta+\nu)\right] \tag{14.18}$$

That is, all determinants of growth in the Swan-Solow model, namely the savings rate (ρ), the capital-output ratio (κ), the rates of KM depreciation (δ) and entry of new workers (ν) are affected by the knowledge externality (β).

In revising the Romer model to the context of KN, one can assume that the knowledge externality prevails with composite capital (K = KM + KN). The estimation of β has been discussed widely – especially in the context of knowledge externalities; for example see: Mankiw et al. (1992); Eaton and Kortum (1996, 1998); and Porter and Stern (2000). However, whilst the overall stock of KM generates a positive externality, the utilisation of KN would also generate significant negative externalities. This matter has been clearly neglected in the growth literature. As argued in Chapter 8 and Thampapillai (2016), the utilisation of KN results in the raising of the entropy of the remaining stock KN and empirical evidence to support this comes from Schneider and Kay (1994).

Suppose that the coefficient for the negative externality associated with KN utilisation is γ. Then the Romer aggregate function in Equation (14.19) can be revised as:

$$Y = \alpha(1-\eta)K^{(\theta+\beta-\gamma)}L^{(1-\theta)} \qquad (14.19)$$

And the rate of economic growth would become moderated by introducing γ besides β and replacing (κ) with ($\bar{\kappa}$) in Equation (14.20) as follows:

$$\frac{\dot{y}}{y} = (\theta+\beta-\gamma)\left[\frac{\rho}{\bar{\kappa}} - (\delta+\nu)\right] \qquad (14.20)$$

The underlying assumption of symmetry in the Romer model is clearly tenuous. Hence several reformulations, as reported for example in Case, Fair and Oster (2014) and Frank, Bernanke and Lui (2015), have generalised the endogenous growth model providing latitude to introduce a broader set of variables. These include the knowledge externalities as well as the effects on KN.

The Generalised Endogenous Growth Model

The generalised model is based on the definition of (Y/N) as an identity expression with reference to two influential variables, namely the average productivity of labour (Y/L) and the employment rate of labour (L/N):

$$\frac{Y}{N} \equiv \left(\frac{Y}{L}\right)\left(\frac{L}{N}\right) \qquad (14.21)$$

The factors that influence these two variables are summarised in Figure 14.3 and are drawn from the literature on development/growth economics. Here, several variables are internalised into the explanation of economic growth.

As one would observe there is a close relationship between these two sets of factors. For example, labour productivity in itself influences the employment rate. The role of these factors is discussed at length in the literature on development economics.

The high value for β in well-performing economies is mainly due to the high levels of attainment of the variables in Figure 14.3. For example, high investments in education, the prevalence of law and order and well-functioning institutions account for the success of economies such as Singapore and South Korea.

The value of γ would be reflected in a range of indicators such as air and water pollution, deforestation and soil degradation. A challenge in contemporary growth theory lies in difficulties of quantifying the variables identified in Figure 14.3, because some of these are clearly intangible. Such challenges are not new to environmental economists and date back to the earlier parts of the last century as reported in Sinden (1967) and Sinden and Worrell (1979).

Environmental Macroeconomics: Long-Run Analysis 243

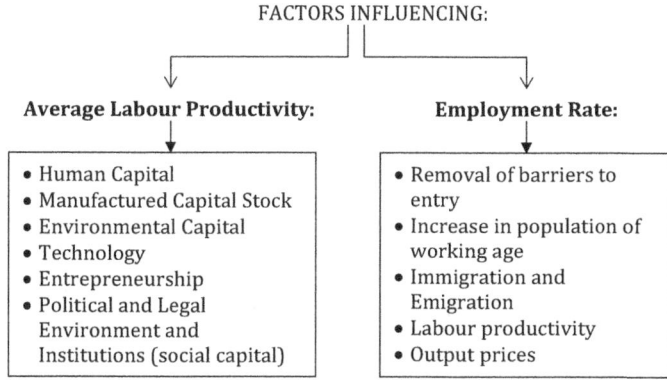

Figure 14.3 Factors Influencing Economic Growth

An Illustration – The South Korean Economy

In this final section, the concepts and tools considered above are illustrated with reference to South Korea in Table 14.1. In 1964, South Korea's per capita Y was around US$120. But in 2016, with per capita Y being in excess of US$30,000, South Korea is classified as a high-income country by the World Bank and is a member of the select group of OECD countries.

Table 14.1 contains information over six years from 2011 to 2016 drawn from the OECD iLibrary over a longer time series (1970–2016). The values given below were derived as point estimates based on the premise that equations (13.1) and (13.12) are valid descriptors of Y. KM stock was derived using the perpetual inventory method over the period 1970–2016. And likewise the utilisation of KN was also estimated for each year of the time domain 1970–2016. This estimation rests on the methodology outlined in Chapter 13, and KN is confined to the air shed. Further, D_{KN} was measured as cost of air pollution abatement with reference to total greenhouse gas (GHG) emissions. These emissions drawn from World Bank (2017) in CO_2 equivalents were valued $100 (United States Dollars) following the Stern (2006) report. The confinement to D_{KN} to air pollution costs does necessarily understate the impact of KN. Nevertheless as illustrated below, the recognition of KN in the analytic frameworks reveals distinct and different outcomes.

The information in Table 14.1 can be grouped into six parts as follows:

I Rows 1–6 display the information on income and resources.
II Rows 7–9 pertain to the coefficients of the H-D model.
III Rows 10–14 contain the coefficients of the Swan-Solow model.
IV Rows 15–16 display the coefficients of the Romer model. Here the knowledge externality coefficient (β) was taken from the study by Evenson and Singh (1997) and the KN externality coefficient (γ) was equated to the average rate of KN utilisation as analysed here over the period 1970–2016.
V Rows 17–22 display the steady state outcomes of the standard and revised Swan-Solow models
VI Rows 22–28 display the target growth rates as dictated by all three sets of models

Table 14.1 Illustration of Model Estimates for South Korea

	2011	2012	2013	2014	2015	2016
Y (10^9 Won)	1,176,786	1,204,372	1,243,786	12,84,928	1,317,890	1,354,922
$Y\text{-}D_{KN}$ (10^9 Won)	1,077,368	1,098,995	1,132,425	1,167,244	1,193,524	1,223,495
KM (10^9 Won)	4,948,833	5,216,256	5,458,407	5,678,313	5,906,810	6,134,083
KN (10^9 Won)	10,772,063	11,594,924	12,278,954	12,968,456	13,636,471	14,424,357
L (10^6 Persons)	24.62	25.06	25.41	25.52	25.73	25.92
k (Won)	200,971,159	208,112,924	214,831,278	222,545,624	229,600,617	236,656,827
ρ	0.269	0.262	0.268	0.273	0.284	0.288
$\frac{\kappa}{\bar{\kappa}}$	4.205	4.331	4.389	4.419	4.482	4.527
	4.593	4.746	4.820	4.865	4.949	5.014
α	2177	2489	2695	3024	3196	3629
θ	0.523	0.515	0.511	0.506	0.503	0.497
η	0.084	0.087	0.090	0.092	0.094	0.097
ν	0.0126	0.0126	0.0126	0.0126	0.0126	0.0126
δ	0.025	0.025	0.025	0.025	0.025	0.025
β	0.300	0.300	0.300	0.300	0.300	0.300
γ	0.176	0.176	0.176	0.176	0.176	0.176
k^* (Won)	634,429,390	575,161,854	599,627,875	627,491,244	678,403,256	693,908,759
$\bar{k}^*$ (Won)	527,279,265	476,172,701	494,933,862	516,661,282	555,758,701	566,601,217
k^{**} (Won)	445,378,159	398,641,120	411,960,584	427,284,477	456,341,466	461,365,312
$\bar{k}_{SE}$ (Won)	751,095,345	687,024,895	720,399,358	758,746,100	826,198,230	852,187,056
k^*_N (Won)	81,901,105	77,531,581	82,973,278	89,376,805	99,417,234	105,235,905
$k^*_{N(SE)}$ (Won)	116,665,955	111,863,041	120,771,483	131,254,856	147,794,975	158,278,298
Growth Rate Standard H-D	0.064	0.061	0.061	0.062	0.063	0.064
Growth Rate Revised H-D	0.059	0.055	0.056	0.056	0.057	0.057
Growth Rate Standard S-S	0.014	0.012	0.012	0.012	0.013	0.013
Growth Rate Revised S-S	0.011	0.009	0.009	0.009	0.010	0.010
Growth Rate Romer	0.022	0.019	0.019	0.019	0.021	0.021
Growth Rate Revised Romer	0.017	0.015	0.015	0.015	0.016	0.016

Note: All Monetary Estimates are presented in Constant 2010 South Korean Won; Primary data source – OECD iLibrary; H-D = Harrod-Domar Model; S-S = Swan-Solow Model.

South Korea's performance in terms of increasing Y is of course impressive. But a closer examination of the time-paths of Y and $(Y - D_{KN})$ in Figure 14.4, reveals that these paths are diverging rather than converging. The reason for this can be partly ascertained from Figure 14.5. That is, the impressive average annual economic growth of nearly 7 per cent over the 45-year period (1970–2016) has been accompanied by the utilisation of KN at an average rate of 9.3 per cent.

As expected, the steady-state equilibrium in the revised version of the Swan-Solow model, $\bar{k}^*$ is reached sooner than in the standard model, that is k^*. Further, the point estimates of the steady-state equilibrium values – revised or otherwise – show a high degree of oscillation, which is absent in the time paths of Y and (Y-D_{KN}). An implication of this observation is that the determinants of the Swan-Solow steady state, namely θ, ρ, ν and δ, fluctuate over time. Also recall the discussion above on the concept of over-capitalisation with reference to the steady-state values. Both the standard and revised versions of the Swan-Solow model reveal that, for the most part, over-capitalisation is not an issue. That is, over most of the time domain the recorded quantities of capital per worker (k) are less than the steady-state values; [(k < k*), and (k < k**)]. Besides the ratio $\left(\dfrac{k^*}{k^{**}}\right)$

does not show signs of convergence towards unity (Figure 14.6). This implies the continued over-utilisation of KN. Further evidence of this over utilisation is shown

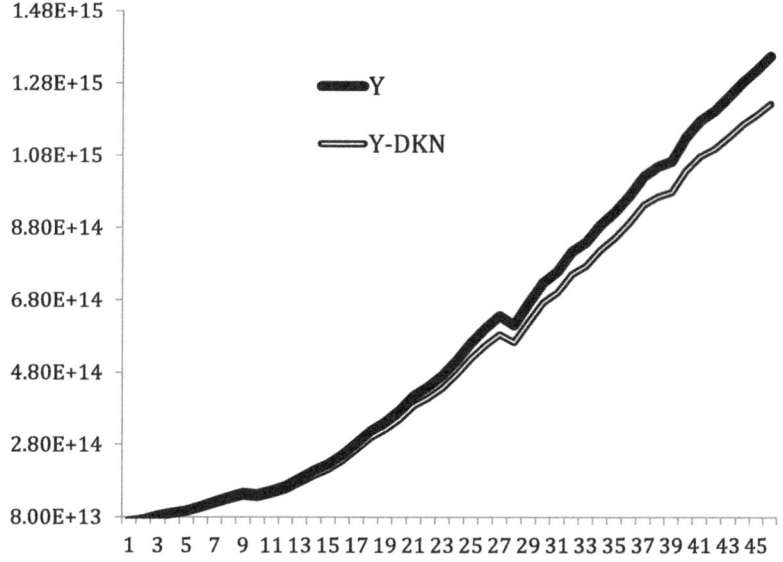

Figure 14.4 Comparison of Income Paths (Y) and $(Y - D_{KN})$ 1970–2016

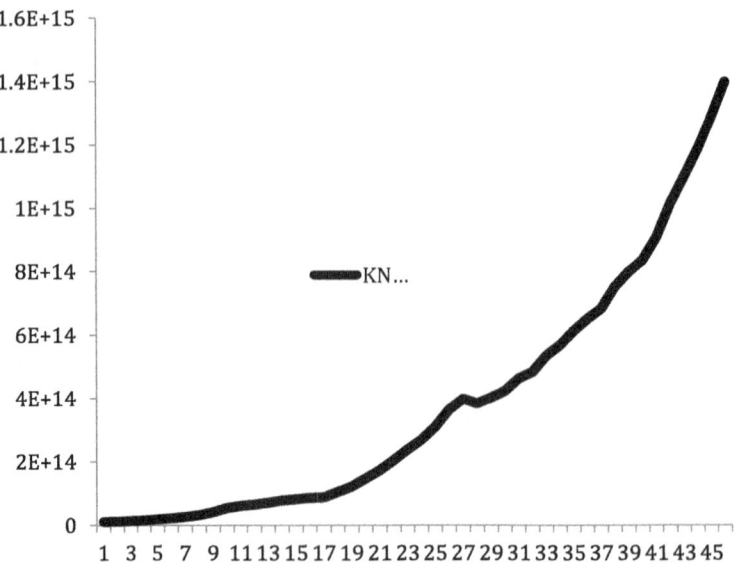

Figure 14.5 KN Utilisation (1970–2016)

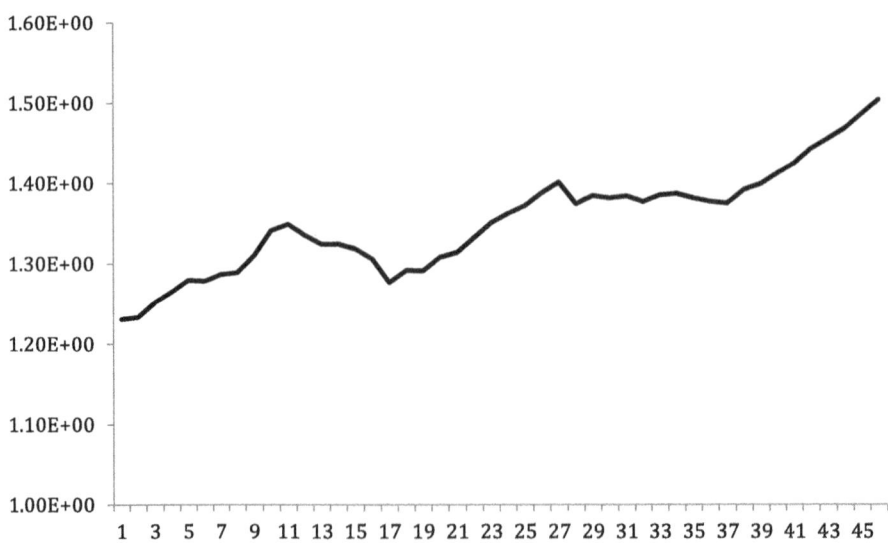

Figure 14.6 The Ratios of KM (k*/k**) for Steady State Equilibrium in Both Standard and Revised S-S Models (1970–2016)

in Figure 14.7, where the quantity of KN utilised following the steady state equilibrium (k*) in the standard model, namely $k_{N(SE)}$, is always in excess of k_N^*, which is the utilisation that follows the steady state equilibrium in the revised model, namely $\bar{k}^*$.

Environmental Macroeconomics: Long-Run Analysis 247

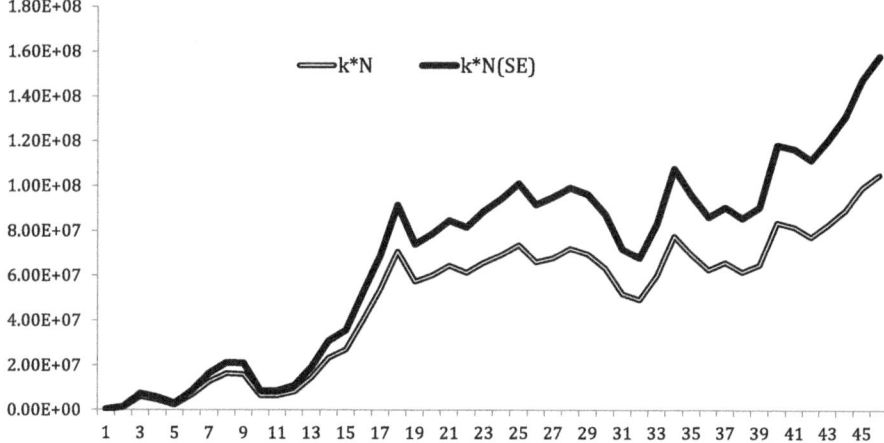

Figure 14.7 KN Utilisation in Both Standard and Revised S-S Models (1970–2016)

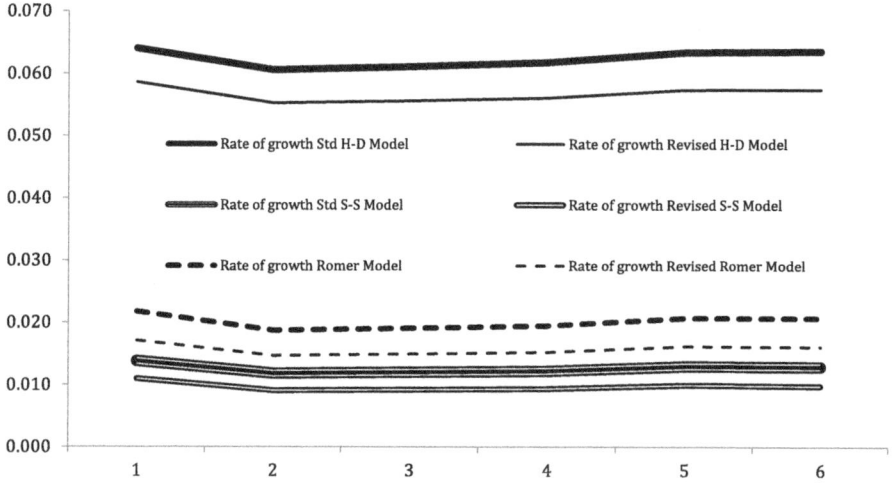

Figure 14.8 Rates of Growth – All Three Models (2011–2016)

The target rates of growth as dictated by the standard versions of all three models remain in excess of those indicated by revised versions of the models.

This is illustrated in Figure 14.8 for the period 2011–2016. During this period, the observed average rate of growth of Y and (Y-D_{KN}) are 3 and 2.6 per cent respectively. It is possible to argue that the target growth rates indicated by the Harrod-Domar model are overestimates on the grounds that economic performance has to be explained by factors beyond savings and capital accumulation. However, the revised versions of both the Swan-Solow and Romer models reveal low rates of growth. At times these rates approach zero. Although the knowledge coefficient used here for the Romer model ($\beta = 0.3$) is a value drawn from the late 1990s, subsequent studies (for example Dutta and Otsuka 2004) reveal that this coefficient might be lower than 0.3. Further, the analysis in terms of KN was confined to the air shed. Had other components of KN such as soil, water and forests been considered, whilst the value of β was raised, the possibility of zero growth could not be ignored.

Concluding Remarks

The main purpose of this Chapter is to explore Environmental Macroeconomics as both a relevant and important theme in economics. Reformulating the three growth models considered above to recognise KN leads to distinct policy conclusions, which centre on moderation rather than expansion. Hence, it is pertinent to argue that had KN been afforded due recognition in macroeconomics, then the inventory of environmental issues that confront contemporary society would be smaller rather than larger. Further, terms 'zero growth' and 'de-growth' as raised by authors such as Victor (2013), might not seem strange, whilst the term 'sustainable growth' would clearly be a physical impossibility for any economy, as argued by Daly (1997).

Review Questions

1 What are the major policy implications of revising the Harrod–Domar framework by internalising KN?
2 Consider the ratios (k^*/k) and (k^{**}/k) that were introduced in the section on Swan-Solow model.

   a What is the significance of this ratio being either equal to 1 or greater than 1 or less than 1?
   b Consider a specific context where (k^*/k) is less than 1, but (k^{**}/k) turns out to be greater than 1. What is the policy implication of this observation?

3 Explain how the internalisation of KN into the Romer model of economic growth affects the rate of economic growth.
4 Consider the definition of per capita income provided in Equation (14.21) above. If KN can be regarded as one of the determinants of (Y/L), is there a need to consider the concept of composite capital, that is, $(K = KM + \eta KN)$ as suggested in the text?

Appendices

Appendix 1: Cobb–Douglas Production Function and the Steady-State Values

We start with a C–D function that displays constant returns to scale. That is:

$$Y = \alpha K M^\theta L^{1-\theta} \tag{14A.1}$$

Divide both sides of Equation (14A.1) by L to get an expression of output per worker (Y/L) as a function of capital per worker {(KM/L):

$$\frac{Y}{L} = \alpha \frac{(KM^\theta)}{L}(L^{1-\theta})(KM^\theta)(L^{-\theta}) = \alpha \left(\frac{KM}{L}\right)^\theta \tag{14A.2}$$

Let $\frac{Y}{L} = y$; and $\frac{KM}{L} = k$. Then Equation (14A.2) can be rewritten as:

$$y = \alpha k^\theta \tag{14A.3}$$

The savings per worker can be explained by multiplying Equation (14A.3) by the savings ratio (ρ):

$$\rho y = \rho \alpha k^\theta \tag{14A.4}$$

Consider now the needs for new workers and the replacement of depreciated capital. When the rate of entry of new workers is v and the rate of depreciation is δ, this can be defined in terms of savings per worker (because this has to come out of savings) as:

$$\rho y = (\delta + v)k \tag{14A.5}$$

The steady state equilibrium is found by equating Equations (14A.5) and (14A.4):

$$(\delta + v)k = \rho \alpha k^\theta \tag{14A.6}$$

Rearranging Equation (14A.6):

$$\left(\frac{k}{k^\theta}\right) = k^{(1-\theta)} = \left[\frac{\rho \alpha}{(\delta + v)}\right] \tag{14A.7}$$

The steady-state value of k is then given by:

$$k^* = \left[\frac{\rho\alpha}{\delta+\upsilon}\right]^{\left(\frac{1}{1-\theta}\right)} \tag{14A.8}$$

Appendix 2: The Rate of Economic Growth in the Swan–Solow Model

Consider Equation (14A.3) where output per worker (y) is defined in terms of capital per worker (k), namely $\{y = \alpha k^\theta\}$. With reference to this expression, the rate of change of (y) per unit of time can be defined as:

$$\frac{dy}{dt} = \dot{y} = \frac{dy}{dk}\frac{dk}{dt} = \theta\alpha k^{(\theta-1)}\dot{k} \tag{14A.9}$$

Dividing Equation (14.A9) by y, we get a definition for the rate of economic growth $\left(\dfrac{\dot{y}}{y}\right)$:

$$\left(\frac{\dot{y}}{y}\right) = \left(\frac{\theta\alpha k^{(\theta-1)}}{\theta\alpha k^\theta}\right)\dot{k} = \theta\left(\frac{\dot{k}}{k}\right) \tag{14A.10}$$

$\dot{k}$ is in fact what remains after the depreciation and new worker needs are subtracted from savings; that is:

$$\dot{k} = \rho y - (\delta+v)k \tag{14A.11}$$

Hence Equation (14.A10) can be redefined as:

$$\left(\frac{\dot{y}}{y}\right) = \theta\left(\frac{\rho y - (\delta+v)k}{k}\right) = \theta\left(\frac{\rho y}{k} - (\delta+v)\right) \tag{14A.12}$$

Given that $\dfrac{k}{y} = \kappa$, the Harrod–Domar ratio, the definition of the growth rate becomes:

$$\left(\frac{\dot{y}}{y}\right) = \theta\left(\frac{\rho}{\kappa} - (\delta+v)\right) \tag{14A.13}$$

Appendix 3: Revised Cobb–Douglas Production Function and Steady-State Values

As above, we start with a revised C–D function that displays constant returns to scale. That is:

$$Y = (1-\eta)\alpha(KM+\eta KN)^\theta L^{(1-\theta)} \tag{14A.14}$$

Divide both sides of Equation (14A.14) by L to get an expression of output per worker (Y/L) as a function of capital per worker {(KM/L):

$$\frac{Y}{L} = \frac{(1-\eta)\alpha(KM+\eta KN)^{\theta} L^{(1-\theta)}}{L} \qquad (14A.15)$$

As before, let $\frac{Y}{L} = y$, and $\frac{(KM+\eta KN)}{L} = \bar{k}$; then Equation (14A.15) can be rewritten as:

$$y = (1-\eta)\alpha\, \bar{k}^{-\theta} \qquad (14A.16)$$

The savings per worker can be explained by multiplying Equation (14A.16) by the savings ratio (ρ):

$$\rho y = \rho(1-\eta)\alpha\, \bar{k}^{-\theta} \qquad (14A.17)$$

As in Appendix 1, the steady-state equilibrium can be defined by equating Equation (14A.17) with the rate of entry of new workers as ν and the rate of depreciation as δ, as:

$$\rho(1-\eta)\alpha\, \bar{k}^{-\theta} = (\delta+\nu)\bar{k} \qquad (14A.18)$$

Rearranging Equation (14A.18):

$$\left(\frac{\bar{k}}{\bar{k}^{-\theta}}\right) = \bar{k}^{-(1-\theta)} = \left[\frac{\rho(1-\eta)\alpha}{(\delta+\nu)}\right] \qquad (14A.19)$$

The steady-state value of composite capital per worker is then given by:

$$\bar{k}^* = \left[\frac{\rho(1-\eta)\alpha}{\delta+\nu}\right]^{\left(\frac{1}{1-\theta}\right)} \qquad (14A.20)$$

Notes

1 The content of this chapter was tested in two different forums, namely Singapore Economic Review Conference (Thampapillai 2017); and Australian Conference of Economists (Thampapillai and Chen, 2018), and was subsequently reported in the Singapore Economic Review (Thampapillai and Chen, 2018).
2 Note that in national income accounting, the definition of GVA is: (GVA = GDP − T). That is, net taxes (T) are subtracted from both the expenditure and income accounts. As indicated in Chapter 13, this definition confines the distribution of national income (Y) to the factors of production.

References

Abel, A. A., Bernanke, B. S. and Croushore, D., *Macroeconomics*, 6th edn, Pearson, Boston, MA, 2008.

Ahmad, Y. J., El Serafy, S. and Lutz, E. (eds), *Environmental Accounting for Sustainable Development*, The World Bank, Washington DC, 1989.

Arrow, K. J., 'The economic implications of learning by doing', *The Review of Economic Studies*, 29(3): 155–173, 1962.

Bouhier, N., Tournadre, J., Rémy, F. and Gourves-Cousin, R., 'Melting and fragmentation laws from the evolution of two large southern ocean icebergs', *The Cryosphere Discussions*, November 2017.

Case, K. E., Fair, R. C. and Oster, S. M., *Principles of Macroeconomics*, 11th edn, Pearson Education Limited, Harlow, 2014.

Cleveland, C. J. and Ruth, M. 1997. When, where, and by how much do biophysical limits constrain the economic process?: A survey of Nicholas Georgescu-Roegen's contribution to ecological economics, *Ecological Economics*, 22(3): 203–223.

Cropper, M. L. and Oates, W. E., 'Environmental economics: A survey', *Journal of Economic Literature*, 30(2): 675–740, 1992.

Daly, H. E., 'Towards an environmental macroeconomics', *Land Economics*, 67(2): 255–259, 1991.

Daly, H. E. 'Sustainable growth: An impossibility theorem', *Development*, 40(1): 121–125, 1997.

Daly, H. E. and Cobb, J. B., *For the Common Good: Redirecting the Economy towards Community, the Environment, and a Sustainable Future*, Beacon Press, Boston, MA, 1989.

Dasgupta, P. and Heal, G. M., 'The optimal depletion of exhaustible resources', *The Review of Economic Studies*, 41(1): 3–28, 1974.

Dasgupta, P. and Heal, G. M., *Economic Theory of Exhaustible Resources*, Cambridge University Press, Cambridge, 1979.

Domar, E. D., 'Capital expansion, rate of growth and employment', *Econometrica*, 14(2): 137–147, 1946.

Dornbusch, R. and Fischer, S., *Macroeconomics*, McGraw-Hill, New York, 1994.

Dutta, D. and Otsuka, K., 'An analysis of knowledge spillover from information and communication technology inn Australia, Japan, South Korea and Taiwan', *33rd Australian Conference of Economists,* Blackwell Press, 2004.

Eaton, J. and Kortum, S., 'Trade in ideas: Patenting & productivity in the OECD', *Journal of International Economics,* 40(3–4): 251–278, 1996.

Eaton, J. and Kortum, S., 'International technology diffusion: Theory and measurement', *International Economic Review*, 40: 537–570, 1999.

Eriksen, M., Lebreton, L. C. M., Carson, H. S., Thiel, M., Moore, C. J., Borerro, J. C., Galgani, F., Ryan, P. G. and Reisser, J., 'Plastic Pollution in the World's oceans: More than 5 trillion plastic pieces weighing over 250,000 tons afloat at sea', *PLoS One*, 9(12), 2014.

Evenson, R. and Singh, L., 'Economic growth, international technological spillovers and public policy: Theory and empirical evidence from Asia', *Center Discussion Paper – 777*, Economic Growth Center, Yale University, 1997.

Fahey, D. W. and Hegglin, M. I., *Twenty Questions and Answers About the Ozone Layer: 2010 Update,* World Meteorological Organization, Geneva, 2011.

Fisher, I., 'Precedents for defining capital', *Quarterly Journal of Economics*, 18(3): 386–408, 1904.

Frank, R. H., Bernanke, B. S. Lui, H.-K., *Principles of Economics*, Asia Global edn, McGraw Hill Education, New York, 2015.

Georgescu-Roegen, N., *The Entropy Law and the Economic Process*, Harvard University Press, Cambridge, MA, 1971.

Glahe, F. R., *Macroeconomics: Theory and Policy*, Harcourt Brace Jovanovich, New York, 1977.

Gowdy, J. and Messener, S., 'The evolution of Georgescu-Roegen's bioeconomics', *Review of Social Economy*, 56(2): 136–156, 1998.

Harrod, R. F., 'An essay in dynamic theory', *The Economic Journal*, 49(193): 14–33, 1939.

Hartwick, J., 'Natural resources, national accounting and economic depreciation', *Journal of Public Economics*, 43(3): 291–304, 1990.

Mankiw, N. G., Romer, D. and Weil, D. N., 'A contribution to the empirics of economic growth, *Quarterly Journal of Economics*, 107: 407–437, 1992.

Marshall, A., *Principles of Economics*, Macmillan, London, 1891.

Mill, J. S. 1848. Principles of political economy with some of their applications to social philosophy, Book IV, Chapter VI, Sec. 2, August M. Kelley Publishers, Fairfield, NJ, 1987.
Munasinghe, M. (ed), *Macroeconomics and the Environment*, Edward Elgar, London, 2002.
Porter, M. E. and Stern, S., 'Measuring the "ideas" production function": Evidence from international patent output', Working Paper 7891, *NBER Working Paper Series*, National Bureau of Economic Research, 2000.
Ramanathan, V., Agrawal, M., Akimoto, H., Auffhammer, M., Devotta, S., Emberson, L., Hasnain, S. I., Iyngararasan, M., Jayaraman, A., Lawrence, M., Nakajima, T., Oki, T., Rodhe, H., Ruchirawat, M., Tan, S. K., Vincent, J., Wang, J. Y., Yang, D., Zhang, Y. H., Autrup, H., Barregard, L., Bonasoni, P., Brauer, M., Brunekreef, B., Carmichael, G., Chung, C. E., Dahe, J., Feng, Y., Fuzzi, S., Gordon, T., Gosain, A. K., Htun, N., Kim, J., Mourato, S., Naeher, L., Navasumrit, P., Ostro, B., Panwar, T., Rahman, M. R., Ramana, M. V., Rupakheti, M., Settachan, D., Singh, A. K., St. Helen, G., Tan, P. V., Viet, P. H., Yinlong, J., Yoon, S. C., Chang, W. C., Wang, X., Zelikoff, J. and Zhu, A., *Atmospheric Brown Clouds: Regional Assessment Report with Focus on Asia*, United Nations Environment Program, Nairobi, 2008.
Romer, P., 'Increasing returns and long-run growth', *The Journal of Political Economy*, 94(5): 1002–1037, 1986.
Ruth, M., *Integrating Economics, Ecology and Thermodynamics*, Kluwer Academic Publishers, Dortrecht, The Netherlands, 251, 1993.
Ruth, M., *Advanced Introduction to Ecological Economics*, Edward Elgar, Cheltenham, England, 2018.
Samuelson, P. A., *Economics*, Mcgraw-Hill, New York, 1948.
Sinden, J. A., 'The evaluation of extra-market benefits: A critical review', *World Agricultural Economics and Rural Sociology Abstracts*, 9(4): 1–16, 1967.
Sinden, J. A and Worrell, A. C., *Unpriced Values: Decisions without Market Prices*, Wiley, New York, 1979.
Solow, R. M., 'A contribution to the theory of economic growth', *Quarterly Journal of Economics*, 70(1): 65–94, 1956.
Solow, R. M., 'Intergenerational equity and exhaustible resources', *Review of Economic Studies: Symposium on the Economics of Exhaustible Resources*, 41(5): 29–45, 1974.
Stavins, R. N., Environmental Economics, Working Paper 13574, National Bureau of Economic Research, Washington, DC, 2007.
Stavins, R. N., 'The evolution of environmental economics: A view from the inside', *Singapore Economic Review*, 62(2): 251–274, 2017.
Swan, T. W., 'Economic growth and capital accumulation', *Economic Record,* 32(2): 334–361, 1956.
Thampapillai, D. J., 'Macroeconomics versus environmental macroeconomics', *Australian Journal of Agricultural and Resource Economics*, 56(3): 332–346, 2012.
Thampapillai, D. J., 'Ezra Mishan's cost of economic growth: Evidence from the entropy of environmental capital', *Singapore Economic Review*, 61(3): 1–10, 2016.
Thampapillai, D. J., 'Environmental macroeconomics – A neglected theme in environmental economics', *Paper presented to the Singapore Economic Review Conference, Singapore*, August 2017.
Thampapillai, D. J. and Chen, Y., 'Environmental macroeconomics – A neglected theme in environmental economics – leave alone economics', *Paper presented to the Australian Conference of Economists, Canberra*, July 2018.
Thampapillai, D. J. and Chen, Y., 'Environmental macroeconomics – A neglected theme in environmental economics – Leave alone economics', *Singapore Economic Review*, October 2018 (https://doi.org/10.1142/S0217590818500327).
Uzawa, H., 'Optimum technical change in an aggregative model of economic growth', *International Economic Review*, 6(1): 18–31, 1965.

15 International Trade and Globalisation

So far, the policy context of our prior analyses has been largely one of achieving sustainability in an individual economy. As illustrated in previous chapters, the premise for such an achievement rests on minimising excessive rates of KN utilisation so that a constant stock of KN can always be maintained. The significant difficulty with a steady state for KN, apart from issues pertaining to the composition of the constant stock, is that KN stocks are not readily delineated by geographic boundaries. Even moderate rates of KN utilisation in one economy could threaten the sustainability of another economy because of the continuity of KN across nation-states and the difference in degree of vulnerability of KN across this continuum. For example, the draining of peat swamp forests for timber harvesting in Indonesia generates significant amounts of GHG emissions due to the mere exposure of peat soils to the sun. The resulting haze does not blanket much of Indonesia but other parts of South-East Asia instead, due to the direction of wind movements during periods of the specific forestry management practices. The issue of trans-boundary pollution is widespread across the world and pertains to all types of KN sinks: atmospheric, as well as surface and sub-surface land, rivers, lakes and oceans. The study of global ecosystems reveals that the emergence of endangered species in one location may be propelled by loss of habitats several thousand kilometres away. Numerous examples of this kind clearly indicate that changes in KN stocks are in fact powerful manifestations of globalisation and global connectivity.

Yet the study of globalisation in economics is primarily focused on international trade and investment. Our starting point then is to illustrate how the internalisation of KN into the theories of trade and investment may change policy outcomes. The approach here is primarily an internalisation of KN-based externalities along the lines covered in Chapter 4. At a broader macroeconomic level, changes in export, import and investment markets lead to changes in the trade balance [Exports (X) – Imports (M)] and in the exchange rate and interest rate. But the question remains: are the mechanisms by which international trade and investment play out able to help sustain KN stocks across geographic boundaries? Of course, not – unless each country is able to identify the externality it would inflict on other countries and alter its pattern of trade and investment accordingly. It is for this reason that several global protocols such as the Kyoto Protocol, the convention on biodiversity, the Convention on International

Trade of Endangered Species (CITES), and the register of heritage-listed items are in force. But, as illustrated in Box 15.1, the forces of international trade and investment can overwhelm these efforts. Further, as we illustrate below, the standard concepts and frameworks pertaining to trade and investment, even after revisions to recognise the role of KN, may fall short of providing policies for achieving the goal of sustainability. Therefore, compliance with the protocols and conventions remains a minimum, necessary, though not sufficient, condition for sustainability.

Box 15.1 Trade as the Priority and KN as an Afterthought

Recent efforts to curb carbon dioxide and other greenhouse gas emissions through the Kyoto Protocol, for example, have resulted in considerable redistribution of emissions profiles, all while seeing emissions rise. For example, research by Glen Peters (2011) of the Center for International Climate and Environmental Research in Oslo and his colleagues provides a consumption-based approach to carbon accounting. This is in contrast to the usual production-based accounting, which keeps track of emissions in a country that are associated with the conversion of fuels and the production of goods and services. Consumption-based emissions accounting keeps track of the emissions attributable to the goods and services that originate outside the country and are imported for consumption, and correct for those emissions that are associated with exports of goods and services, which are allocated to the receiving countries. To accomplish this, Peters et al. draw on a trade-linked global database for CO_2 emissions covering 113 countries and 57 economic sectors from 1990 to 2008. For that time period, they find that the emissions transfers via international trade from developing to developed countries increased fourfold – from 0.4 Gt CO_2 in 1990 to 1.6 Gt CO_2 in 2008 – which exceeds the Kyoto Protocol emission reductions.

Such an outcome should hardly be surprising in a world where trade agreements take on a paramount role in boosting the exchange of materials and energy across the globe for the sake of economic growth. Environmental and social repercussions are an afterthought. Small, local businesses are destroyed along the way, making reliance on long-distance supplies imperative, thus setting off a vicious cycle in which not only KN becomes undermined, but also the resilience of economies, as recent disruptions to shipping routes illustrate – be it because of a single container ship getting lodged in the Suez Canal, for example, or pirates attacking vessels off the coast of Africa.

Source: Peters, G.P., J.C. Minx, C.L. Weber, and O. Edenhofer (2011)

Comparative Advantage and Specialisation

In the case of a set of trading partners, the theory of comparative advantage dictates that the welfare of all trading countries would increase if individual countries opted for specialisation in only certain items. In other words, each country would enjoy higher welfare by being an exporter and an importer than by being solely an exporter – even if it had absolute advantage in terms of all items that are traded. This principle is illustrated in the two frameworks we will display below. The first stems from the production and factor-utilisation framework, while the second is based on the market model. Both frameworks assume that each country is a price-taker and that world prices are competitively determined. In practice, such price determination takes place through auction bids on spot markets. However, in real-world settings the price-taker condition is marred by the fact that there are powerful buyers and sellers in these auctions. Nevertheless, we will retain the price-taker assumption and confine our treatment to two goods: X, the export good, and M, the import good. We will denote the world prices of these two goods as P_X and P_M. Further, we will define the exchange rate (e) as the amount of world price currency units (WCU) per local currency unit (LCU); that is, $\left[e = \dfrac{WCU}{LCU} \right]$. Initially we will regard e as given, and illustrate its determination subsequently.

Given world prices P_X and P_M and the exchange rate e, the domestic prices are simply (P_X/e) and (P_M/e). Note that, in the context of the price-taker assumption, even if exchange rates change and cause changes in the domestic prices faced by exporters and importers, the relative price ratios will remain the same.

The Factor Endowment and Trade Framework

The Standard Framework

1 The country's production possibility frontier, described by $[Q_M = m\,(Q_X, KM, L)]$, is an efficiency frontier depicting the complete utilisation of resources, namely KM and L. Because the resource set (KM, L) is fixed during a given time period, there can only be one production possibility frontier for that time period.
2 Society's consumption preferences described by indifference curves $[Q_M = U(Q_X)]$ explain utility maximising and substitution behaviour in terms of X and M. Following our discussion in Chapter 7, one could envisage the existence of an infinite number of indifference curves of the form $[Q_M = U(Q_X)]$ in $Q_X - Q_M$ space. The axiom of utility maximisation dictates that consumers would prefer to be on the highest possible indifference curve that is possible for any given endowment.
3 Iso-revenue lines of fixed gradient, namely the ratio of prices $[P_X/P_M]$, describe the relative prices faced by the country. It is possible to envisage the existence of an infinite number of such lines in $Q_X - Q_M$ space. The lines closer to the origin display lower fixed revenue compared to those further away from the origin.

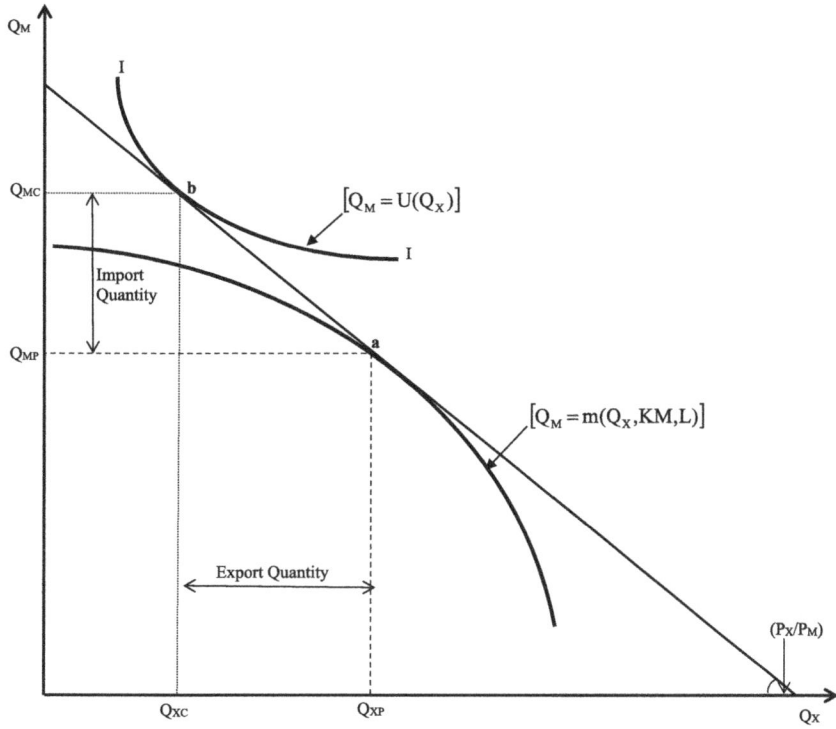

Figure 15.1 Factor Endowment and Trade – Standard Framework

Production decisions within the frontier are inefficient because they lead to inefficient use, or under-utilisation, of resources, while any production decision outside the frontier is infeasible, given the particular technology in use. Note that the production capability is overstated because KN has been excluded – but we will consider this aspect with the revised framework below. Given that $[Q_M = m(Q_X, KM, L)]$, production is optimised by the tangency of the frontier with the highest attainable iso-revenue function. This point of tangency is represented by point **a** in Figure 15.1 leading to the production decision (Q_{XP}, Q_{MP}).

Consumption decisions are determined by the tangency of the highest attainable iso-revenue line and the indifference curve. Highest consumption possibility is displayed by the tangency of the indifference curve with the iso-revenue line that is also tangent to the production possibility frontier. Should the tangencies of all three functions – production possibilities, indifference and iso-revenue – coincide, then production and consumption decisions would be identical and trade would not be warranted. The illustration in Figure 15.1 displays a consumption decision governed by the tangency at point **b** leading to the consumption set (Q_{XC}, Q_{MC}).

It is now evident that this country has a surplus of X and a shortage of M. This provides the basis for trade by exporting $(Q_{XP} - Q_{XC})$ and importing $(Q_{MC} - Q_{MP})$.

The production and consumption decisions, both chosen in terms of the same iso-revenue line, enable the achievement of trade balance; that is

$$\text{Export revenue} = \{P_X(Q_{XP} - Q_{XC})\} = \text{Import expenditure} = \{P_M(Q_{MC} - Q_{MP})\} \quad (15.1)$$

The literature on trade and development subscribes to the expansion of consumption possibilities by recourse to the expansion of production possibilities. A range of options considered in the previous chapter, such as savings and capital accumulation, human capital development and knowledge accumulation can enable this. However, these possibilities would inevitably be limited when the constraints surrounding KN are considered. We turn to this issue next.

The Revised Framework

The recognition of KN results in the production possibility frontier being now defined as

$$[Q_M = n(Q_X, KM, L, KN)] \quad (15.2)$$

Recall from Chapter 8 the observation that the recognition of KN leads to the production function shifting downwards, because output is now attributed to three factors instead of two. So, as illustrated in Figure 15.2, the revised three-factor production possibility frontier falls within the frontier determined with reference to two factors. In Figure 15.2, the conceptual tools pertaining to the revised framework are displayed with dashed lines. A comparison of the outcomes of the standard and revised frameworks reveals that both consumption and production decisions have to be reduced by the following quantities:

Consumption: $(Q_{XC} - Q_{XC'})$; $(Q_{MC} - Q_{MC'})$
Production: $(Q_{XP} - Q_{XP'})$; $(Q_{MP} - Q_{MP'})$

Also note that the production possibility frontier can further contract in the context of environmental regulations. Consider, for example, the comparative advantage enjoyed by China with reference to a range of consumer goods, the manufacture of which relies on the utilisation of KN stocks. The imposition of restrictions on KN utilisation, for example discharge of production residues into rivers, oceans and atmosphere, would have the effect of shifting the production possibility frontiers further inwards.

The subject of comparative advantage in the context of trade in natural resources warrants deeper scrutiny. Countries such as Australia and Canada, being endowed with large deposits of energy resources such as coal and other fossil fuels, have a comparative advantage and hence export these resources to other countries that

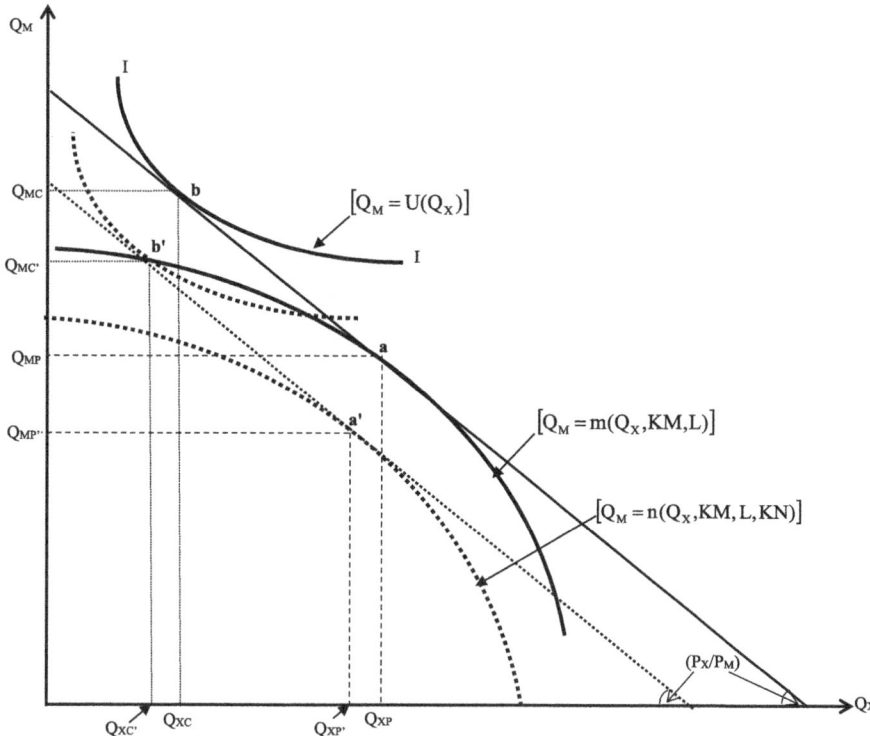

Figure 15.2 Factor Endowment and Trade – Revised Framework

require them. While the extraction and export of these resources intensify, there are reports of correlations between fossil fuel extraction and seismic activity (Yerkes and Castle 1976; Van Eijs et al. 2006). The cognisance of these correlations could prompt the appreciation that the extractable resources are components of much larger KN stocks that afford the earth surface greater stability. The net effect of such appreciation is that the production possibility frontier involving an export item such as coal or shale oil and shale gas would have a quantity domain that would be much smaller than one would expect. Similar considerations would apply to trade in forestry products and the cognisance of biodiversity losses.

A difficulty with the production and factor-endowment framework is that export and import goods are not necessarily substitutable and the effects of changing consumption preferences due to considerations pertaining to KN may not be readily captured. For example, the awareness of increased frequency of earthquakes caused by mineral and fuel extraction, or the loss of biodiversity caused by logging and land conversion, would depress consumer demand for these items. The framework based on the market model provides a basis for recognising such effects.

Trade Framework Based on the Market Model

This framework is illustrated in Figure 15.3a for exports and Figure 15.3b for imports. As shown, the production and consumption decisions are made with reference to the effect of global demand and supply on the domestic market. Global demand and supply are represented by world market prices. A surge in global demand and/or a fall in global supply would push world prices upwards, while a surge in global supply and/or a fall in global demand would have the reverse effect. With exports, the world market price remains above the fundamental market equilibrium price and the reverse applies to the case of imports. That is, when the world price of a good or service is in excess of its fundamental market equilibrium price, then the country has the potential to specialise in the production of this good or service towards exporting it. This potential stems (at least in part) from the country's capability to produce goods and services cost-effectively. Such capability is manifested in a relatively elastic supply (marginal cost) schedule. In general, countries that lack comparative advantage with reference to the production of a specific good or service display domestic supply schedules that are relatively less elastic compared to those of countries that are endowed with the comparative advantage.

The effects of recognising KN can be considered at two levels: global and domestic. Although we do not illustrate these effects in Figure 15.3, they may be inferred from the discussion that follows.

At the global level, awareness of the roles and values of global ecosystems and the adverse ecological impacts of producing certain goods should diminish global demand and push world prices downwards. Such awareness has prompted

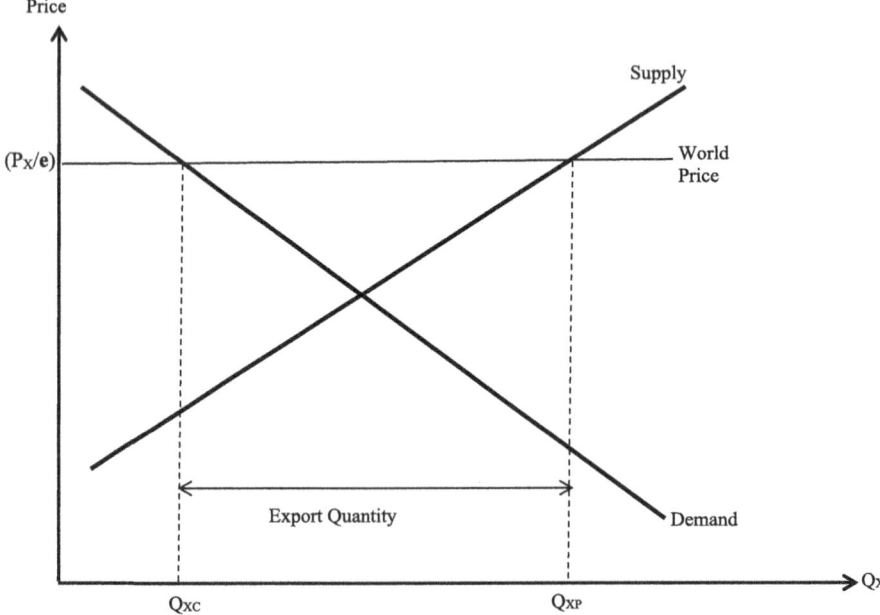

Figure 15.3a Market for an Export Good

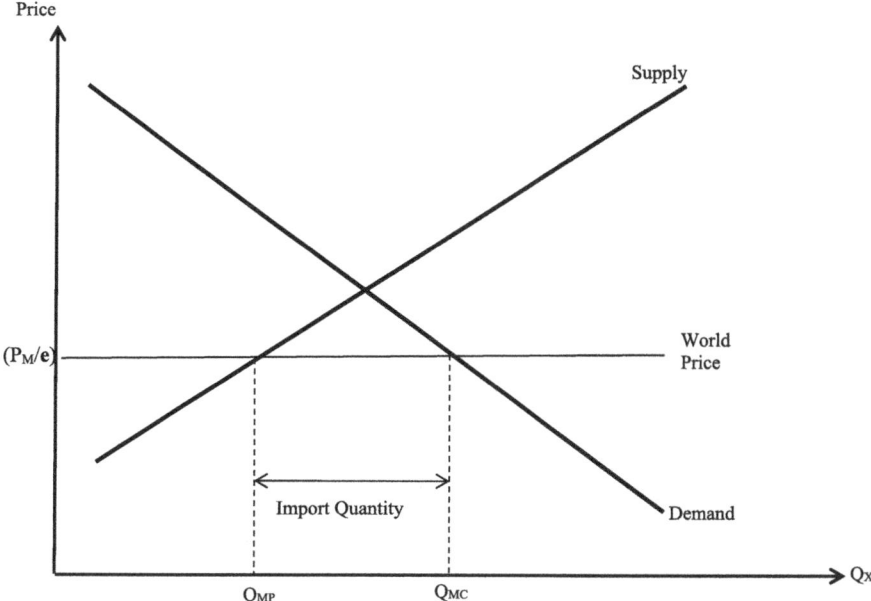

Figure 15.3b Market for an Import Good

producers of some goods and services to make explicit their methods of production through labelling. For example, several consumer items carry labels that convey compliance with ecological stewardship and production methods approved by the International Standards Office (ISO). Yet the global demand for energy resources such as oil, gas and coal – extraction and utilisation of which inflict significant global externalities – grows unabated, propelling upward shifts in the world market price. The proliferation of exploration and mining activities worldwide is in fact the response to such world market signals.

At the domestic level, the internalisation of KN into the production function would raise the marginal cost of production and thereby diminish the extent of comparative advantage. Recall from Chapter 8 that the recognition of KN within the factor-utilisation framework of a firm causes the marginal (and average) cost schedules for producing any good or service to resemble those of a non-renewable resource. Therefore, the internalisation of KN should, at least in theory, lead to the contraction of production and export domains. On the demand side, the awareness of KN externalities could prompt the shift of domestic demand inwards to the left. In some instances, such shifts could have a perverse effect on the world market by creating extra export capacity in exporting countries. For example, the demand for asbestos has been virtually non-existent in Canada since the 1970s. Yet, asbestos mines operated in the province of Quebec until as late as 2011 and all quantities extracted were exported to countries such as India.

In theory, the contraction of export domains in exporting countries due to compliance with environmental standards would propel world market prices upward. Such escalation of prices would act as an incentive for the importing countries to search for cheaper alternatives. For example, China imports significant quantities of coal and gas from countries such as Australia and Canada. The increase in global demand for coal from other emerging economies such as India has led to the steady increase in the world market price of coal. China's initial response was to explore and extract its own coal reserves. However, more recently China has diversified its energy portfolio by investing in wind and solar power. Until about 2009, Germany was hailed as the largest producer of solar panels in the world, but by 2011 China had surpassed Germany. The province of Xinjiang in China has been almost exclusively dedicated to the utilisation of wind power development owing to its specific geographic location.

Diversification and specialisation are two related aspects of trade. As Box 15.2 illustrates, however, the argument that comparative advantage lies behind trade must be understood with much more nuance, as well as a healthy dose of scepticism.

Box 15.2   Seemingly Nonsensical Trade Patterns

The UK Interdependence Report (Simms et al. 2006) claims that between the United Kingdom (UK) and the rest of the world, 'identical products are being shipped backwards and forwards with heavy environmental costs'. A couple of illustrative examples are as follows:

- The UK exports 17,600 tons of chocolate waffles, whilst also importing 17,200 tons of the same item.
- For Boneless Chicken, UK's export and import quantities (measured in tons) are respectively 51,000 and 44,000.

Why are the imports and exports indicated above nearly identical? One reason may be that, for example, consumers in England prefer the taste of chocolate waffles and boneless chickens produced in other countries over their own, and that other countries prefer those from Britain. So, why then not exchange the recipes among countries and/or details on production methods, and as a result cut down on the unnecessary material and energy use associated with tens of thousands of shipments?

The argument that near-equal quantities of exchange of the same products among different countries are attributable to differences in preferences and tastes can readily be debunked by data on products that are highly homogeneous, even extremely narrowly defined. The following data from the Bureau of the Census (2011) provides two examples from the USA World Trade Statistics, and many such examples can be found for virtually all countries. For example:

- The export and import quantities (measured in tons) of Vitamin B12 for the USA are respectively 327 and 328
- For Tetrachloroethylene, such quantities (in tons) are 54,032 and 54,011

Vitamin B12 is vitamin B12. It is used as a dietary supplement, and clearly defined by its chemical composition. Similarly, what constitutes tetrachloroethylene (used, for example, in dry-cleaning products) is unambiguously defined. Differences in tastes and preferences cannot distinguish between imports and exports of these products. Rather, the fact that in a particular year, the USA imported almost exactly as much of either of these two substances (and the same holds for many others), is related to a host of factors, among them seasonal differences in domestic capacities to produce products relative to domestic demand, differences in demand over the course of the year, differences in fuel and other costs of transport, as well as exchange rate differentials that foster arbitrage and speculation. In a world where financial markets are global and adjust nearly simultaneously to market signals, trade flows follow suit. As the examples from both tables showcase, the result are large amounts of materials that are exchanged over long distances for the economic benefit of a few, at high costs of emissions and environmental impact, borne by all.

The Market for Foreign Exchange

Thus far, we have regarded the exchange rate as given. Because volumes and directions of trade can be affected by changes in the exchange rate, it is pertinent to consider the mechanisms underlying the determination of exchange rates. In essence, the exchange rate is the amount of another country's currency that can be purchased with a unit of the local currency. As indicated above, we will define the exchange rate (**e**) as the amount of world price currency units (WCU) that can be purchased with one local currency unit (LCU); that is, $\left[e = \dfrac{WCU}{LCU} \right]$. Some dominant WCUs in operation are the Unites States Dollar; the United Kingdom Pound, the Euro and the Japanese Yen. The basic determinant of **e** is the interaction between demand and supply with reference to LCUs. This is illustrated in Figure 15.4.

Demand for LCUs (D_{LCU}) is displayed by foreign entities that wish to:

- Purchase local goods and services that are on offer for export and/or
- Invest in local assets such as bonds, stocks and shares and real estate.

Supply of LCUs (S_{LCU}) is provided by local entities that wish to:

- Import foreign goods and services and/or
- Invest in foreign assets.

When **e** is determined by the free interplay of demand and supply, the currency market is labelled as a free float. However, most foreign exchange markets represent managed floats. That is, most monetary authorities permit **e** to fluctuate within a band defined by upper and lower limits, and would only intervene if **e** displayed

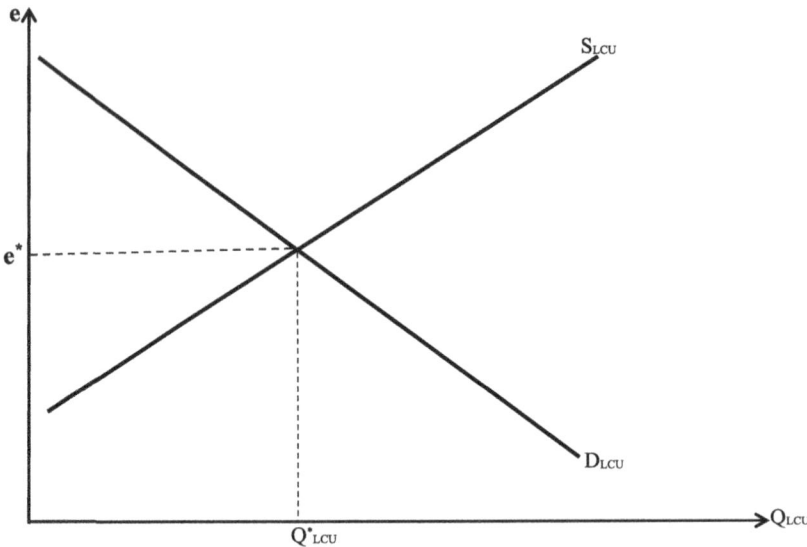

Figure 15.4 Market for Foreign Exchange

signs of going outside this band. For example, if **e** exceeded the upper limit of the band, then monetary authorities would inject more LCUs and make S_{LCU} shift to the right and thereby bring **e** within the band. The reverse of this – buying more LCUs to prompt a D_{LCU} shift to the right – would happen if **e** fell below the lower limit.

Leaving the subject of interventions aside, our purpose here is to illustrate the effects of recognising KN on the determination of **e**. Such effects would ultimately depend on how trade and investment volumes are affected by the recognition of KN. Our major conclusions from the foregoing text on the effects of affording recognition to KN could be summarised as follows:

1. The domestic production of both X and M will contract by the mere recognition of the factor-utilisation function involving three factors (KM, L, KN) instead of that involving two factors (KM, L). This observation implies that the rates of return to investment would be lower than those envisaged in terms of the two-factor function.
2. The effect, on export and import volumes, depends on the extent to which the awareness of KN-related matters has penetrated into the global and local communities.

   - If the global community has recognised KN-related vulnerabilities with reference to X and M, then the world prices P_X and P_M will both fall.
   - If such recognition has also pervaded the local community, then the change in preferences and demand will lead to a contraction of domestic consumption of X and M.

The most likely net effect of these considerations is that there would be a reduction in the volume of trade. Hence, both D_{LCU} and S_{LCU} in Figure 15.4 could shift to the left. The effect on the exchange rate would depend on which shift was more significant. However, it is clear that the volume of transactions (Q^*_{LCU}) in the foreign exchange market will be smaller than when KN goes without due recognition.

Global Issues and the Need for a Global Paradigm

The main outcome of internalising KN into the trade and exchange rate frameworks is that markets would contract. This conceptual result follows the line of reasoning advanced with KN and intergenerational externalities in Chapters 5–9. However, in reality, several obstacles prevent the realisation of such outcomes. We consider some of these below.

World trade is driven primarily by a myriad of economic entities in different countries engaged in the exchange of goods and services across national boundaries. The primary goal of organisations such as the World Trade Organisation (WTO) and the United Nations Commission on Trade and Development (UNCTAD) is to promote and facilitate such exchange and remove barriers to trade. In pursuing this mission of trade promotion and facilitation, the protection of KN does (inadvertently) get neglected. Such neglect is despite the presence of various international agencies such as the United Nations Environment Program (UNEP) and the Intergovernmental Panel on Climate Change (IPCC). There is indeed an explicit need for institutional cross-compliance across various agencies so that the protection of KN becomes a clear article in the various protocols pertaining to trade.

Whilst global institutions such as the WTO and UNCTAD assist and facilitate global trade, several countries have drawn up bilateral free trade agreements (FTAs) of their own. For example, Singapore has 22 FTAs with various countries, whilst this number for Canada is around 15 with several more under negotiation. The basic idea of a FTA is to create much larger markets compared to the domestic market. For example, Canada has a population of approximately 36.5 million people. By entering into the North American Free Trade Agreement (NAFTA), Canada has entered into a market of nearly 500 million people.

In the context of proliferation of multiple FTAs, keeping track of KN protection would inevitably be tenuous. Although KN protection has surfaced in some FTAs, the restoration KN does not appear clearly. This is because the restoration of KN is indeed a difficult area in terms of undoing centuries of damage and the potential of losing valuable commercial revenue. For example, China, Sri Lanka and Kenya are the world's leading exporters of tea. Each country earns on average nearly $1.5 billion in export revenue. Tea plantations surfaced on the mountainous landscape of these countries some 100 to 150 years ago. The prior state of the mountainous terrain was dominantly forests comprising of tall trees. Reforestation of at least part of this terrain may help towards mitigation against climate change. Although reforestation of commercially viable land areas would prove exceedingly difficult, it is pertinent to consider, at least in general, as to how the restoration of lost KN endowments could make an appearance in FTAs – especially in those pertaining to agreements between developing and developed countries.

All FTAs and WTO protocols are clearly definable because the individual trading entities are nation-states with clearly defined boundaries. In contrast, KN does not have national boundaries – it is the global ecosystem – and some would argue that the KN's boundary is the planetary boundary. In the preceding chapters, we adopted the premise of the KN endowments of a nation being the foundational capital that supports the economy in terms of the 'source' and 'sink' functions. But in reality, these KN endowments are part of a broader collective, namely the global ecosystem. The water-logged peat soils of South East Asia as illustrated in Box 15.3 is one example (amongst many) that shows the connectivity of KN endowments across the world.

Box 15.3 Palm Oil and Green House Emissions from South East Asian Peat Soils

South East Asia carries near 30 million hectares of water-logged peat forests. The soils in these forests display the unique characteristic of rolling peat mounds being submerged in water. These soil mounds are the preferred sites for palm oil plantations. Nearly 6 million hectares of these sites have been so far cleared for palm oil production. Indonesia and Malaysia together produce nearly 85 per cent of the world's palm oil production.

The dominant pattern of palm oil plant cultivation appears to be one where the terrain is drained of its water and seedlings are planted on the naturally occurring mounds of peat. Such mounds are perceived as significant sources of nutrients for the palm oil plant seeds/seedlings to grow.

However, the mere exposure of peat to sunlight triggers its combustion and the generation of significant volumes of GHG. The so-called slash and burn practices in the cultivation of palm oil plants contribute to an ancillary factor that simply exacerbates the GHG emission volumes. A misconception prevails with reference palm oil cultivation. In response to recognising the importance of having peat soils persistently covered by water to prevent combustion on exposure to sunlight, some have sought to flood the previously drained peat soils. Apart from severe hydrological disequilibria – this is also flawed and even more damaging. Microbiologists at the National University of Singapore (Mishra et al. 2014) have shown that non-combustion of peat soils requires not only submergence by water but also the presence of specific microbial activity. These microbes are destroyed when the water-logged peat soils are drained for cultivation purposes.

The solution is for the water-logged peat soils to be preserved and left intact. Such acts of preservation will significantly reduce GHG emissions but cost revenue losses.

But, what is connection between the South East Asian water-logged peat forests and the global ecosystem? If the clearing of water-logged peat forests continues GHG emissions would rapidly escalate and global temperatures will rise. This raising of temperatures could in turn lead to melting of ice that covers the Arctic peat amongst other events including sea level rise.

Trade theorists claim that perfect free trade in the global economy guided by the theory of perfect competition will maximise global welfare. However, such maximisation will be infeasible if the foundational capital of the global economy, namely the global ecosystem is not adequately cared for. Important drivers of KN damage that are not readily curtailed by global protocols are war and conflict. In such contexts, the arms industry has displayed rapid expansion and proliferation in the world market. KN damages ensue from the both the manufacture and utilisation of the arms. Of course, the use of long-range ballistic missiles inflict far greater KN damage compared to small scale arms. Mcfarlane and Volcovici (2023) report that just one year of the Ukraine conflict could have led to net increase of GHGs by about 120 million tons of CO_2 equivalent. Note that conflict appears to have permeated several corners of the globe. The display and use of arms in impoverished countries like the Sudan and Papua New Guinea shows that trade in arms (legal or otherwise) is universal. KN damage that stems from the use of arms is further exacerbated by the human misery that unfolds due to conflict. Clearly, peace and negotiations are vital tools of KN preservation and sustainability.

Concluding Comments

If there were worldwide recognition of the importance of KN, then the volumes of global trade and investment will contract. This outcome is consistent with the main result of Chapter 9. That is, when sustainability is also recognised as an additional condition besides the standard ones that depict perfect competition, then 'less' becomes preferable to 'more'.

At present, the recognition afforded to KN on the global spectrum is partial; only some countries have complied with global protocols and conventions, and even where they have, they may have fallen short of the hopes to reduce adverse economic impact on the life support functions on which their economies rely. Such countries, overall, tend to be high-income countries. Yet even within these countries, the stewardship towards KN is not widespread. Further, the consumption baskets of the compliant countries inevitably include raw materials and items produced by non-compliant countries. Firms and corporations of the so-called compliant countries also relocate their operations to countries where compliance is not required. They do so in order to maximise their profits. Additional drivers behind trade, and the ways they undermine KN quality, as indicated above, pertain to war and conflict.

As long as there is no change in the global mindset, protocols and conventions will remain a necessity. Such changes in mindset are difficult to achieve when multilateral agencies such as the World Bank, the Asian Development Bank (ADB) and the International Monetary Fund (IMF) fail to grasp the importance of the global ecosystem perspective on KN. Such failure is evidenced, for example, by the construction of the Three-Gorges Dam in China – a project sponsored by the World Bank, and the deepening of the Mekong River for river trade between Thailand and Yunan Province in China – a project financed by the Asian Development Bank.

With reference to trade theory, it is pertinent that we replace the concept of 'comparative advantage' with one of 'comparative and KN advantage'.

Review Questions

1 Critically evaluate the following claim made in this chapter: 'Even moderate rates of KN utilisation in one economy could threaten the sustainability of another economy because of the continuity of KN across nation states and the different degrees of vulnerability of KN across this continuum'.
2 Explain how the recognition of KN could alter a nation's comparative advantage. (This could be approached at two levels, namely a system of KN endowments at the national level and KN in terms of a global ecosystem.)
3 Discuss the effects of internalising KN in trade theory frameworks on exchange rate determination.

References

Bureau of the Census, *US World Trade Statistics*, Government Printing Office, Washington, DC, USA, 2011.

Mcfarlane, S. and Volcovici, V. Accounting for war – Ukraine's climate fallout, Reuters News Report, 6 June 2023. https://www.reuters.com/world/accounting-war-ukraines-climate-fallout-2023-06-06/

Mishra, S., Lee, W. A., Hooijer, A., Reuben, S., Sudiana, I. M., Idris. A. and Swarup, S., 'Microbial and metabolic profiling reveal strong influence of water table and land-use patterns on classification of degraded tropical peatlands', *Biogeosciences*, 11, 1727–1741, 2014.

Peters, G. P., Minx, J. C., Weber, C. L. and Edenhofer, O., 'Growth in emission transfers via international trade from 1990 to 2008', *Proceedings of the National Academy of Sciences of the United States of America*, 108(21): 8903–8908, 2011.

Simms, A., Moran, D. and Chowla, P., The UK Interdependence Report: *How the world sustains the nation's lifestyles and the price it pays*, New Economics Foundation, Brighton, England, 2006.

Van Eijs, R. M. H. E., Mulders, F. M. M., Nepveu, M., Kenter, C. J. and Scheffers, B. C., 'Correlation between hydrocarbon reservoir properties and induced seismicity in the Netherlands', *Engineering Geology*, 84(3–4): 99–111, 2006.

Yerkes, R. F. and Castle, R. O., 'Seismicity and faulting attributable to fluid extraction', *Engineering Geology*, 10(2–4): 151–167, 1976.

Part IV
Valuation

16 Valuation of KN

Microeconomic Basis

The valuation of KN remains an important, and yet controversial, subject in policy analysis. It is important because the value assigned to KN can in several instances make the difference between the preservation of KN and the loss of KN. It is controversial because there is often little consensus on the validity of the values assigned to KN. In this chapter, we deal with valuation at the microeconomic level. We will consider its valuation at the macroeconomic level in the next chapter.

The value of any form of capital is usually ascertained by recourse to the valuation of the services it provides. As illustrated in this chapter, the valuation of KN is no exception to this. Although applications of methods for the valuation of KN and its goods and services date back as far as the 1930s, concerted efforts to use these values in policy formulation are a more recent development. The object of this chapter is to present a concise review of the various valuation methods that are pertinent to KN. As indicated, the conceptual tools considered in Chapters 3–9 would now become useful for a better appreciation of some of these methods.

The Basis for Valuation

The literature on environmental economics considers three types of value: *existence value*, *option value* and *use value*. The first of these, existence value, is the value that is attributed by individuals to the mere existence of KN endowments, even if they do not directly utilise these endowments. For example, many individuals are familiar with the existence of Lake Baikal in Siberia and could be willing to pay for its existence, even when Siberia is an unlikely place for them to visit. An important caveat is in order. The value of KN endowments extends much further beyond value in terms of visitations and/or extraction of specific services. Being part of the global ecosystem, these endowments, could in a preserved (protected) state of existence, offer greater stability to the earth and the planetary system. Nevertheless, the literature on valuation generally proceeds on the basis of use value.

Option value refers specifically to the preservation of KN endowments and its services, which have no current use value but potential future use value. It captures the right that individuals would like to exercise in the future in terms of enjoying these endowments and their services. So, with the Lake Baikal example, one

272 Valuation

simply gains value from the awareness that he/she may in fact visit this ancient lake in Siberia sometime in the future.

Use value deals with various forms of KN endowments that are in current use. With the Lake Baikal example, the Russian Government does promote it as an important tourist destination. Given its remote location, only about 150,000 people visit the lake area each year and that too the Southern parts of the lake, which can be accessed through the Trans-Siberian Railway. If the value of the Lake Baikal were to be confined to use value, then such value will clearly be an underestimated one because the vast richness of one of the world's largest and deepest lakes cannot be captured by the visits of some tourists. Yet, as illustrated below, most methodological developments in Environmental Economics are centred on use value.

Following standard microeconomic theory, the area below the demand curve and/or the area below the supply (marginal cost) curve define the value of a good or service. The former measures value in terms of either willingness to pay (WTP) or willingness to accept (WTA), while the latter equates value to the opportunity costs of providing the good. Based on this conceptual premise, the methods of KN valuation can be classified into two broad categories, as shown in Figure 16.1. As shown below, these methods attempt to elicit the value of KN by either improvising markets or making use of actual transactions that occur in related markets. Each method of valuation is considered in turn under its relevant heading.

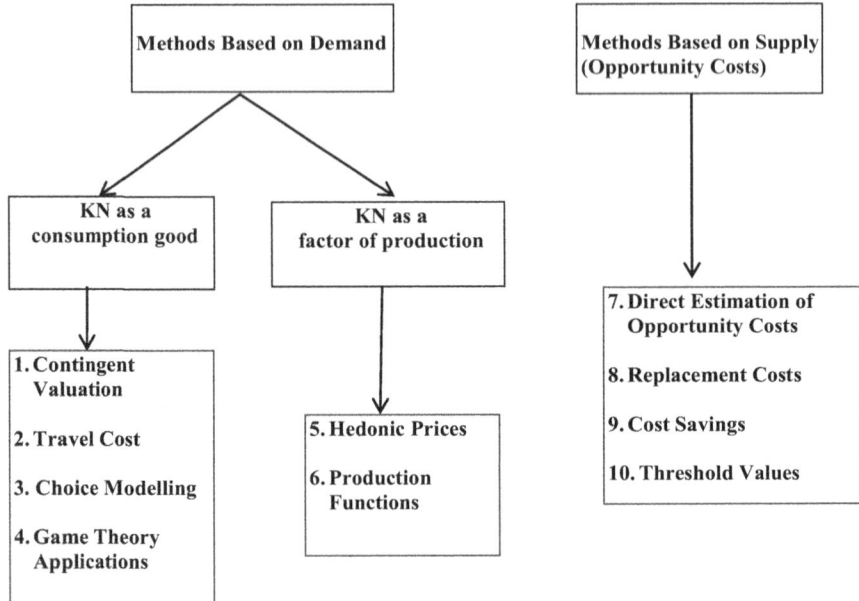

Figure 16.1 An Overview of the Valuation Methods for Environmental Outcomes

Methods Based on Demand: KN as a Consumption Good

Most of the applications that have been reported in this category attempt to value KN as a good that is consumed. That is, KN enters directly into a consumption utility function. For this reason, some authors describe these methods as those based on stated preferences. Therefore, either direct or indirect estimation of consumption utilities is an essential feature of these methods.

Contingent Valuation Method (CVM)

This method has been used mostly to estimate the value of wilderness areas in terms of either preservation or the losses inflicted by damaging economic activities. It is claimed to be relevant when the wilderness area in question has several important environmental features. It is not suggested that this method can readily capture the value of all these features. Rather, the implication is that with the exception of perhaps choice modelling described below, most of the other methods of valuation that are described in this Chapter are not capable capturing the diverse and complex attributes of KN endowments. The method has been applied widely, but not without controversy. Examples are the valuation of the effects of the Exxon-Valdez oil spill in Alaska (Carson et al. 1992); the value of preserving the Kakadu National Park in Australia (Wallace 1992); rural water supply decisions in Nigeria (Whittington et al. 1990); preserving the Kariba lake shore in Zambia (Thampapillai et al. 1992); and preserving wildlife habitats (Kontoleon and Swanson 2003; Bandara and Tisdell 2004; Surendran and Sekar 2010; Jin et al. 2010).

To illustrate the method, consider estimating the value of preserving KN at a specific location, such as a lake or a wilderness area. In its simplest form, the method involves the direct questioning of a sample of individuals on their WTP for preservation. As a first step, the person being questioned is made aware of the characteristics of KN as exhaustively as possible. The respondent is then asked whether he/she would be willing to pay a certain amount, say $500 per year, for preserving KN. If the response to this initial question is yes, then no further questions need be asked, and the person's WTP is recorded as $500. However, if the response is negative, then the person is asked whether he/she would be willing to pay either more or less than $500. If the person is willing to pay more than $500, then the person is bid up until he/she says 'no more'. Alternatively, if the person is willing to pay less than $500, then he/she is bid down until he/she says 'no less'. When this method has been repeated in this manner with various individuals, an average value of WTP can be derived for the sample of individuals who have been questioned. The method derives its name from the fact that the value of WTP that is derived for each person is contingent upon the initial information that has been provided. The value of KN can be determined by multiplying the average value of WTP by an appropriate population factor.

However, several issues emerge in the context of applying the CV method. These pertain to the types of biases that can prevail among those carrying out the survey as well as among those responding to the survey (Pearce and

Turner 1990), and to the lack of consistency between hypothetical transactions and actual economic commitments (Neill et al. 1993). Response biases were evident in the survey of individuals in Lusaka for the estimation of WTP for the preservation of the Kariba Lake Shore in Zambia (Thampapillai et al. 1992). While, at one extreme, some people were willing to contribute in excess of fifty per cent of their income for preservation, at the other extreme, there were people who were willing to contribute none. Further, some argue that the public good characteristics of KN may prompt respondents to overstate their WTP, while others argue the reverse by recourse to experimental economics (for example, see Cummings and Harrison 1992).

Knetsch (1994) and, subsequently, Numes and Nijkamp (2011) provide a wide range of examples to clearly illustrate two significant difficulties with the CVM. These are *anchoring* and *embedding*. Anchoring is evidenced in the observation that most responses centre around the initial bid value that is proposed. In some instances, when individuals were given unrelated information containing irrelevant numerical values, a spurious correlation between these irrelevant numbers and the average bid value was observed. Embedding is illustrated by the fact that most individuals set aside a certain amount of their income for altruistic (non-self) purposes. The value assigned for KN is very often embedded in the WTP or WTA for altruistic purposes. Hence, when questions are not properly formulated, it is the value for altruism rather that for KN that is ascertained. For instance, when respondents face a sequence of questions, it is likely that the response to the first question is the intrinsic non-self-centred value that people uphold. Knetsch (1994) illustrates this by selecting samples of similar socioeconomic background and then changing the first question on WTP for each sample. Despite the initial question being different across the samples, there was a significant similarity in the WTP value recorded as the first response.

Despite the above difficulties, CVM continues to be a method that is much preferred because it is easy to apply and its applicability to a diverse set of contexts. Frykblom (1997) and Szabo (2011) illustrate some specific improvements in terms of calibrating the responses in CVM surveys for consistency. Further, some of the inconsistencies and disparities in CVM responses have been minimised by recourse to a dichotomous (binomial) response method (Petrolia and Kim 2009). With the binomial choice methods, the larger sample is divided into several smaller subsamples, each of equal size. The response sought is that of a 'yes' or a 'no' for a WTP or WTA question concerning a KN endowment. The first subsample is asked about its WTP or WTA in excess of a trivially small amount, such as $1. As one would expect almost all respondents are likely to deliver a 'yes' response for such a trivially small amount. The WTP (or WTA) amount is then gradually increased across the subsamples, and the final sample is exposed to a relatively large amount, say $5,000. In the context of a significantly large amount ($5,000), it is likely that almost all respondents would register a 'no'. It is now possible to search for the relationship between the proportion of persons saying 'yes' in each subsample and the WTP or WTA amount that is offered to each subsample. As one would expect, this will be a downward-sloping curve and will be a proxy for the demand curve and could be fitted with a functional form using a statistical technique such as logistic regression. However, an issue remains, namely that similar responses may be elicited for any good. This is because the number of persons responding with a 'yes' will be always very high for a trivially small WTP or WTA amount, and the reverse would hold for a significantly high WTP or WTA amount. Therefore, in the response surveys, it is very important to clearly articulate the important attributes of KN goods in order to differentiate them from non-KN items.

Choice Modelling (CM)

CM is in fact a further calibration of CVM beyond the dichotomous choice method by focusing on the attributes of KN goods and services. The application of the method follows a sequence of steps similar to that adopted in Cost-Benefit Analysis. So the method begins with a clear description of the issue/problem. Very often the need to preserve and/or expand KN is at the centre of the issue. CM enables the identification of a very large number of alternative ways of dealing with the issue and a value for KN results from a comparison of the alternatives.

In order to generate the alternatives, the KN endowment is clearly defined in terms of its attributes such as for example, biodiversity, water quality and scenic beauty. Different levels are also specified for the attributes (for example, high/low). This enables the analyst come up with several alternatives by recourse to different combinations of the attributes. For example, if there are two attributes and three levels, then we can have by factorial design, (3^2) namely nine possible combinations. Respondents are then presented with the various combinations where each of the combinations would have different levels of attainment of the attributes – and then asked to rank or choose amongst the alternatives. Respondents' choices of their preferred alternatives demonstrate their willingness to trade-off one attribute against another. If one of the attributes is a monetary outcome, it is possible to estimate respondents' WTP to obtain additional units of the KN benefits described by other attributes.

Applications of CM can be found in Bennett et al. (2004), van Bueren and Bennett (2004), Morrison and Bennett (2004), and Hanley et al. (2001).

Some of the challenges inherent in the application of CM pertain to the need to identify different levels for attributes – especially for KN – to be realistic. However, this requirement results in one having a very large number of alternatives for the respondents to evaluate. For example, Hanley et al. (2001) generated 3,072 alternatives by defining four levels for four of the attributes amongst others. Although these were subsequently reduced to 36 alternatives, respondents are faced with task of making meticulous comparisons for the analyst to ascertain reasonable estimates of value. Given the large number of alternatives and hence responses, analysts make use of specially designed software that primarily perform logistic regressions that yield the relationship between WTP (or WTA) and the quantity estimate of the attribute. Besides the surveys are likely to cost a significant amount of both time and money.

276 *Valuation*

Travel-Cost Method (TCM)

The origins of this method go back to a letter written by Hotelling (1949) to the United States Department of the Interior. This method is applicable to the valuation of sites such as forests and national parks that have recreational potential. The underlying assumption is that the value of KN is equal to the value of recreational benefits that are provided by the site. The method has been applied and illustrated widely; for example, see Knetsch (1963, 1964), Clawson and Knetsch (1966), Sinden (1973), Sinden and Worrell (1979), Pearce and Turner (1990), Farr et al. (2011) and du Preez and Hosking (2011). The method, in its simplest form, involves asking a sample of visitors (recreationists) at a given site two questions. These are:

How far do you have to travel to visit this site?
How often do you visit this site?

By grouping the respondents into areas that are equidistant from the site, it is possible to find, for each group, an estimate of the average number of visits per year. The 'distance travelled' is then translated into a cost that includes travel and other related recreational expenditures. This cost, which is referred to as the 'travel-cost', is taken as a proxy for the price of recreation, while the average number of visits per year is taken as a proxy for the quantity of recreation.

Because those who live closer to the park visit the site more frequently than those who reside further away, the relationship between the travel-cost and the number of visits is usually downward-sloping, as shown in Figure 16.2. This relationship is in fact the demand for recreation and, by virtue of the assumptions that have been made it is also the demand for KN. The value of KN at the site can now be found by measuring the area below the demand curve.

In some instances, the travel-cost method can be modified to include some features of the CV method. For example, it is possible to estimate the value of

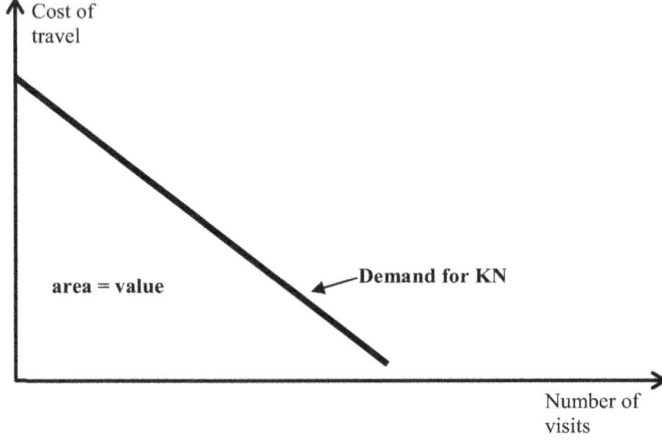

Figure 16.2 The Demand for the KN by TCM

wilderness areas that offer facilities for tourism by bidding respondents with tour package prices in terms of the number of visits they are likely to make per year. The main question asked of respondents is: 'How many times per year will you visit the wilderness area for X days, if you have to pay $Y?'

Tisdell (1991) provides a comprehensive list of deficiencies that are associated with this method. These include the non-homogeneity of population attributes, multiple purposes that can be associated with recreational visits, and the fact that the economic value of a natural area cannot confined to the number of visits alone. Further, given that this method is applicable to only those sites that have recreational facilities, it is not readily relevant to the valuation of remote wilderness areas such as those contested for mining leases.

Game Theory Methods

Game theory methods involve individual interviews about specific aspects of KN and the interviews are structured on a game theoretic model. For convenience, we present the details of the game theory application in the Appendix of this Chapter and confine the narrative here to the salient features of the method. The interview responses are then used to derive an individual's indifference map. This map is then used in turn to derive the individual's demand curve for KN. Society's demand for KN is derived by first deriving a sample of individual demand curves, and then aggregating them through vertical aggregation. The value of KN is then the area below the aggregate demand curve so derived. Although a variety of game theoretic models can be used, we illustrate the Ramsey model (shown in the Appendix) following the applications by Sinden (1974), Sinden and Wyckoff (1976) and Thampapillai (1985).

For example, Thampapillai's (1985) application of the Ramsey model involved the elicitation of a surrogate decision-maker's indifference map for two policy goals: income maximisation (IM) and the preservation of KN. The application of the Ramsey model to the individual resulted in the display of a family of utility curves for his/her utility towards preserving KN at different levels KN_1, KN_2 and so on, as illustrated in Figure 16.3(a). Each utility curve in this family of curves explains the variations in the individual's utility for KN at a fixed level of IM. For example, in the lowermost curve of Figure 16.3(a), IM is fixed at IM_1. Consider now the cross-section across the family of utility curves at utility level U_1. The points on this cross-section represent various combinations of IM and KN that give the same level of utility U_1. Hence these points form the basis for a single indifference curve as shown in Figure 16.3(b). Several other cross-sections, such as the one at U_1, enable the demonstration of the individual's indifference map.

Deriving the demand for KN involves nominating a budget and an appropriate price for IM, say, the return per dollar of investment. By arbitrarily setting a range of prices for KN, a series of budget lines (BL), as illustrated in Figure 16.3(b), can be derived. Following standard microeconomic theory (as illustrated in Chapter 5), the various points of tangency of the budget line with the indifference curves results in a price consumption curve (PCC), and from this curve the demand for KN can

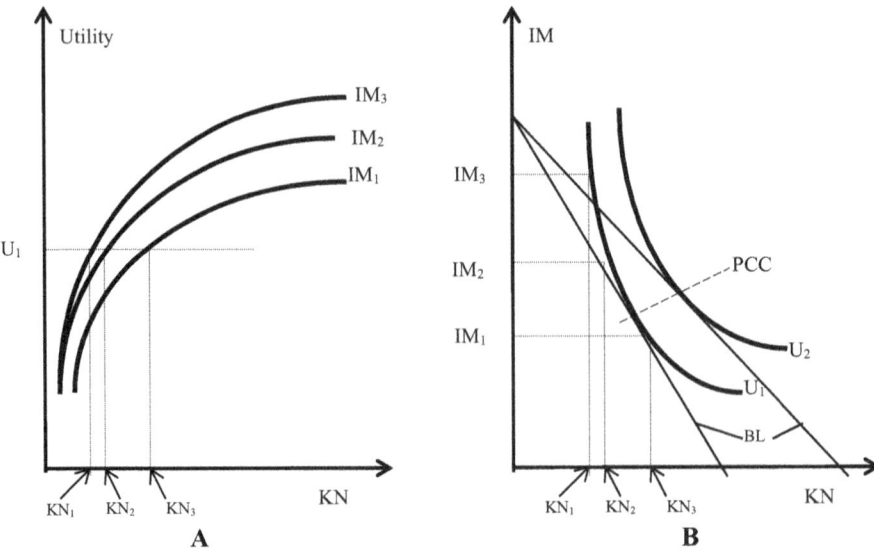

Figure 16.3 Indifference Map from a Game Theory Model

be derived. The main difficulty with this method is that it deals with hypothetical transactions, and is therefore inevitably inadequate in portraying true economic commitment. Besides, the method is time-consuming to administer.

Methods Based on Demand: KN as an Input in Production

In all methods reported in this category, a production function is either explicit or implicit and the demand for KN is derived in terms of the concept of marginal value product. Hence, the value of KN enters consumption utilities indirectly through goods and services that rely on KN for their provision. Hence, the elicitation of demand here is based on revealed preferences.

Hedonic Prices

This method is usually applied to estimate the value of KN with reference to an environmental amenity or disamenity. For example, consider the polluting activities of firms. These activities do clearly cause a disamenity – say reduced air quality and inflict losses in the value of KN. Such losses are often also reflected in variations in residential property value (PV) with reference to proximity to the polluting activity. That is, PVs are likely to be lower for residences that are closer to the source of pollution than those that are more distant from it. Therefore, the method approximates the value of KN to the differentials in PV. The metric for KN is usually an index, such as an air quality index. An increase in the value of the index reflects a gain in KN and a decrease the reverse. In most applications, a relationship between PV and the KN metric is estimated through a regression analysis. In some instances, a stepwise regression procedure is used to narrow and focus on the relationship between PV and KN. Should the relationship between PV and KN display significance, then the demand for KN is defined in terms of the rate of change in PV with respect to KN. For example, suppose that [PV = f(KN)] displays diminishing marginal returns. Then the derived demand for KN will be a downward-sloping function, namely $\{[dPV/dKN] = g(KN)]\}$; that is, the marginal value of PV.

While it is true that changes in KN can prompt variations in PVs, the hedonic method cannot be applied over a wide geographic area and has to be confined to a narrow geographic area, such as the vicinity of a polluting industry or an urban park. This is because, over a broader spatial area, several other variables such as proximity to transport and schools could cloud the influence of KN. Tisdell (1991) argues that differentials in PVs can at times be poor indicators of the value of preserved natural areas. This may happen, for example, when PVs closer to the natural areas are sometimes lowered, for example because of restrictions that are imposed on property owners to protect the attributes of the natural area. The tighter these restrictions, the lower the property value may be. Examples of such restrictions are limitations on the methods of waste disposal and the holding of domestic pets.

Dose-Response Methods

The principles of these methods are similar to those outlined above. The value of KN is approximated to differentials in the value of output. For example, in several studies in agricultural economics, agricultural output (Y) has been related to the extent of soil conservation, which becomes a measure of KN; (Walker 1982; Walker and Young 1986; van Vuuren 1986; Sinden and Yap 1987; Sinden and King 1988). Mendelsohn (1992) demonstrates similar differentiation in output with reference to differences in air pollution levels.

The essential features of dose-response methods are illustrated in Figure 16.4. Consider two firms employing a similar level of factors – labour (L) and manufactured capital (KM) – to produce the same type of output. We denote the combined input of L and KM as (LKM). However, the main variation between the two firms is that one is endowed with a relatively higher level of KN, namely (KN_H), whilst the other with a lower level of KN, that is (KN_L). For example, in an agricultural context, KN may be measured as the depth of topsoil or the concentration of organic matter. One may observe that utilising a similar level of L and KM, say $[(LKM)_1]$, by each firm yields different outcomes. In other words, the size of Y from the firm endowed with KN_H will be higher than that of the firm endowed with KN_L.

In Figure 16.4, the value of ($KN_H - KN_L$) could be represented by the value of output differential $\{P(Y_H - Y_L)\}$, where P represents the market price of output. Extensions of the principles underlying this method involve the search for a range of proxies, such as the value of workday losses due to ill health caused by pollution and the added costs of mitigating pollution.

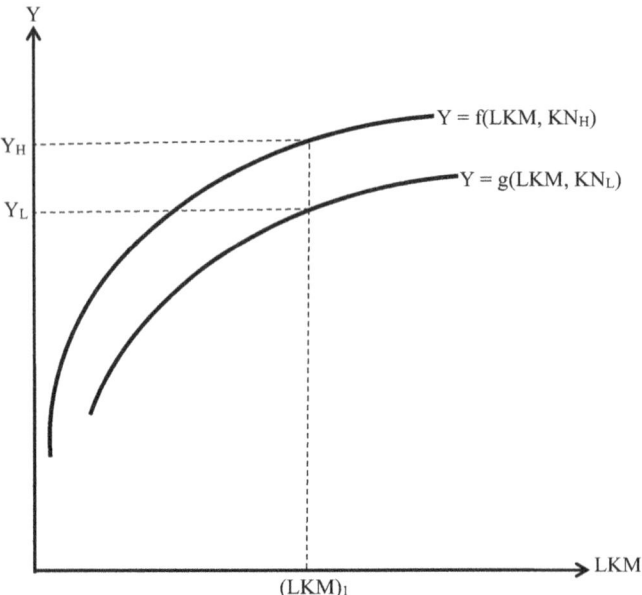

Figure 16.4 Illustration of the Dose-Response Method

Methods Based on Opportunity Costs (OC)

The methods reported in this category equate the value of KN to the value of income benefits that have to be forgone. As indicated below, while this treatment is explicit in some methods, it is implicit in others. When these methods are used for the estimation of benefits, the acknowledged premise is that the estimates so derived represent a minimum value. That is, the benefits are *at least equal to* the opportunity costs. Generally, this method has been associated with valuing the preservation of KN, which would otherwise have been used for generating income from some form of commercial activity.

Direct Estimation of OCs

The OC of preserving KN at a specific location is equal to the *highest* net income that has to be forgone from an alternative income-generating activity that could be undertaken at that location. Therefore, to calculate the true OC of preservation, one needs to identify all feasible alternative activities and estimate the net income of each activity. In practice, such a procedure is not feasible, and the estimation of OC is invariably limited to a select few activities. Further, the use of OCs as a basis for value is always easy when a decision has been made. For example, when the Australian government decided to preserve the Kakadu National Park and prohibit any form of mining activity in 1993, the OC value of preservation was conveniently narrowed to the value of mining income that had to be forgone.

The concept of OC has also been central to the discussion of KN irreversibility; for example, see Krutilla and Cicchetti (1972) and Fisher and Krutilla (1985). The argument is that, should a KN endowment be unique, then it is irreplaceable and so the OC of exploiting it approaches infinity. Further, when a development activity that is intended for a given natural area is subject to technological change, then the OC of preservation is most likely to diminish over time. For example, should there be cost-effective advances in using solar energy, then the OC of preserving natural areas with energy resource deposits such as coal and uranium, instead of mining them, could become very low. Hence, following Krutilla and Cicchetti (1972), a basic definition for the OC value of preservation is:

$$(B_d - C_d) - (B_a - C_a) \tag{16.1}$$

where

B_d and C_d represent respectively the benefits and costs of energy from mining
B_a and C_a represent the benefits costs of energy from an alternative technology or method.

Should B_a and B_d be equal, then, the OC value of preservation is given by $(C_a - C_d)$. The argument, therefore, is that, when several alternative methods of income generation are explored, the cost of preserving the natural area may not turn out to be far too expensive.

Box 16.1   The OC of Preserving Old-Growth Forests

As an example of the OC principle for environmental decision-making, consider the controversy that emerged in 1987 in Australia. Harris-Diashowa, a woodchip exporter, had applied for a renewal of its logging and export licence for a period of 17 years. The main issue here was the logging of old-growth forests spanning an area of nearly 60,000 hectares around the coastal township Eden in South Eastern New South Wales. These forests supposedly harboured some unique species. Hence, the decision problem was one of choosing between preserving the forest and logging it.

The argument in favour of logging centred on the claim that the export of woodchips generated an annual income of approximately $40 million for Australia. On the basis of this information, one would conclude that the OC of preservation was $40 million.

However, Harris-Diashowa is a fully foreign-owned enterprise, and therefore the income that Australia would have had to sacrifice in the context of preservation was not export revenue, but the tax on export revenue.

> For convenience, if we assume a tax rate of 30 per cent, this would have amounted to $12 million; that is, the OC preservation in terms of tax revenue forgone would have been $12 million per year.
>
> Consider now another option: that of getting raw material for the chip mill from a more distant location, say from some plantation forests. Due to the higher costs associated with this option, suppose that export revenue reduces to $35 million per year. The tax earnings to Australia, if this option were feasible, would be $10.5 million. In the context of this alternative option for logging, the OC of preserving the old-growth forests reduces to $1.5 million per year. In present value terms, this OC is equal to $15.67 million (assuming a 17-year period and a 7 per cent discount rate).
>
> It may appear that, had these OC values been seriously considered in 1987, the forests around Eden might have been preserved. The then Minister for the Environment, Senator Peter Cook, recommended the renewal of the logging lease on the recommendations of a committee headed by Professor Henry Nix of the Australian National University. This committee's primary focus was the search unique species and it concurred that the forest area did not carry any unique species.

Consider the example in Box 16.1. The OC of preserving the old-growth forests spanning over several thousand hectares reduces to $1.5 Million. Even then in the 1980s, homes in Sydney's exclusive suburbs fetched market prices above $1 Million. Hence the question posed here is could we sacrifice $1.5 Million to save nearly 60,000 hectares of old-growth forests, which were perhaps older than 200 years. Hence, in instances such as the forests in Eden, the OC principle can guide decision-making by recourse to the notion of 'acceptable sacrifice'. That is, the pertinent question, way back in 1989, should have been: 'Can we give up $1.5 Million to preserve 60,000 hectares of old growth forests?'

Replacement Costs

This method is based on the premise that the value of KN that is committed to development should be at least equal to the value of restoring a site to its near original state from the altered state due to economic activity such as mining or housing development. The method is reasonably relevant if the site that is proposed for economic activity is either a forest or a wetland. Then the value of KN can be equated to the cost of either reforestation or regeneration – assuming, of course, that the damages done previously by the economic activity are reversible, which is rarely, if ever, the case. Similarly, in the context of an infestation by algal blooms, the value of KN would equal the cost of removing the blooms and treating the water supplies to restore quality levels that make it suitable for consumption. The method has been applied in the context of reforestation by

Water Resources Engineers (1970). When replacement includes the rehabilitation of wildlife, the costs tend to be prohibitively high. No doubt the cost of replacing unique environments will be infinity.

Consider an example where the value of KN was estimated for natural areas bearing coal deposits that were to be strip-mined (Thampapillai 1988). The value of the KN at any given time was equated to the sum of two items:

- The OC of preservation; that is, the value of the resource left without being mined
- Cost of restoring the mined-out area to its original state

As indicated in the method, the cost of restoration becomes a determinant of the extent of irreversibility. To illustrate this procedure, suppose that H hectares of land bearing a uniformly distributed mineral deposit of size Q, is considered for clearing and mining. Let V_i denote the value of KN after h_i hectares of land have been mined. Following a definition offered by the Water Resources Engineers (1970),

$$V_i = (C_i + B_i) = (C_i) + P(Q_i) \tag{16.2}$$

where

C_i = Cost of restoring h_i hectares to their near original state
B_i = Income forgone by preventing the remaining $(H-h_i)$ hectares from being mined
P = Profit per unit (tonne) of the mineral
Q_i = quantity of mineral remaining in $(H-h_i)$ hectares.

If the mineral is uniformly distributed across the H hectares of land and the mining firm is a price-taker, the relationship between B_i and the area mined (h hectares) will be linear decreasing function. Note that, if (h = 0), then the entire mineral deposit remains intact and (B = PQ). The relationship between C_i and h will depend on the degree of irreversibility of changes to the environment. At this point a qualification is in order. Mining is in essence, always irreversible because there is typically no refilling of mines with the original ore. The aspects of mining that may be reversible are the impacts on ecosystems from tailings and other wastes. It is this type of reversibility that we consider here; for example, reforestation by filling the mined-out area with bulk that remains after the ore has been separated. Should the rate of increase in C_i be higher than the rate of decrease in B_i, as h increases, then KN is irreversible relative to the opposite situation, as illustrated in Figure 16.5. As indicated in Thampapillai (1988), in some locations, the bulk that remains after the separation of the ore does expand and thereby remove the expenses of searching suitable fill materials including topsoil. For such locations, the rate increase in C_i could be less than the rate of decrease of B_i. Valuation methodology of this type could facilitate choices between preservation and economic activity. In other words, as an initial step, miners can decide to mine those sites where their actions are relatively reversible.

284 Valuation

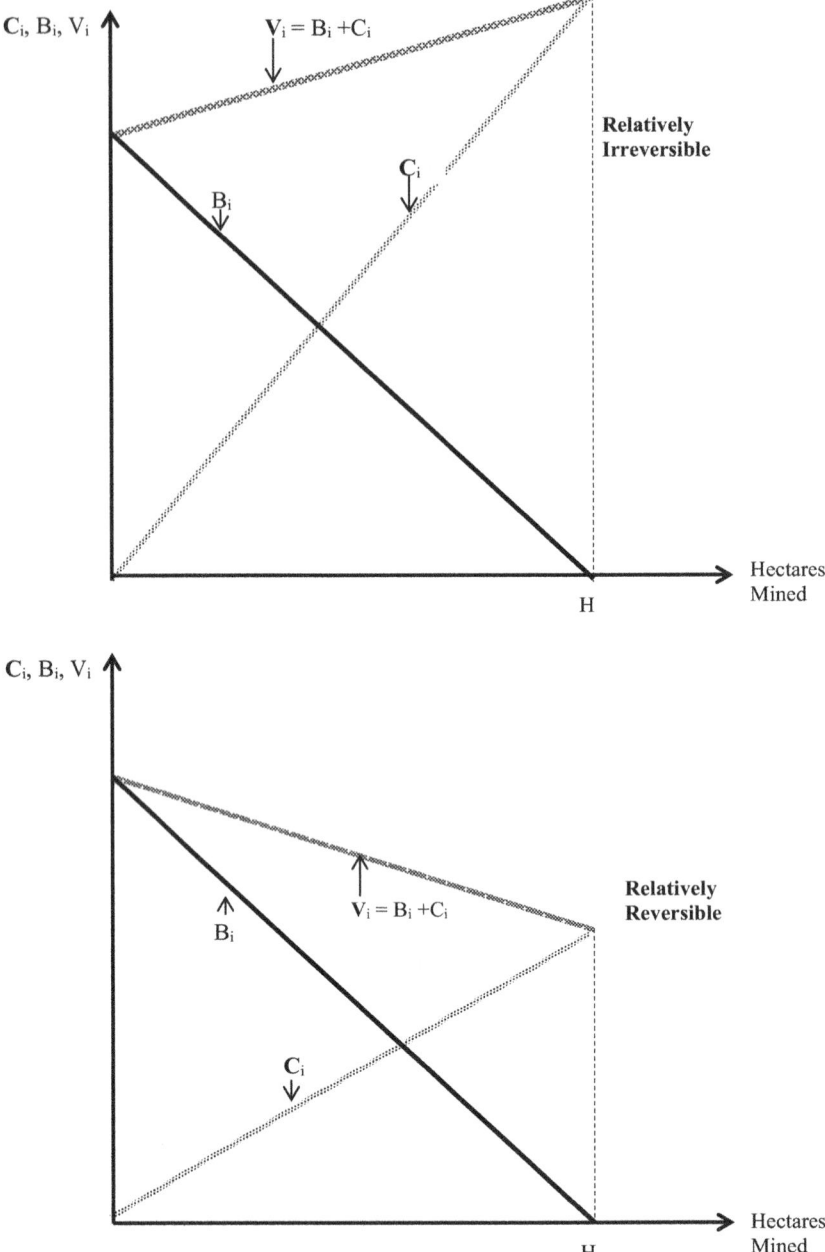

Figure 16.5 The Value of KN in Varying Degrees of Irreversibility

With reference to mining, there is yet another observation that policy makers need to come to terms with. That is, the popular belief is that a mineral is worthless, as long as it remains beneath the ground. Hence the economic rationale would be extract the mineral and generate revenue. However, from a geophysical point of view, leaving the mineral intact beneath the ground without being mined out could enhance earth surface stability. If this view is upheld, the minimum value of contribution to earth surface stability would be (B = PQ).

Cost Savings

In this method, the value of KN is equated to the value of costs saved when the KN is left undamaged in its original state. Greig and Devonshire (1981) provide an illustrative demonstration of the method. Here, the value of KN is approximated to the value of tree cover with reference to the impact of planting trees on salinity of soils. The method proceeds in two steps, as follows. In the first step, a functional relationship between tree cover and salinity is derived by recording salinity levels of soil samples taken from sites with different magnitudes of tree cover. This relationship is illustrated on the right-hand side of Figure 16.6. That is, the magnitude of salinity increases as the size of tree cover decreases.

In the second step, the study area is surveyed to find the functional relationship between household (and farm) costs and salinity. As expected, these costs increase as salinity increases. This is because, for example, increases in salinity prompt more frequent replacement of appliances and the added expenses of soil treatment

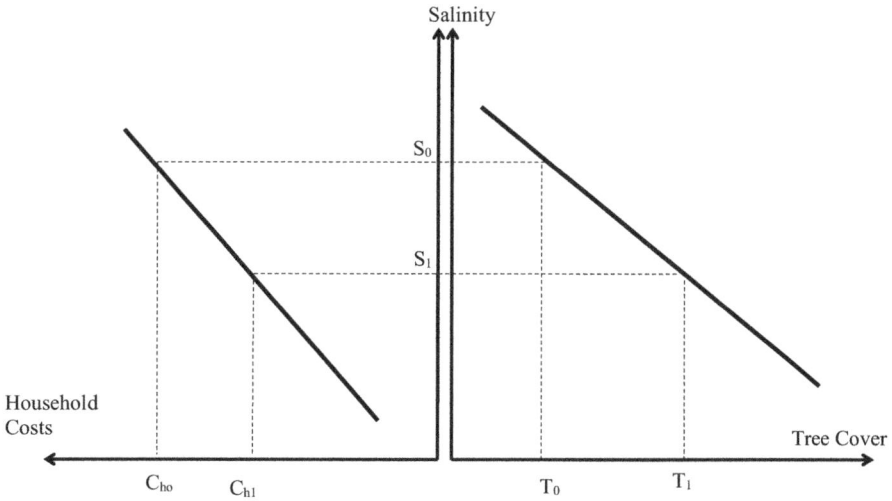

Figure 16.6 An Illustration of the Cost-Saving Method

on farms. The relationship between household costs and salinity is shown on the left-hand side of Figure 16.6. For illustrative purposes, both relationships have been drawn as straight lines. These two relationships can now be used to estimate the value of tree cover in terms of costs that can be saved. For example, consider the reduction of tree cover from T_1 to T_0 in Figure 16.6. This results in an increase in salinity from S_1 to S_0, which in turn creates an increase in household expenditure from C_{h1} to C_{h0}. Hence, the value of tree cover (and, by assumption, the value of KN) is approximated to the distance between C_{h1} and C_{h0}.

The concept of cost savings has been used in several contexts. For example, the benefits of improving KN by reducing pollution have been estimated in terms of savings in health and medical expenses. The benefits of improved public transport facilities can be estimated in terms of savings in road maintenance costs, as commuters may substitute public transport for private transport. Similarly, the benefits of flood mitigation can be estimated in terms of expenditure savings from the reduced number of emergency operations and charity payments.

Threshold Values

Krutilla and Cicchetti (1972) introduced this concept for the valuation of preserving unique and irreversible KN. In this context, the threshold value has three properties:

1 It is the minimum value of the benefit from the preservation of KN in the initial year.
2 This initial year benefit is expected to grow at some specific rate over time.
3 The present value of the benefits of preservation is at least equal to the highest net present value (NPV) of the alternative to preservation.

That is, the threshold value (TV) is the minimum value for the initial year's benefit in the temporal stream of preservation benefits that would render preservation just as desirable as its alternative.

If the TV can be estimated, then its relevance has to be subjectively appraised by recourse to the principle of 'acceptable sacrifice' that was mentioned earlier. The estimation of TV requires the following information:

- Complete data on the alternatives to preservation
- The growth rate for the benefit of preserving KN.

The latter may be determined by scientific inquiry or the use of pertinent data; for example, the rate at which demand for recreation is expected to grow or the rate at which biodiversity is expected to grow. The former is usually found in business proposals that target natural areas for economic activities such as the construction of infrastructure and buildings.

Valuation of KN: Microeconomic Basis 287

The definition of the TV is embedded in the simple calculus of cost-benefit analysis and can be explained as follows:

- Consider two alternatives at a specific location, such as commercial development and preserving the KN at the location. Suppose that these two alternatives are mutually exclusive. That is, the preservation of KN and commercial development cannot coexist.
- Let the net present value (NPV) from commercial development be $\$B_d$ and the value of preservation in the initial year be $\$b_{0p}$. Further, suppose that the preservation benefits grow at r% and that the discount rate is i%.
- The NPV of preserving KN can be defined as:

$$b_{0P} + \left\{\frac{b_{0P}(1+r)}{(1+i)}\right\} + \left\{\frac{b_{0P}(1+r)^2}{(1+i)^2}\right\} + \ldots + \left\{\frac{b_{0P}(1+r)^T}{(1+i)^T}\right\} \quad (16.3)$$

(16.3) can be re-written as:

$$b_{0P}\left[1 + \left\{\frac{(1+r)}{(1+i)}\right\} + \left\{\frac{(1+r)^2}{(1+i)^2}\right\} + \ldots + \left\{\frac{(1+r)^T}{(1+i)^T}\right\}\right] \quad (16.4)$$

If NPV of preservation exceeds the NPV of commercial development, then:

$$b_{0P}\left[1 + \left\{\frac{(1+r)}{(1+i)}\right\} + \left\{\frac{(1+r)^2}{(1+i)^2}\right\} + \ldots + \left\{\frac{(1+r)^T}{(1+i)^T}\right\}\right] \geq B_d \quad (16.5)$$

Note that the term within square brackets on the left-hand side of Equations (16.4) and (16.5) can be defined as follows:

$$\left[1 + \left\{\frac{(1+r)}{(1+i)}\right\} + \left\{\frac{(1+r)^2}{(1+i)^2}\right\} + \ldots + \left\{\frac{(1+r)^T}{(1+i)^T}\right\}\right] = \binom{\text{The PV of \$1 growing}}{\text{at r\% discounted at i\%}}$$

Hence Equation (16.5) for recommending the preservation decision can be rewritten as:

$$b_{0P}\left[\text{The PV of \$1 growing at r\% discounted at i\%}\right] \geq B_d \quad (16.6)$$

It is now possible to define the amount that the initial year's KN preservation benefit must at least equal to, so that the preservation of KN just as desirable as development. That is:

$$b_{0P} \geq \frac{B_d}{\left[\text{The PV of \$1 growing at r\% discounted at i\%}\right]} \quad (16.7)$$

288 *Valuation*

The right-hand side of Equation (16.7) is the TV because, if b_{0p} exceeds the value estimated from the right-hand side, then the preservation of KN would be the preferred option. To reiterate, the denominator of the right-hand side in the last statement above is the present value of $1 growing at the same rate as the benefits of KN preservation. Therefore, a formal definition of the threshold value for the preservation of KN (in the context of such preservation being mutually exclusive to a well-defined income-generating activity) is as follows:

$$TV = \left[\frac{\text{NPV of Income Generating Activity}}{\text{PV of \$1 growing at the same rate as the benefits of Preserving KN}}\right]$$

The use of TVs inevitably requires a subjective assessment of what is 'large' or 'small'. Nevertheless, in some instances, such subjectivity can be guided by readily observable benchmarks such as house prices and average incomes. Should the TV turn out to be trivially small, then, decision-making becomes simplified. That is, a small value of preservation benefit can permit the NPV of preservation to exceed that of economic activity. Besides, should the income-generating activity have alternatives of its own – for example, a windmill as an alternative to a coal power station – then the TV can be even smaller. Let us return to the example in Box 16.1, dealing with wood chipping in Eden in Australia. It was indicated that the present value of the OC of preservation could have been $15.67 million, when an alternative logging option were available and we had assumed a discount rate of 7 per cent. Assuming this same discount rate, the present value of $1 growing at a very small rate per year, say 0.0005 per cent, over a 17-year period is $10.447. In this context, the threshold value of preservation is:

$$(\$15.67 \text{ million})/10.447 = \$1.5 \text{ million}$$

That is, had it been possible to demonstrate that the benefits of preservation in 1987 could have been more than $1.5 million, then preservation would have taken precedence over forest logging. Even way back in 1987, the prices of single dwellings in exclusive suburbs of suburbs of Sydney did approach $1 million. Hence, justifying the initial value of 600,000 hectares of old growth forest as $1.5 million would not have been difficult.

Box 16.2 deals with yet another Australian example. The State Government of New South Wales (NSW) had by 2010 issued mining exploration licences to two mining giants BHP-Billiton and Shenhua Watermark at the price of nearly $500 million per licence. In order to strengthen their claims to mine the coal, the two firms had been actively buying up farm properties at prices ranging between $8,000 and $10,000 per hectare.

But, as argued in Box 16.2, preservation of the ground water resource system and mining are mutually exclusive. An economic analysis of the preservation of the groundwater system reported in Thampapillai (2011) reveals the threshold value of

Box 16.2 The OC Preserving Sensitive Environmental resources in Mining – Liverpool Plains in New South Wales, Australia

The Liverpool Plains is perhaps one of the most fertile agricultural land areas in Australia and generates annual export revenues in excess of $330 million. Located some 400 kilometres North-West of Sydney, this area also carries very large deposits of black coal. The size of the deposit is so large that two mining giants – BHP-Billiton and Shenhua Watermark – had plans for extracting some 10 million tonnes of coal per year over a period 20 years. Both mining firms have significant foreign ownership. Shenhua Watermark is 70 per cent owned by Shenhua, which is a Chinese government subsidiary. The issue is not one of foreign ownership but one of ecological sensitivity.

The subterranean ecosystem in the Liverpool Plains includes Australia's second largest and perhaps most pristine groundwater systems. According to hydro-geologists Acworth and Timms (2003), this system is a mixture of aquifers and aquitards. The former consist of porous matter, such as sand and gravel through which water flows, while the latter consist of hardened materials (clays and rock) that confine the water.

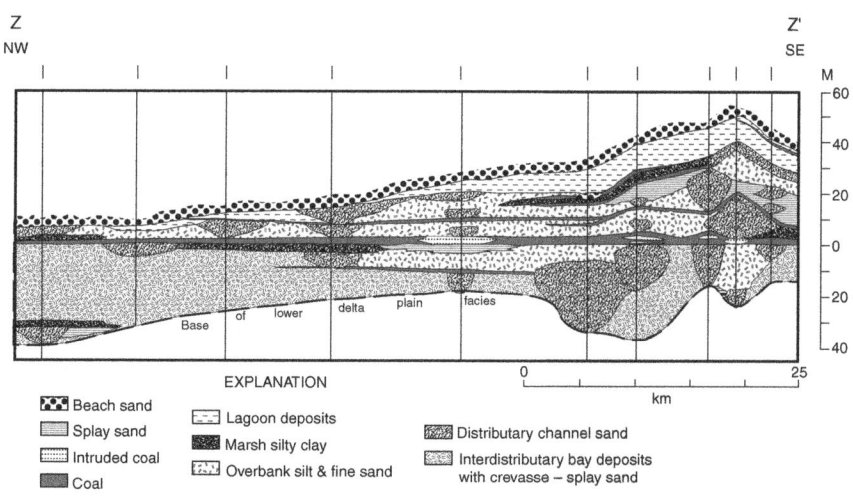

Figure 16.7 A Specific Strike Section Through Lower Delta-Plain Facies in the Liverpool Plains

Reproduced from Hamilton (1985, p. 55).

> A specific geological section taken by Hamilton (1985) is illustrated in Figure 16.7. It is clearly evident that the coal seams of varying thickness are dispersed amid the aquifers and the aquitards. Therefore, mining and the preservation of the groundwater system are mutually exclusive. Agriculture, which has been practised on the Plains since the 1800s, seems not to have affected the groundwater system.
>
> By 2021, both BHP-Billiton and Shenua-Watermark sold back their mining leases to the government of NSW. As explained by Lodhi and Thampapillai (2024), these mining ventures have been a costly exercise, notwithstanding the community disruptions they caused. Had mining eventuated the damage to the groundwater resource system would have been immeasurable. Fortunately, the NSW government managed to prevent the damage.

preservation to range between $4,000 and $5,500 per hectare. The purchase prices paid by the mining firms ($8,000–$10,000) can be regarded as indicators of the capitalisation of the net present value of mining incomes. However, such purchases suggest that the benefits of preservation outweigh those of mining in terms of the threshold value principle. This is because the smaller land values ($4,000–$5,500) can lead to a stream of benefits that is equivalent to the present value of mining incomes.

Concluding Remarks

The review of the various methods of valuation presented above reveals that almost all methods contain varying degrees of deficiencies. For example, CVM can theoretically deal with all attributes of KN. Nevertheless, the wide variations in expressed values and the inconsistencies between hypothetical transactions and economic commitments, or inconsistencies that are observed between WTP and WTA render the method inefficient. Similarly, CM and game theory methods, apart from being time-consuming, are also limited by the hypothetical nature of the transactions. As noted earlier, the travel-cost method does not adequately deal with all features of KN and cannot be applied unless the specific KN endowment offers recreational facilities. The hedonic price method, too, is limited by spatial restrictions and the influences of other socio-economic variables. The success of methods involving production functions and dose-responses will depend on the availability of data on KN, at least on a cross-sectional basis. At the other end of the spectrum, the methods based on opportunity costs will always tend to provide underestimates of the value of KN.

There is an emerging tendency among decision-makers to merely seek a number for the value of environmental outcomes in order to portray the image of having demonstrated environmental awareness. Further, there is also the danger of controversies due to conflicting values presented by different interest groups, especially in the context of sensitive policy decisions; for example, the determination of compensation payments by Exxon-Valdez for the Alaskan oil spill. Therefore, there is a clear need to move towards less contentious methods of valuation. The use of threshold values in conjunction with other methods may reduce the extent of these difficulties.

Review Questions

1 An environmental economist collected the following data from seven samples of respondents in applying the binomial approach to contingent valuation in order to estimate the monetary value of an environmental endowment. The size of each sample – thirty respondents – was the same.

Sample Number	WTP Value	Number of Persons Willing to Pay
1	$1	30
2	$5	27
3	$10	24
4	$15	25
5	$20	15
6	$25	9
7	$30	3

Illustrate how the above data could be used in estimating the value of the endowment. Also indicate whether the binomial approach to contingent valuation could overcome the problems of 'anchoring' and 'embedding'.

2 A popular beach area is also the site for a sewage treatment plant's (STP) ocean outfall. A decline in the amenity value of the beach area has been attributed to the age and outdated technology of the STP. Suppose that the upgrading of the STP and its infrastructure will result in the appreciation of property values in the vicinity of the beach area. Illustrate how the hedonic pricing method could be used to value the environmental benefits of upgrading the STP.

3 Comment on the following statement: 'As a valuation methodology, contingent valuation presents more issues than solutions'.

4 Provide three examples where the OC and TV principles could be applied to resolve environmental decision-making problems.

Appendix
The Ramsey Model of Utility Estimation

To illustrate, we choose an application provided in Thampapillai (1985). The person interviewed – whose utility function is to be estimated – is offered two prospects. Each prospect contains a two-element combination yielding four possible outcomes (x_1, x_2, x_3 and x_4) with neutral probabilities, as illustrated in Table 16A.1.

The notations (i)–(iv) refer to positions in the matrix and x_1 to x_4 are values of outcomes. As illustrated by Sinden (1974), positions (i) and (iii) pertain to the preservation of KN, while positions (ii) and (iv) pertain to a priced alternative, namely a specific level of income. In this illustration, KN was quantified as the number of trees preserved.

The neutral probabilities and the assumption of additive linear utilities enable the indifference between Prospect 1 and Prospect 2 to be defined as:

$$0.5\,[u(x_1)] + 0.5\,[u(x_2)] = 0.5\,[u(x_3)] + 0.5\,[u(x_4)] \qquad (16A.1)$$

This is the same as:

$$[u(x_1)] + [u(x_2)] = [u(x_3)] + [u(x_4)] \qquad (16A.2)$$

$$[u(x_2)] - [u(x_4)] = [u(x_3)] - [u(x_1)] \qquad (16A.3)$$

The interval $\{[u(x_3)] - [u(x_4)]\}$ can be arbitrarily assigned a utility value of five utiles (a cardinal measure for utility).

Consider now the first game, where the interviewee was given two prospects as shown in Table 16.A2.

Table 16A.1 The Ramsey Model

Probability of Occurrence	Prospect 1	Prospect 2
0.5	(i) x_1	(iii) x_3
0.5	(ii) x_2	(iv) x_4

Table 16A.2 The Ramsey Model – Game 1

Probability of Occurrence	Prospect 1	Prospect 2
0.5	(i) 0 Trees Preserved	(iii) γ Trees Preserved
0.5	(ii) $3 Million Income	(iv) $0 Income

Table 16A.3 The Ramsey Model – Game 2

Probability of Occurrence	Prospect 1	Prospect 2
0.5	(i) 300 Trees preserved	(iii) γ Trees Preserved
0.5	(ii) $3 million income	(iv) $0 Income

The interviewee was asked to specify a value for γ such that there is indifference between the prospects. The assignment of γ = 300 by the interviewee enabled Equation (16A.3) to be restated as:

$$[u(\$3M)] - [u(\$0)] = [u(300\,\text{Trees})] - [u(0\,\text{Trees})] = 5\,\text{Utiles} \qquad (16A.4)$$

The second game was then played with the indifference value of 300 trees being placed in position (i), as shown in Table 16.A3.

The value of 500 trees for γ in this game results in the following statement:

$$[u(\$3M)] - [u(\$0)] = [u(500\,\text{Trees})] - [u(300\,\text{Trees})] = 5\,\text{Utiles} \qquad (16A.5)$$

These two games reveal that the first 300 trees yielded 5 utiles, while an additional 200 trees also yielded a further 5 utiles. Additional games along this line will enable a utility curve to be constructed. Note that the differential in income level in these games had been fixed at $3 million. Further sets of games by fixing the income differentials at different values ($4 million, $2 million, $1 million and so on) can yield a family of utility curves conceptualised in Figure 16.3A.

References

Acworth, R. I. and Timms, W. A., 'Hydrogeological investigation of mud-mound springs developed over a weathered basalt aquifer on the Liverpool Plains, NSW, Australia', *Hydrogeology Journal*, 11(6): 659–672, 2003.

Bandara, R. and Tisdell, C. A., 'The net benefit of saving the Asian Elephant: A policy and contingent valuation study', *Ecological Economics*, 48(1): 93–107, 2004.

Bennett, J., van Bueren, M. and Whitten, S., 'Estimating society's willingness to pay to maintain viable rural communities', *The Australian Journal of Agricultural and Resource Economics*, 48(3): 487–512, 2004.

Carson, R. T., Mitchell, R. C., Hanemann, W. M., Kopp, R. J., Presser, S. and Ruud, P. A., *A Contingent Valuation Study of Lost Passive Use Values Resulting from the Exxon Valdez Oil Spill: A Report to the Attorney General of the State of Alaska*, November 1992, 3.

Ciriacy-Wantrup, S. von, 'Soil conservation in European farm management', *Journal of Farm Economics*, 20(1): 86–101, 1938.

Clawson, M. and Knetsch, J. L., *Economics of Outdoor Recreation*, Johns Hopkins University Press, Baltimore, MD, 1966.

Cummings, R. G. and Harrison, G. W., 'Was the Ohio Court informed in their assessment of the accuracy of the contingent valuation method?', Paper B-92–07, College of Business Administration, University of South Carolina, Columbia, South Carolina, 1992.

Du Preez, M., Lee, D. E. and Hosking, S. G., 'The recreational value of beaches in the Nelson Mandela Bay area, South Africa', *Journal for Studies in Economics and Econometrics*, 35(3): 85–102, 2011.

Farr, M., Stoeckl, N. and Beg, R. A., 'The efficiency of the environmental management charge in the Cairns management area of the Great Barrier Reef Marine Park', *Australian Journal of Agricultural and Resource Economics*, 55(3): 322–341, 2011.

Fisher, A. C. and Krutilla, J. V., 'Economics of nature preservation', in A. V. Kneese and J. L. Sweeney (eds), *Handbook of Natural Resource and Energy Economics*, North Holland, Amsterdam, 1985.

Frykblom, P., 'Hypothetical question modes and real willingness to pay', *Journal of Environmental Economics and Management*, 34(3): 275–287, 1997.

Greig, P. J. and Devonshire, P. G., 'Tree removals and saline seepage in Victorian catchments: some hydrologic and economic results', Australian Journal of Agricultural Economics, 25(2): 134–148, 1981.

Hamilton, D. S., 'Deltaic depositional systems, coal distribution and quality, and petroleum potential Permian Gunnedah Basin, NSW, Australia', *Sedimentary Geology*, 45(1): 35–75, 1985.

Hanley, N., Mourato, S. and Wright, R. E., Choice modelling approaches: A superior alternative for environmental valuation?' *Journal of Economic Surveys*, 15(3): 435–462, 2001.

Hotelling, H., Letter quoted in *Economics of Public Recreation: An Economic Study of the Monetary Evaluation of Recreation in National Parks Service*, National Parks Service, United States Department of the Interior, 1949.

Jin, J., Indab, A., Nabangchang, O., Thuy, Truong, D., Harder, D. and Subade, R. F., 'Valuing marine turtle conservation: A cross-country study in Asian cities', *Ecological Economics*, 69(10): 2020–2026, 2010.

Knetsch, J. L., 'Outdoor recreation demands and benefits', *Land Economics*, 39: 387–396, 1963.

Knetsch, J. L., 'Economics of including recreation as a purpose of eastern water projects', *Journal of Farm Economics*, 46: 1148–1157, 1964.

Knetsch, J. L., 'Environmental valuation: Some problems of wrong questions and misleading answers', *Environmental Values*, 3(4): 351–368, 1994.

Knetsch, J. L., 'Asymmetric valuation of gains and losses and preference order assumptions', *Economic Inquiry*, 38(1): 138–141, 1995.

Kontoleon, A. and Swanson, T., 'The willingness to pay for property rights for the Giant Panda: Can a charismatic species be an instrument for nature conservation?' *Land Economics*, 79(4): 483–499, 2003.

Krutilla, J. V. and Cicchetti, C. J., 'Evaluating benefits of environmental resources with special application to the Hells Canyon', *Natural Resources Journal*, 12: 1–29, 1972.

Lodhi, I. A. and Thampapillai, D. J., 'A Knetsch proposition and a due process for the evaluation of mining projects in ecologically sensitive areas: An illustration through two Australian case studies', *Singapore Economic Review*, January 2024 (https://doi.org/10.1142/S0217590823450017).

Mendelsohn, R., 'Measuring hazardous waste damages with panel models', *Journal of Environmental Economics and Management*, 22(3): 259–271, 1992.

Morrison, M. and J. Bennett, 'Valuing new south wales rivers for use in benefit transfer', *Australian Journal of Agricultural and Resource Economics*, 48(1): 591–612, 2004.

Neill, R. H., Cummings, R. G., Ganderton, P. T., Harrison, G. W. and McGuckin, T., 'Hypothetical surveys and real commitments', Paper B-93-01, College of Business Administration, The University of South Carolina, Columbia, South Carolina, 1993.

Numes, P. and Nijkamp, P., 'Economic valuation, values and the contingent method: An overview', *Regional Science Inquiry Journal*, 3(1): 95–116, 2011.

Pearce, D. W. and Turner, K., *Environmental and Natural Resource Economics*, Wheatsheaf and Harvester, New York, 1990.

Petrolia, D. R. and Kim, T.-G., 'What are barrier islands worth? Estimates of willingness to pay for restoration', *Marine Resource Economics*, 24(2): 131–146, 2009.

Sinden, J. A., 'The evaluation of extra market benefits: A review', *World Agricultural Economics and Rural Sociology Abstracts*, December 1967.

Sinden, J. A., 'A utility approach to the valuation of recreational and aesthetic experiences', *American Journal of Agricultural Economics*, 56: 61–72, 1974.

Sinden, J. A. and King, D. A., 'Land condition, crop productivity, and the adoption of soil conservation measures', Paper presented to the Australian Agricultural Economics Society Conference, Melbourne, February 1988.

Sinden, J. A. and Worrell, A. C., *Unpriced Values—Decisions without Market Prices*, Wiley, New York, 1979.

Sinden, J. A. and Wyckoff, J. B., 'Indifference mapping—An empirical methodology for economic valuation of the environment', *Regional Science and Urban Economics*, 6: 81–103, 1976.

Sinden, J. A. and Yap, T., 'The opportunity cost of land degradation in New South Wales: A case study', Paper presented to the Australian Agricultural Economics Society Conference, Adelaide, February 1987.

Surendran, A. and Sekar, C., 'An economic analysis of willingness to pay (WTP) for conserving the biodiversity', *International Journal of Social Economics*, 37(7–8): 637–648, 2010.

Szabo, Z., 'Reducing protest responses by deliberative monetary valuation: Improving the validity of biodiversity valuation', *Ecological Economics*, 72(1): 37–44, 2011.

Thampapillai, D. J., 'Trade-offs for conflicting social objectives in the extraction of finite energy resources', *International Journal of Energy Research*, 9: 179–192, 1985.

Thampapillai, D. J., 'The value of natural environments in the extraction of finite energy resources: A method of valuation', *International Journal of Energy Resources*, 12: 527–538, 1988.

Thampapillai, D. J., 'Value of sensitive in-situ environmental assets in energy resource extraction', *Energy Policy*, 39(12): 7695–7701, October 2011.

Thampapillai, D. J., Maleka, P. T. and Milimo, J., Quantification of the trade-offs between environment, employment, income and food security, *Working Paper No. 229*, International Labour Office, Geneva, 1992.

Tisdell, C. A., *Economics of Environmental Conservation*, Elsevier, Amsterdam, 1991.

Van Bueren, M. and Bennett, J., 'Towards the development of a transferable set of value estimates for environmental attributes', *The Australian Journal of Agricultural and Resource Economics*, 48(1): 1–32, 2004.

Van Vuuren, W., 'Soil erosion: The case of market intervention', *Canadian Journal of Agricultural Economics*, 33: 41–62, 1986.

Walker, D. J., 'A damage function to evaluate erosion control economics', *American Journal of Agricultural Economics*, 64: 690–698, 1982.

Walker, D. J. and Young, D. L., 'The effects of technical progress on erosion damage and economic incentives for soil conservation', *Land Economics*, 62: 83–93, 1986.

Wallace, N. W. (ed.), *Natural Resource Management: An Economic Perspective*, Australian Bureau of Agricultural and Resource Economics, AGPS, Canberra, 1992.

Water Resources Engineers Inc., *Wild River Method of Evaluation*, United States Department of the Interior, Office of Water Resources, Washington DC, 1970.

Whittington, D., Okorafor, A. and Okore, A., 'Strategy for cost recovery in the rural water sector: A case study of Nuskka District, Anambra State, Nigeria', *Water Resources Research*, 26(9): 1899–1913, 1990.

17 Valuation of KN
Macroeconomic Basis – I

The previous chapter dealt with the valuation of specific environmental capital (KN) endowments and their services at the microeconomic level. In this chapter and the next, we consider the valuation of KN at the macroeconomic (aggregate) level. This chapter will focus on the utilisation of KN, whilst the next chapter deals with stock estimates.

The three types of values discussed in Chapter 16, namely *existence*, *option* and *use values*, are relevant in the macroeconomic context as well. The estimation of existence and option values relies primarily on the application of the contingent valuation method (CVM). Costanza et al. (1997) illustrate such valuation by employing mainly CVM, and estimate the average annual value of global ecosystems to be around US$33 trillion (constant 1994 values). In a subsequent analysis involving CVM, Costanza et al. (2014) illustrate that the value of global ecosystem services in 2011 had significantly appreciated to $125–$145 trillion per year (2007 USD). The procedure for estimating existence and/or option values of KN endowments at the national (aggregate) level using CVM would essentially be the same as that described in Chapter 16. That is, the application would require carefully prepared narratives of the endowments; questionnaires on willingness to pay/accept; and the choice of samples that are representative of society as a whole. As indicated in Chapter 16, the use of CVM is not without controversy. Therefore, we focus on *use value* at the macroeconomic level. As illustrated below, the estimation of use value relies on the principles of environmental accounting, considered in Chapter 11, and the principles of the conceptual framework introduced in Chapter 13.

Use Value of KN in Macroeconomics

The determination of use value for KN in macroeconomics can be approached in two steps, as the estimation of the following two items:

1 D_{KN}, namely the depreciation of KN, which is the size of KN utilised as illustrated in Chapters 13 and 14.
2 The share of national income (Y) that accrues to KN and thereby the price of KN.

DOI: 10.4324/9781003408574-21

Valuation of KN: Macroeconomic Basis – I

Recall from Chapters 11–15 the distinction we made with reference to the equilibrium between national income and expenditure on the basis of D_{KN}. That is, a more sustainable equilibrium ($Y \equiv GVA - D_{KN}$), compared to one that is unsustainable ($Y \equiv GVA$). The rationale here is that, if D_{KN} is not recognised, then we mistakenly think that a higher amount of Y is available and will hence expedite expenditures that are unsustainable. In the analyses reported in Chapters 13–14, it is evident that the size of KN utilised depends on the size of D_{KN} and relatedly the ratio of D_{KN} to GVA, namely η and [η = (D_{KN})/GVA]. For example, consider the definition for the size of KN utilised provided in Equation (13.14) in Chapter 13. For convenience, this equation is reproduced here:

$$KN = KM\left[\left(\frac{1}{\eta(1-\eta)^{(1/\theta)}}\right) - \frac{1}{\eta}\right] \tag{17.1}$$

The following observations become apparent from this definition. If (η → 0), then (KN → 0); and if (η → 1), then (KN → ∞).

In Chapters 13 and 14, we conceptualised, purely for illustrative purposes, KN as consisting of only the airshed. As a result, D_{KN} was confined to the cost of air pollution abatement. Had KN been regarded in terms of all environmental sinks, the estimates of D_{KN} would have been much higher and hence the estimates for the size of KN utilised would also have been correspondingly higher. As illustrated below, the size of KN utilised affects the share of Y that accrues to KN and as a result affects the value (price) of KN.

In order to define the share of Y that accrues to KN, consider the definition of factor-utilisation provided in Chapter 13, Equation (13.13). For convenience, this definition is also reproduced here:

$$Y = (1-\eta)\alpha[KM + \eta KN]^{\theta} L^{\lambda} \tag{17.2}$$

Recall that KM and KN are measured on the same scale, such that [(KM + ηKN_t) = K] and η accounts for the resilience of KN. In standard macroeconomic analyses, the share of Y accruing to a factor is determined in terms of the marginal product of that factor. The premise is that the reward for a unit of the factor would be its marginal product. Factor income would then be the result of multiplying factor quantity utilised by its marginal product. The division of factor income by Y would then reveals the factor's share of income.

The marginal product of KN and KM from Equation (17.2) would be:

$$\frac{\partial Y}{\partial KN} = \left(\frac{\partial Y}{\partial K}\right)\left(\frac{\partial K}{\partial KN}\right) = \frac{\left(\theta\eta(1-\eta)\alpha L^{\lambda}\right)}{\left((KM + \eta KN)^{(1-\theta)}\right)} \tag{17.3a}$$

298 *Valuation*

$$\frac{\partial Y}{\partial KM} = \left(\frac{\partial Y}{\partial K}\right)\left(\frac{\partial K}{\partial KM}\right) = \frac{\left(\theta(1-\eta)\alpha L^\lambda\right)}{\left((KM+\eta KN)^{(1-\theta)}\right)} \quad (17.3b)$$

Note that, given the nature of the factor-utilisation function in Equation (17.2) the marginal product of KN will also be η times that of KM. That is:

$$\frac{\partial Y}{\partial KN} = \eta\left(\frac{\partial Y}{\partial KM}\right) \quad (17.4)$$

The amount of income that is allocated respectively to KN, KM and K, namely Y_{KN}, Y_{KM} and Y_K, can be defined as:

$$Y_{KN} = \left(\frac{\partial Y}{\partial KN}\right) KN \quad (17.5)$$

$$Y_{KM} = \left(\frac{\partial Y}{\partial KM}\right) KM \quad (17.6)$$

$$Y_K = \left(\frac{\partial Y}{\partial K}\right) K \quad (17.7)$$

The division of Equation (17.5) by Equation (17.2) would reveal the factor share of Y that accrues to KN. Similarly, the division of Equation (17.6) by Equation (17.2) and Equation (17.7) by Equation (17.2) would reveal respectively the factor shares of Y that accrue to KM and K. Note that the share of Y that accrues collectively to KM and KN, namely K is θ. And as explained in Chapter 13, point estimates of θ are derived as the ratio (GOS/GVA) from the national accounts. Denoting the share of Y accruing to KN and KM respectively as (θ_{KN}) and (θ_{KM}), it can be shown that:

$$\theta_{KN} = \theta\left[\frac{\eta KN}{(KM+\eta KN)}\right] \quad (17.8)$$

$$\theta_{KM} = \theta\left[\frac{KM}{(KM+\eta KN)}\right] \quad (17.9)$$

The use value of KN can now be defined as a unit price (P_{KN}) by simply dividing the factor income accruing to KN by the utilised quantity of KN:

$$P_{KN} = \left[\frac{(\theta_{KN})Y}{\eta KN}\right] \qquad (17.10)$$

If one were to now define the unit price of KM, namely P_{KMt}, along the same lines of Equation (17.10), it would become apparent that:

$$P_{KM} = \left[\frac{(\theta_{KM})Y}{KM}\right] \qquad (17.11)$$

The definitions given above are the result of the specific nature of the functional form assigned for factor-utilisation. The assignment of this functional form in turn was based on the premises advanced in Chapters 8 and 13, namely that KM and KN constitute a composite item of capital K and, further, that the contribution to Y stems from not just KM alone, but KN as well.

To summarise, the use value of KN in macroeconomics can be expressed at two levels. The first is the amount of national income that is distributed towards KN for its utilisation. The second is the unit price as shown in Equations (17.10) and (17.11). These two value measures enable the appreciation of the role played by KN in economic performance. However, as discussed in the next section, the time trends of these value measures do not necessarily address the issue of KN scarcity.

Illustration of the Use Values of KN and the Scarcity of KN

The values of KN as discussed above, namely Y_{KN} and P_{KN}, are displayed for Australia in Table 17.1 for the period 2011–2020. This table also displays the additional information that is pertinent to the estimation of these values. The time trends of these values for the period 1990–2020 are presented in Figures 17.1 and 17.2. It is important to note that the estimation of KN utilisation is based on D_{KN} being confined to the costs of abating GHG emissions. Hence, as in the illustrations made in the previous chapters, KN has been narrowed to the air shed.

It is evident from Figure 17.1 that the value of KN in terms of factor income (Equation 17.5) has been increasing over time. However, the value of KN at unit cost level (Equation 17.10), displays a modest increase with tendency to fall in the later years of the dataset. Unit price of KM (Equation 17.11) displays a slight, yet clear, downward trend. These are shown in Figure 17.2.

As discussed below, prices can serve as indicators of scarcity. However, interpreting the observed trends of P_{KN} may not be so straightforward. Therefore, we will consider first the subject of scarcity with reference to KN. The discussion of scarcity often entails the consideration of the efficiency of resource utilisation.

Table 17.1 The Valuation of KN in Australia

	2011	2012	2013	2014	2015	2016	2017	2018	2019	2020
KN Depletion	1.17E+13	1.20E+13	1.24E+13	1.29E+13	1.36E+13	1.41E+13	1.48E+13	1.54E+13	1.61E+13	1.66E+13
KM Stock (Accumulation)	6.60E+12	6.80E+12	7.03E+12	7.27E+12	7.49E+12	7.70E+12	7.90E+12	8.09E+12	8.29E+12	8.46E+12
θ_{KN}	5.37E-02	5.29E-02	4.60E-02	4.70E-02	4.73E-02	4.55E-02	4.93E-02	4.88E-02	4.87E-02	4.97E-02
θ_{KM}	5.56E-01	5.58E-01	5.60E-01	5.54E-01	5.43E-01	5.37E-01	5.23E-01	5.15E-01	5.05E-01	4.99E-01
P_{KN}	1.10E-02	1.09E-02	9.55E-03	9.63E-03	9.57E-03	9.20E-03	9.87E-03	9.83E-03	9.79E-03	9.75E-03
P_{KM}	1.24E-01	1.26E-01	1.26E-01	1.23E-01	1.20E-01	1.18E-01	1.15E-01	1.13E-01	1.11E-01	1.08E-01
Y_{KN}	7.93E+10	8.15E+10	7.27E+10	7.59E+10	7.79E+10	7.68E+10	8.53E+10	8.70E+10	8.89E+10	9.07E+10
KN/KM	1.77E+00	1.76E+00	1.76E+00	1.78E+00	1.81E+00	1.83E+00	1.88E+00	1.91E+00	1.94E+00	1.96E+00
KN/Y	1.47E+02	1.47E+02	1.70E+02	1.70E+02	1.74E+02	1.84E+02	1.74E+02	1.77E+02	1.81E+02	1.83E+02

Notes: All Monetary Estimates in Constant 2019 AUD.

Valuation of KN: Macroeconomic Basis – I   301

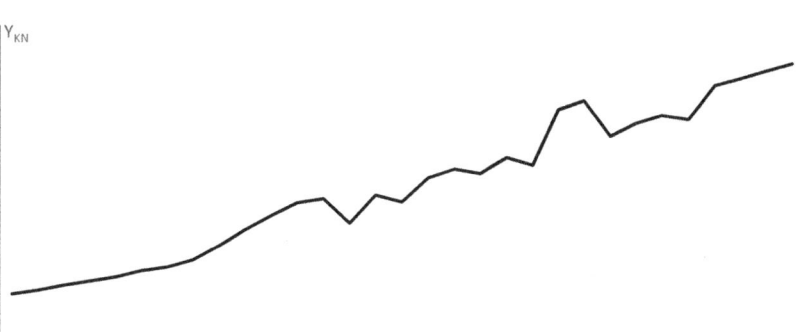

Figure 17.1 Time Trend of Y_{KN} in Australia (1990–2020)

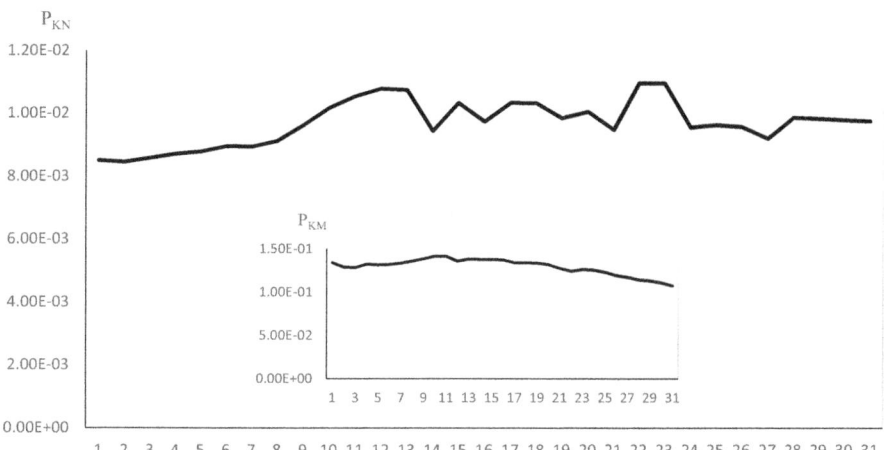

Figure 17.2 Time Trends of P_{KN} and P_{KM} in Australia (1990–2020)

Valuation

In other words, the general argument is that improved efficiency of resource use can offset resource scarcity. For this reason, we will consider the efficiency of KN utilisation as well.

The Relative Scarcity of KN

The scarcity debate of the 1970s (Forrester 1971; Meadows et al. 1972) invoked the analysis of natural resource prices. The main hypothesis in these analyses was the premise that, if natural resources were becoming scarce (due to progressively higher rates of extraction and depletion), then their relative prices must rise. Most analyses had rejected the scarcity hypothesis on the grounds of falling natural resource prices (Nordhaus 1973; World Bank 1992). For example, recall from Chapter 1 the statement by the World Bank (1992, p. 37) with reference to non-ferrous metal prices between the period 1900 and 1991: 'Declining price trends also indicate that many non-renewables have become more, rather than less, abundant'. The observed falls in price are attributed to technological improvements in resource utilisation (Samuelson and Nordhaus 1990; Mankiw 2004). Such improvements can reduce the demand for certain resources because production can continue with less of the resource. Alternatively, efficient methods of extraction can also reduce the costs of resource use.

If one were to directly interpret the observations displayed in Figure 17.2 using the above rationale, then one would conclude that the scarcity of KN has a clear tendency to fall in Australia. This, however, is not the case, for the following reason. The downward trend of P_{KN} during the latter segment of the period 1990–2020 also coincides with increased KN utilisation. As a result, the denominator of Equation (17.10) has become progressively larger over the years. The utilisation of larger quantities does not imply that KN has become abundant – as implied by the World Bank's (1992) statement cited above. Hence a true assessment of KN scarcity must be based on the value of KN per unit of KN stock that *remains* subsequent to utilisation. Recall the discussion in Chapter 8 of the distinction between the accumulation and the utilisation of KN. We provide a simple methodology for estimating the size of KN stock that remains in the next chapter.

The size of KN estimated in the analyses reported in this text so far refers to quantities that are utilised from an undefined stock. Accordingly, the size of KN that remains is necessarily undefined. To illustrate the relevance of this argument to the subject of scarcity, suppose that KN_{At} represents an extremely large number that has been arbitrarily assigned to the stock size of KN that exists in its entirety at the beginning of a period. Given that the quantity of KN utilised during a given period, namely KN_t, can be estimated – for example, the application Equation (17.1) – the quantity of KN stock that remains (KN_{Rt}) is simply $[KN_{At} - KN_t]$. If KN_{Rt} should be the denominator of Equation (17.10), P_{KN} is likely to display an upward trend, confirming the rising scarcity of KN. The only exception would be the context where the net additions to KN stocks offset the utilisation of KN stocks. However, as indicated in Chapter 8, the time lag for stock additions must be recognised.

The Efficiency of Utilisation with Respect to KN

One of the important arguments in the resource scarcity debate has been the claim that improvements in technology would offset the constraints imposed by scarcity because those improvements would enable greater efficiency of resource utilisation. In this context, the ratios (KM/KN) and (Y/KN) can be used to measure the efficiency of KN utilisation. Whether, however, efficiency gains occur at higher rates than the rates of increased demand for KN (e.g. because of the growth of the economy) is another matter, and it is quite likely that efficiency improvements are used to further fuel the economy, thus negating KN-saving tendencies.

(KN/KM) measures the quantity of KN required for the formation of one unit of KM. Hence the smaller the ratio, the greater is the efficiency. Similarly, (KN/Y) measures the amount of KN required for the formation of a unit of Y. If these ratios decline over time while KM and Y are both on the increase, the inference is that efficiency gains exist. This is because smaller amounts of KN are being progressively utilised to generate one unit of KM or Y.

This is not the case in Australia. Both (KM/KN) and (KN/Y) display an upward trend as shown in Figures 17.3 and 17.4. Further, note that, if the concept of KN is broadened to include a wider range of KN sinks, the size of utilised KN will of course be higher than those presented here and the upward trends may be more pronounced.

A related test for efficiency would be to examine the ratio $\left[\dfrac{\partial Y_{KN}}{\partial KN}\right]$. This ratio can also be regarded as a proxy price for the utilisation of KN and is denoted in

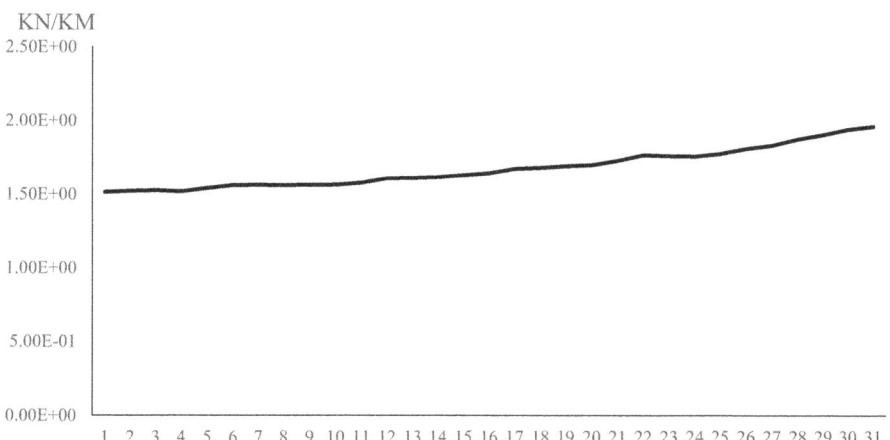

Figure 17.3 (KN/KM) in Australia (1990–2020)

304 *Valuation*

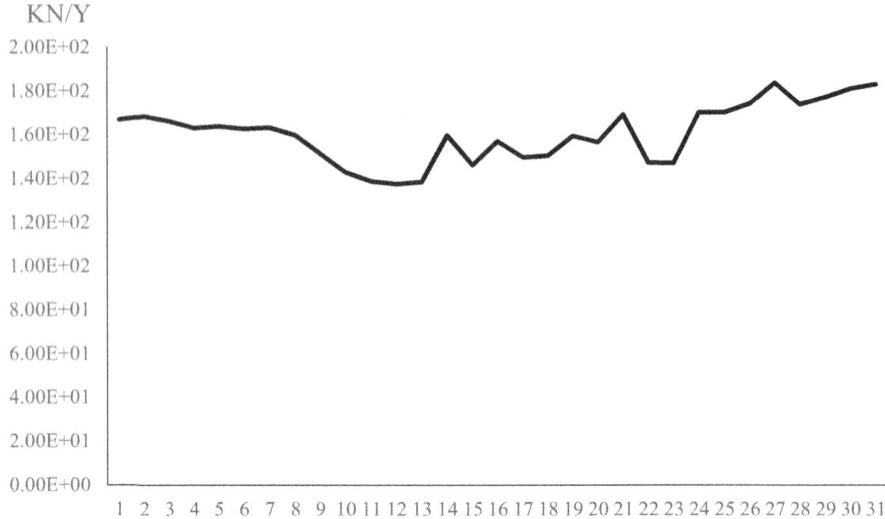

Figure 17.4 (Y/KN) in Australia (1990–2020)

Table 17.2 An Alternative Efficiency Indicator for Utilised KN in Australia

	1990–1998	1999–2007	2007–2013	2014–2020
dY_{KN}/dt ($\$10^9$)	1	2	2	3
dKN/dt ($\$10^9$)	200	300	500	600
P_{KN} (Alt)	0.005	0.0067	0.004	0.005

Table 17.2 as [P_{KN} (Alt)]. As illustrated in this table, it is the result of dividing $\left[\dfrac{\partial Y_{KN}}{\partial t}\right]$ by $\left[\dfrac{\partial KN}{\partial t}\right]$ and defines the additional factor income that accrues to KN from the utilisation of an additional unit of KN. The basis for efficiency is as follows. If the rate of KN utilisation per unit of time is more than the rate of income earned per unit of time, then the ratio will be less than one, implying poor efficiency. In Table 17.2, this ratio has been estimated for the time period 1990–2020, as well as for discrete subsets of four decades within the 30-year period. The ratio of less than 1 in all cases implying a lack of efficiency. This is because the rate of KN utilisation per unit of time is in excess of the rate of income accrual per unit of time.

Concluding Remarks

We have presented here a simple alternative framework for the valuation of KN at the aggregate level. As illustrated, the aggregate level valuation enables the assessment of the scarcity of KN and the efficiency with which KN is utilised. Furthermore, as indicated, a decline in KN-scarcity is generally not possible when the criterion for assessment is *the size of KN stocks that remains after utilisation*. The policy implication of this observation is the need for focusing on KN investments that would add to KN stocks, despite the length of time taken for the addition to take effect. Another implication is the need for finding various ways by which KN endowments can be conserved and protected – such as in the case of Lake Tissø in Denmark, on which we reported in Chapter 8.

The analysis of efficiency ratios reveals that much effort needs to be expended on the area of KN efficiency. As indicated, although efficiency gains have been very significant in the utilisation of KM, these have been lacking with respect to KN. Therefore, there is an explicit need to search for technical improvements that directly target KN. These could include the development of cost-effective methods of waste management and a transition to the utilisation of renewable KN from non-renewable KN. This also implies that, during periods of stagnation, the enhancement of KN could become a vehicle for economic recovery. Recall that this inference was drawn in Chapters 12 and 13 as well.

It is pertinent to conclude this chapter with the words of Alfred Marshall (1891):

Man does not create things. He only rearranges matter.

This statement is in no uncertain terms the first law of thermodynamics. When matter is rearranged, the second law of thermodynamics takes effect. It is therefore difficult to ignore the issue of KN scarcity because economic activity inevitably involves the rearrangement of KN and an increase in its entropic value and hence scarcity. It is pertinent to consider the policy initiatives that would mitigate such changes. We will do so Chapter 19. However, prior to a consideration of policy initiatives, we consider a method to determine the stock size of KN as a macroeconomic estimate in the next chapter.

Review Questions

1 Critically assess the concept of 'use value' of KN in macroeconomics that was advanced in this chapter.
2 Consider the following statement by the World Bank (1992) as cited in the text: 'Declining price trends also indicate that many non-renewables have become more, rather than less, abundant'. Critically review the validity of this statement.
3 Explain the difficulties associated with the measurement of the relative scarcity of KN.

References

Costanza, R., d'Arge, R., de-Groot, R., Farber, S., Grasso, M., Hannon, B., Limburg, K., Naeem, S., O'Neil, R., Paruelo, J., Raskin, R., Sutton, P. and van den Belt, J., 'The value of the world's ecosystem services and natural capital', *Ecological Economics*, 25(1): 3–15, 1997.

Costanza, R., d'Arge, R., de Groot, R., Sutton, P., van der Ploeg, S., Anderson, S. J., Kubiszewski, I., Farber, S. and Turner, R. K., 'Changes in the global value of ecosystem services', *Global Environmental Change*, 26: 152–158, 2014.

Forrester, J. W., *World Dynamics*, Wright-Allen Press, Cambridge, MA, 1971.

Mankiw, N. G., *Principles of Macroeconomics*, 3rd edn, Thomson South-Western, Mason, OH, 2004.

Meadows, D. H., Meadows, D. L., Randers, J. and Behrens, W.W., *The Limits to Growth: A Report of the Club of Rome's Project on the Predicament of Mankind*, Universe Books, Earth Island, NY, 1972.

Nordhaus, W. D., 'World dynamics: Measurement without data', *The Economic Journal*, 83(332): 1156–1183, 1973.

Samuelson, P. A. and Nordhaus, W., *Economics*, Mcgraw-Hill, New York, 1990.

World Bank, *World Development Report*, Oxford University Press, New York, 1992.

18 Valuation of KN
Macroeconomic Basis – II

So far, our treatment of KN pertained to the concept of utilisation, which is tantamount to net depletion. An important question that was raised in the last chapter was whether KN is becoming increasingly scarce. Towards resolving this question, we considered various indicators of scarcity. As indicated in the previous chapter, the 'size of KN stock that remains' is a better indicator of KN scarcity than various other measures that are usually derived from the utilisation of KN. Therefore, in this chapter, we provide a proxy method to estimate the stock size of KN so that an estimate of the 'size of KN that remains' after utilisation is possible. An important caveat is in order. The stock size we display here is not an absolute number. Instead, it is a reference value for a stock size that could have existed at the start of a nominated time period. The estimation of this reference value is based on the utilisation and resilience properties of KN and follows the methodology for estimating the initial stock size for KM. As with the previous chapters (13, 14 and 17) our illustration pertains to KN as an air shed.

Recall from Chapters 1 and 11, that capital theory in economics owes its origins to the conceptualisation of nature as capital. For example, Fisher's (1904) seminal paper, which is perhaps a cornerstone for the literature on the economics of capital, builds on the premise that nature is a durable stock that generates flows of services – the utilisation of which results in degradation. The three essential properties of capital, namely durability, generation of service flows, and depreciation – all originate from the conceptualisation of nature as capital – abbreviated as KN throughout this text.

Stocks and Flows of KN

In this chapter, we distinguish between stocks and flows with reference to KN. In order to make this distinction, we denote the size of the stock as KN_S. This stock is the natural endowment that generates service flows. Such flows are indeed complex. For example, in terms of the air shed, there are at least two basic flows, amongst others. These are the services of the air shed as a source as well as that of a sink. That is, the air shed acts as a source for the air we breathe, whilst it is also a sink for the emissions we discharge into the atmosphere. When the emissions are not adequately dispensed with and they accumulate, the size of the air shed

DOI: 10.4324/9781003408574-22

diminishes. Such diminution is mainly due to the volume of KN that is utilised towards the formation of national income (Y). The aggregate measurement of this utilisation of KN is denoted hereafter as KN_U. That is, the definition offered in Equation (17.1) in the previous chapter pertained to KN_U. Hence, KN_{Ut} is a depletive flow that stems, in time t, from the stock, KN_{St}, that prevailed at time t. That is, KN_{Ut} results in the contraction of KN_{St}. However, should KN be in a steady state, by retaining its resilience properties, say due to human effort, then KN_U might not result in a reduction of KN_S. For example, Bush (2024) and Graham (2024) provide examples of how community action resulted in the restoration of highly degraded water resource systems, namely creeks and rivulets. It may be possible to extract some services from such restored ecosystems up to some threshold level without diminishing KN_S. As indicated, our illustration in this Chapter is confined to the air shed that gets affected by GHG emissions. Recall our discussion in Chapter 6, especially on measures to achieve net-zero emissions, as efforts that could restore some form of balance in the atmospheric system. However, as indicated in Chapter 6 and subsequent ones, the attainment of net-zero emissions status faces significant challenges. Hence, our main premise in this Chapter is that emissions, represented by KN_U, would result in a net reduction of KN_S. The aim of this chapter is to present a simple method to measure the stock size, namely KN_S, from which the utilisation occurs.

The literature on environmental and ecological economics abounds with methods that detail the valuation of *services* provided by environmental capital, but not the *size of stock* that generates the services. For example, besides our review of methods in Chapter 16, see Bennett (2011a, 2011b); Costanza et al. (1997, 2014); Mishan and Quah (2007); Quah and Boon (2003); and Sinden and Worrell (1979). As indicated in Chapter 17, Costanza et al. (1997) estimated the value of global ecosystem services as $33 trillion per year (1995 USD) and then, in a subsequent analysis [Costanza et al. (2014)], illustrated that the value of global ecosystem services in 2011 had significantly appreciated to $125–$145 trillion per year (2007 USD). Such appreciation of the monetary values does in fact validate the premise that KN stocks, which provide the ecosystem services, are becoming increasingly scarce.

The size of certain KN endowments such as forests and minerals can be ascertained by recourse to surveys of geographic information systems and geological records. Nevertheless, the estimation of the stock size of other endowments such as air sheds and oceans has remained outside the calculus of KN size valuation. This relegation is at least partly due to the mistaken and widespread belief that these endowments are infinite in their availability.

The proposed framework draws on an accounting method employed with reference to the accumulation of manufactured capital stock (KM) in macroeconomics. This method, the perpetual inventory method (PIM), enables the quantification of the stock size of KM that could have existed at the initial time period of a time series dataset. Hence, in adapting the PIM to the context of KN, the question posed is: *What could have been the stock size KN_S at the start of a specific time period?* As shown below, the subsequent changes to the stock will depend on the withdrawal as well as the resilience of KN. The metric employed is the same as that for KM in the national accounts, namely, constant base year prices.

The next section provides a brief description of the PIM. This is followed by its adaptation to the estimation of KN_S and an empirical illustration with reference to the air shed of the Australian continent.

The Perpetual Inventory Method (PIM)

In the statement of national accounts, estimates of KM are generally displayed in inflation-adjusted constant dollars, thereby enabling diverse items of KM, such as infrastructure and machinery, to be aggregated into a single number. A comprehensive review of the various approaches to PIM can be found in Berlemann and Wessehöft (2014), and Dey-Chowdhury (2008). In this chapter, we employ a basic approach as illustrated by Nehru and Dhareshwar (1993) that combines a geometric depreciation at a constant rate alongside the assumption of an economy being in a steady state. Recall our discussion concerning the steady state with reference to the Swan-Solow model in Chapter 14.

The PIM defines the size of KM at the start of a given time period, in terms of the stock that prevailed at the start of the previous period. This previous period stock is then adjusted for investments (I) and depreciation (D) that would have occurred during the previous period. For example, the stock of KM in time t would be described as:

$$KM_t = KM_{t-1} + I_{t-1} - (\delta_{KM})KM_{t-1} \tag{18.1}$$

In Equation (18.1), δ_{KM} usually represents a fixed rate of depreciation of the aggregate KM stock. That is, $\{D = [(\delta_{KM})KM_{t-1}]\}$. The process of change in stock size is one of net accumulation of KM with (I > D).

Consider now the initial time period (t = 0) and the period immediately following it, that is (t = 1). The application of Equation (18.1) with reference to (t = 0) and (t = 1) will be as follows:

$$KM_1 = KM_0 + I_0 - \delta_{KM}KM_0 \tag{18.2}$$

And when rearranged, Equation (18.2) could be written as:

$$KM_1 - KM_0 + \delta_{KM}KM_0 = I_0 \tag{18.3}$$

In the PIM, one assumes that the difference $(KM_1 - KM_0)$ can be approximated to a value by recourse to the average annual rate of growth of investment in the time series dataset. Suppose that this rate of growth is σ. Then $(KM_1 - KM_0)$ equates to (σKM_0), and then the stock of KM that would have existed in the initial period becomes:

$$KM_0 = \frac{I_0}{(\sigma + \delta)} \tag{18.4}$$

310 *Valuation*

In Equation (18.4), the coefficient σ is usually estimated on the premise that investments grow at an exponential rate over time. That is $I_t = Ae^{\sigma t}$ and hence the rate of growth is $\{[dI_t/dt]/I_t\} = \sigma$. The coefficient δ_{KM} is typically based on the assumption that KM stocks have an average life of 30–40 years and is hence assigned a fixed value ranging between (1/30) and (1/40); that is the constant depreciation rate mentioned above. In complex analyses (Berlemann and Wessehöft 2014; Dey-Chowdhury 2008), the KM stock is differentiated into various categories on the basis of asset life. The simple approach based on a constant rate is considered here so that the adaptation of the method towards the estimation of KN stocks would be relatively easy. A more nuanced approach based on "asset life" could be applied as well.

Adapting the PIM for KN

A simple adaptation of the PIM to estimate KN_S would necessitate the measurement of the depreciation in KN stocks. As indicated, such depreciation is the net withdrawal of KN, namely (KN_U) from the KN stock, and the methodology underlying the estimation of KN_U was illustrated in Chapter 13- Equation (13.14). We reproduce this equation again now, with a change in notation for the annual net utilisation as KN_U:

$$KN_{ut} = KM_t \left[\left(\frac{1}{\eta_t (1-\eta_t)^{(1/\theta_t)}} \right) - \frac{1}{\eta_t} \right] \qquad (18.5)$$

To recapitulate from previous chapters, the determinants of KN_U are:

a η, is the ratio (D_{KN}/GVA), where D_{KN} is the allowance for mitigating the depreciation of KN,
b φ, which is KN's resilience coefficient and is related to η by the expression [φ = 1 − η],
c θ, which is the share of Y accruing to composite capital stock (K), and this was defined as $\{[KM+((1-\varphi)KN)] = [KM + \eta KN]\}$, and
d KM_t – the accumulated stock of manufactured capital.

We use the term 'net utilisation' for KN_U with reference to Equation (18.5) because the resilience properties of KN as identified in Chapter 2 have already been incorporated. The study and analysis of resilience is complex; for example, see Atkinson and Dietz (2009), Brand and Jax (2007), and Singh and Perwez (2015). However, we adopted a simple premise as outlined in Chapter 13. The context of complete resilience, (φ = 1), coincides with (η = 0), whilst complete non-resilience is dictated by the context of (φ = 0) and D_{KN} being as high as GVA – that is (η=1). So, with reference to Equation (18.5) above, if (η → 0), then (KN → 0); and if (η → 1), then (KN_U → ∞). Hence, we were able to use the coefficient η, as also an indicator of resilience besides depreciation of KN, namely (D_{KN}).

Further, recall that the utilisation of KN was expressed as a determinant of national income (Y) alongside labour (L) and KM. In explaining Y, the distinction between KN and KM is as follows. The utilisation of KM is based on the net accumulation of stock as outlined in the PIM above. In contrast, it is reasonable to assume that with most KN endowments, the net changes in stock during any given time period is one of net depreciation rather than accumulation.

Consider now the adaptation of PIM in Equation (18.2) above to the context of KN stocks. The definition of KN stocks in a given year (t), namely (KN_{St}) would be made with reference to KN stocks of the previous year (KN_{St-1}):

$$KN_{St} = KN_{S(t-1)} - KN_{U(t-1)} \tag{18.6}$$

Therefore, in the context of complete resilience of KN_{Ut}, ($KN_{St} = KN_{S(t-1)}$). The expression in Equation (18.6) could be now restated with reference to the initial year and the subsequent one, namely (t = 0) and (t = 1) as:

$$KN_{S1} = KN_{S0} - KN_{U0} \tag{18.7}$$

Note that KN_{U0} in Equation (18.7) is the volume of KN withdrawn in (t = 0) as per Equation (18.5) above. Rearranging Equation (18.7) one gets:

$$KN_{S0} - KN_{S1} = KN_{U0} \tag{18.8}$$

As done with the PIM, the lefthand side of Equation (18.8) can be approximated to an expression in terms of the initial stock size, namely KN_{S0}, by recourse to the average rate of withdrawal of KN. Should this rate be δ_{KN}, then Equation (18.8) could be rewritten as:

$$\delta_{KN} \cdot KN_{S0} = KN_{U0} \tag{18.9}$$

However, in the case of KN, because the change in stock is that of net depreciation, instead of net accumulation, it is pertinent to compare the initial year with the terminal year. Therefore, as an extension of Equation (18.9) above, consider the comparison of the initial year (t = 0) with the terminal year (t = T). This comparison could be explained as follows:

$$[\delta_{KN} KN_{S0}] = KN_{S0} - KN_{ST} = \sum_{t=0}^{T} KN_{Ut} \tag{18.10}$$

T represents the number of years between the initial year and the terminal year. As illustrated in the next section, T is the time period over which annual emissions data is collected. Hence the minimum stock size of KN that could have existed in (t = 0) is:

$$KN_{S0} = \left\{ \frac{\sum_{t=0}^{T}(KN_{Ut})}{\delta_{KN}} \right\} \quad (18.11)$$

Hence, the accumulated withdrawal of KN and the average annual rate of withdrawal of KN become the sole determinants of the stock size of KN at the beginning of the time period.

Empirical Illustration

The method is illustrated with reference to Australia's air shed as the KN endowment. As in previous chapters, the cost of abating greenhouse gas (GHG) emissions was nominated as a proxy for D_{KN} and the estimation of these costs is as described in Chapters 13 and 14.

The estimates of KN_S for 1990 and 2020 are illustrated in Table 18.1.

As indicated in Table 18.1, the size of KN (air shed) that could have existed in 1990, in national income accounting terms, is approximately $8,021 trillion and this value dropped to nearly $7,736 trillion by 2020. The average annual loss of the air shed during the 30-year period (1990–2020) was about $9.2 trillion.

Recall the discussion about the price of KN displayed in Figure 17.2 of the previous chapter. This price is better defined as 'utilisation price' because it was based on the income accruing to KN (Y_{KN}) being divided by the volume of KN utilised, namely KN_{Ut}. We suggested there that the appropriate indicator of scarcity would stem from dividing Y_{KN} by the stock of KN that remains. We present this information in Figure 18.1. As expected, P_{KN} based on the size of KN stock that remains, shows a clear upward trajectory indicating the presence of scarcity.

As mentioned above, the change in stock sizes as well as the size of the initial and terminal stock are all contingent on the assumptions pertaining to the resilience coefficient and the length of time period (T) for the analysis. Recall our assertion at the beginning of this Chapter, namely that the stock size will not be an absolute value – instead it will be reference value. Such reference value will indeed change

Table 18.1 Estimation of KN Stock

ΣKN_{Ut} (1990–2020)	*$51.12 trillion*
δ_{KN}	0.0376
$\Phi = (1 - \eta)$	Reduces from 0.955 in 1990 to 0.946 in 2015
T	31
$KN_{S(1990)}$	$8,021 trillion
$KN_{S(2020)}$	$7,736 trillion
Stock depreciation	$285 trillion

Valuation of KN: Macroeconomic Basis – II 313

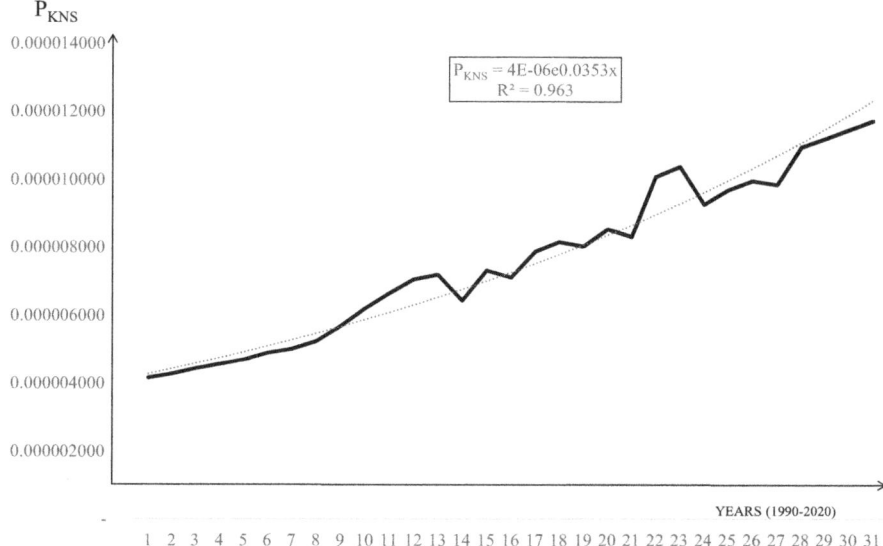

Figure 18.1 Price of KN in Australia (1990–2020) – Based on KN_S

in magnitude depending on the assumptions. For example, the state of complete non-resilience could be reached even before (D_{KN} = GVA), that is, η being less than 1. Nevertheless, the analysis illustrated here bears policy relevance. For example, a study of Figure 18.1 reveals that the loss of KN_S is increasing at the rate of around 3.5 per cent each year. Thus a search for KN stabilisation measures will be in order.

Concluding Remarks

The illustration above rests on the assumption that the Australian air shed is an isolated entity; that is, the emissions are dispersed only within the Australian continent. This assumption is made purely for illustrative purposes. Yet, the methodology advanced here can be extended to broader geographic areas beyond sovereign borders and rendered more robust with scientific data. Alternatively, we may presume that each county should be responsible for their own emissions, in which case the delimitation of system boundaries as done here could be defended on ethical as well as practical grounds.

There is a widespread belief that endowments such as air sheds and oceans are infinite. This mistaken belief propels the lack of stewardship to protect nature. Several nation states take KN sinks such as air sheds and oceans for granted and fail to adequately regulate discharges into these sinks. The likelihood of stewardship towards such sinks could increase if policy makers had some quantitative appreciation of the stock size of KN sinks. That is, an appreciation that the KN sinks would some day cease to exist. In this regard, an important policy directive would be to search for and invest in technologies to reduce emissions and discharges and at the same time also enhance the resilience of KN. We consider the policy directives that follow from this logic in the next chapter.

Review Questions

1. Critically review the adaptation of the perpetual inventory method to the context of estimating the stock size of KN.
2. Suggest potential proxies that may be developed for the estimation of the resilience coefficient j.

References

Atkinson, G. and Dietz, S., 'Progress in the measurement of sustainable development', in Ng, Y-K., and Wills, I. (eds), *Welfare Economics and Sustainable Development*, EOLSS Publishers, Oxford, 2009.

Bennett, J., *The International Handbook on Non-Market Environmental Valuation*, Edward Elgar, Cheltenham, 2011a.

Bennett, J., 'Valuing Australia's environment', *Australasian Journal of Environmental Management*, 18(1): 21–32, 2011b.

Berlemann, M. and Wesselhöft, J.-E., 'Estimating aggregate capital stocks using the perpetual inventory method – A survey of previous implementations and new empirical evidence for 103 countries', *Review of Economics*, 65: 1–34, 2014.

Brand, F. S. and Jax, K., 'Focusing the meaning(s) of resilience: Resilience as a descriptive concept and a boundary object', *Ecology and Society*, 12, article 23, 2007. Available from URL: http://www.ecologyandsociety.org/vol12/iss1/art23/

Bush, J., Making merry: How we brought Melbourne's Merri Creek back from pollution, neglect and weeds, *The Conversation*, 7 May 2024.

Climate Change Authority, *Reducing Australia's Greenhouse Gas Emissions – Targets and Progress Review (Final Report)*, Commonwealth of Australia, Canberra, 2014. Available from URL: https://www.climatechangeauthority.gov.au/annual-reports

Costanza, R., d'Arge, R., de Groot, R., Farber, S., Grasso, M., Hannon, B., Limburg, K., Naeem, S., O'Neill, R. V., Paruelo, J., Raskin, R. G., Sutton, P., and van den Belt, M., 'The value of the world's ecosystem services and natural capital', *Nature*, 387: 253–260, 1997.

Costanza, R., d'Arge, R., de Groot, R., Sutton, P., van der Ploeg, S., Anderson, S. J., Kubiszewski, I., Farber, S. and Turner, R. K., 'Changes in the global value of ecosystem services', *Global Environmental Change*, 26: 152–158, 2014.

Dey-Chowdhury, S., 'Perpetual inventory method', *Economic and Labour Market Review*, 2: 48–52, 2008.

Fisher, I., 'Precedents for defining capital', *Quarterly Journal of Economics*, 18: 386–408, 1904.

Graham, S., This group rid one Australian river of its privet problem – and strengthened community along the way, *The Conversation*, 7 May 2024

Mishan, E. J. and Quah, E. T., *Cost-Benefit Analysis*, Routledge, New York, 2007.

Nehru, V. and Dhareshwar, A., 'A new database for physical capital stock: Sources, methodology and results', *Revista de Analisis Economico*, 81: 37–59, 1993.

Quah, E. and Boon, T. L., 'The economic cost of particulate air pollution on health in Singapore', *Journal of Asian Economics*, 14: 73–90, 2003.

Sinden, J. A. and Worrell, A. C., *Unpriced Values: Markets without Prices*, Wiley, New York, 1979.

Singh, G. and Perwez, A., 'Estimation of assimilative capacity of the air shed in iron ore mining Region of Goa', *Indian Journal of Science and Technology*, 8: 1–7, 2015.

Part V
Policy

19 Policies for Sustainable Development

The content of this text has emphasised to the following basic principle:

> The sustainability of environmental capital (KN) is an essential prerequisite for the sustainability of an economy.

Recall the explanation of economic sustainability as explained by Hicks in Chapter 1 (Box 1.1), namely the ability to maintain the same level of spending in real terms over time. As illustrated in the chapters that followed, the maintenance of real spending depends on the maintenance of a steady stock of KN. To properly understand and model changes in KN, we have attempted the following:

- Invoked the laws of thermodynamics and ecological resilience in Chapter 2
- Explained the role of markets and their failure in Chapters 3–7
- Revised production functions in Chapter 8
- Modified the theory of perfect competition in Chapter 9
- Detailed the relevance of residual externalities introduced in Chapter 4 to the macroeconomy in Chapters 10–18
- Elaborated on the role of KN depreciation (D_{KN}) and resilience in Chapters 11–18

The concept of global ecosystems considered in Chapter 15 suggests that the responsibility for maintaining 'above-threshold' levels of KN stocks extends beyond national geographic boundaries. It is therefore pertinent to the discussion of policies in this chapter to centre on the requirement of maintaining KN not only in the national economy but also the global economy. This is indeed a challenge, because the same way as the principles of conventional economics are premised on the axioms of self-interest and present gains, policy formulation in general also follows the lines of promoting national interest and the welfare of current generations.

Further, we can relate the types of policies that are required in contemporary societies, to the analogy drawn in Chapter 1 for the damages inflicted on KN. Recall the plight of the prisoner who has been receiving 100 lashes a day. As a result, the prisoner suffers considerably. Reducing the number of lashes to 50 or even ten per day will not help heal the prisoner's wounds. The lashes must stop completely, and

DOI: 10.4324/9781003408574-24

treatment must be offered for immediate relief and long-term recovery. Following the lines of this analogy, policies for achieving sustainability fall into three categories. These are policies that:

1 Strive to reduce damages to KN
2 Attempt to minimise KN damages towards eliminating them
3 Overarch the above 2 and without which, any form of sustainability would be difficult to achieve

Table 19.1 displays the taxonomy of these policies. The first category of policies strives to reduce damages. However, such efforts will not suffice because the cumulative toll of damages will continue to increase despite the reduction in damages. Hence, the second category of policies attempts to eliminate damages to KN. However, any human activity will inevitably entail disturbances to KN systems. Hence the second category is better described as those that minimise as much damage as possible. In this category, we include strategies to restore and expand KN assets. The third category of policies overarches the first two as a precondition. This is because without these any safeguards for KN as prescribed by the first two categories would fail. We emphasise the importance of social capital (KS) and need to remove conflict and maintain peace and harmony. An important part of this KS is the knowledge – and acceptance of that knowledge – that KN is vital to social and economic interests. This acknowledgement in the context of our current 'post-truth' and 'fake news' era is vital. It is also a motivator for writing a book like this in the first instance.

Table 19.1 The Policy Mix

I Policies Aimed at Reducing Damages to KN
1 Taxes and subsidies
2 Regulation, tradable permits, net zero emissions
3 Prices: markets for goods and services; capital, labour and global trade markets
4 Education, changing attitudes and lifestyles
5 Population management
6 Property rights
7 Improving productivity of KN
II Policies Aimed at Minimising KN Damages Towards Zero
8 Closed-loop production systems: waste management; energy use and management
9 Bio-mimicry
10 Changes to business models
11 Restore and expand KN assets
III Overarching Policies – Social Capital (KS) – A Precondition
12 Enhance social cohesion
13 Remove corruption – good governance
14 Social conscience

Policies for Sustainable Development

The various policies considered below rely on conceptual premises presented in the preceding chapters. Further, earlier chapters also contained a discussion of many of these policies – but in the context of specific conceptual frameworks. The purpose here is to provide a synthesis of these policies and highlight the synergies and/or the lack of synergies among them.

Policies Aimed at Reducing Damages

Note that, although the policies considered in this section mainly reduce damages, some of them have the potential to progress towards the second category; that is, minimising towards zero damages.

Taxes and Subsidies

Taxes generally represent a set of disincentives designed to reduce the extent of damages inflicted on KN. As illustrated in Chapters 4–9, such damage reductions can involve reductions in production and consumption. For example, consider Figure 4.3 in Chapter 4. Here, the internalisation of a negative externality raises the marginal cost of production and, as a result, the market equilibrium is resolved at a higher price and a lower level of production and consumption than when the externality remains without being internalised. The price and cost differentials that stem from such internalisation can become the basis for a tax. The internalisation of externalities across various markets would lead to a lower level of aggregate demand (AD) and aggregate supply (AS), as illustrated in Chapters 12–15. However, recall our reference to residual externalities that accumulate.

Some taxes target specific externalities. For example, a carbon tax aims to reduce greenhouse gas emissions. This tax may result in higher energy prices, especially when energy utilities rely mainly on fossil fuel resources for power generation. But tax revenues may also be used to incentivise investment in new technology and efficiency improvements, thus lowering the impact of taxes on prices. In the longer run, higher prices prompt consumers to search for more cost-effective solutions, such as solar power, and energy utilities to adopt cleaner methods of power generation. These developments may usher in a transition in the energy sector to lower carbon fuels, improved air quality and public health, reduced dependence on imports of fossil fuels, and a host of other 'co-benefits' – aside from the targets improvements in KN quality, which lead to higher standards of living and quality of life. Being able to create and quantify co-benefits from improvements in KN – and correcting the benefits and co-benefits for potential costs that arise as well, such as when wetlands are rehabilitated and with them mosquitoes and the diseases, they may spread become more prevalent – serve as an important contribution to building public support for KN policies can (Ruth et al. 2015).

Some commentators argue that the surge of renewable energy options such as solar panels and windmills in the European Union could have been the result of carbon pricing and taxes. The opposition to carbon taxes stems mainly from those who have stakes in maintaining the status quo with reference to current practices that are carbon intensive. These include energy utilities that have invested in the mining and combustion of fossil fuel resources.

A resource rent tax (RRT) targets specific activities such as mining, and rests on the premise that the ultimate owner of KN resource stocks is the society that harbours these resources. Given the global connectivity of KN stocks, this can be a tenuous premise. Nevertheless, the formulation of RRT usually proceeds without reference to global connectivity. The RRT is the difference between total revenue earned from mineral sales and all legitimate costs associated with mineral extraction. A further important premise is that mining firms undertake the task of extraction on behalf of the resource owner for a fee, which is labelled normal profit. Such profits, in turn, must cover the dividends that mining firms have to pay their shareholders. When the premises underlying the RRT are not clearly articulated, especially those pertaining to ownership, mining firms mistakenly assume propriety of the mineral assets. Estimates of normal profits usually fall between 10 and 15 per cent of capital and exploration costs associated with mining ventures. In the study concerning the Liverpool Plains in Australia (Box 16.2 in Chapter 16), the annual estimates of normal profits to BHP-Billiton and Shenhua-Watermark were respectively $120 million and $100 million. On this basis, each mining firm would have had to pay from their proceeds of mining an RRT ranging between $350 million and $450 million per year. In the absence of the RRT, each mining firm would envisage its profit as ranging between $470 and $550 million per year. When the RRT is in force, mining firms can expect to receive only their profits indicated above. Hence it is plausible that the RRT could even prompt mining firms to seek alternative investment ventures involving renewable energy options. As a side note, the Federal Government of Australia abolished the RRT in 2013. Nevertheless, as of 2021, BHP-Billiton had dropped its plans for mining on the Liverpool Plains, and the State Government of New South Wales has purchased back the mining lease from Shenhua-Watermark. The RRT is successfully in force in Norway and Chile. In these countries, the RRT is reinvested in the community to build up social capital and alternative economic activities that are viable in the long run.

Subsidies are incentives designed to prompt the adoption of activities that are less damaging to KN relative to others. An important consideration with subsidies is the subject of cross-compliance raised in Chapter 6. This means that, while environmental taxes aim to curb the extent of damages to KN, subsidies should not counter such efforts. For example, returning the RRT as a subsidy to promote further exploration and mining would not comply with the aim of reducing damages on KN. Yet, such returns of tax revenues to the government tend to make it back to the firms inflicting damages to KN, sometimes in subtle ways, which when these firms are allowed to write off from their tax liabilities as (failed) exploration efforts.

Recall the discussion in Chapter 15 of the extractable resource stocks being components of larger global resource stocks that afford the Earth greater stability in terms of seismic activity. An important policy challenge is the return of environmental taxes as KN investments. As illustrated in Chapter 8, the returns from KN investments are time-consuming and, as considered in Chapter 17, such returns are difficult to quantify and value. Yet some studies (Thampapillai et al. 2010) show that environmental taxes, when returned as KN investments in Vietnam – especially for reforestation and cleaner methods of sewage treatment – would reduce D_{KN} and elevate the time path of sustainable income (Y_S). Recall the discussion of such time paths in Chapters 12 and 13.

Regulation, Tradable Permits and Net Zero Emissions

As discussed in Chapter 6, the basis for regulation typically stems from scientific information. Recall the description standards set by various environmental agencies and the International Standards Office. An extension of this discussion to the context of cumulative damages would be that, over time, the stringency of the regulatory limits should increase. The ultimate level of stringency is the requirement of zero damage, and where possible the reversal of past damages. Porter and van der Linde (1995) use the success of the Dutch flower industry to argue that stringent regulations can drive firms to innovate:

> This static view of environmental regulation, in which everything except regulation is held constant, is incorrect.... Properly designed environmental standards can trigger innovations that lower the total cost of a product or improve its value. Such innovations allow companies to use a range of inputs more productively — from raw materials to energy to labour — thus offsetting the costs of improving environmental impact and ending the stalemate. Ultimately, this enhanced resource productivity makes companies more competitive, not less.

A persuasive argument in the literature on regulation is that the setting of standards alone would not suffice. Effective regulation requires the presence of monitoring and enforcement, and these in turn require institutional capacity. These make regulation a costly exercise. Another extension of this subject is the possibility of 'self-regulation', which could avert the need for costly intervention. This means that firms would not wait for the imposition of regulations – instead, they would foresee emergent issues and set their own standards and codes of practices. However, the emergence of this type of behaviour would be contingent on other areas of policy intervention, such as changes in attitudes and education. We will consider these below.

Although some consider implementing regulation costly, the costs stemming from the absence of regulation can be phenomenally high. For example, recall our illustration in Chapter 4 of the potential origins of Covid-19 and the global pandemic that ensued. As argued in Thampapillai et al. (2021), China's animal husbandry markets did not have adequate regulation, and as such could have been an important factor contributing to the origins and spread of the virus. Those pursuing 'free markets' often argue against regulatory barriers on the basis of welfare maximisation under perfect competition. However, perfect competition does also embody the criterion of 'perfect information', and one needs to recognise that 'free markets' are unlikely to be perfect markets.

322 *Policy*

As illustrated in Chapter 6, taxes are likely to be more effective when combined with regulation than without it. This is because, for example, polluters could simply pay the taxes and continue to pollute – and even pollute more than before. As also illustrated in Chapter 6, tradable permits represent an alternative way of recouping environmental taxes by improvising a market for the utilisation of KN services within a regulated context. To recapitulate, the users of KN services (for example, polluters) purchase permits that restrict their use of KN within a regulated limit. The expectation is that efficient users would accumulate a surplus of permits, which the inefficient users could then purchase. Such an arrangement, often referred to as the cap-and-trade system, can incentivise inefficient users to change their practices and gain greater efficiency. Once again, in the context of recognising cumulative damages, the limit would have to become progressively more stringent. Issues pertaining to monitoring, enforcement and institutional capacity are relevant for this instrument as well. In a review of carbon trading markets, Pearse and Böhm (2014) indicate that the major shortcoming of these markets is their inability to reduce emissions. Not all permit markets have not been confined to GHG emissions. Water trading markets by recourse to transactions of permits aim to limit excessive abstractions from river systems. Vinoli Thampapillai (2009) illustrates that not all permit-holders are willing sellers.

As indicated in Chapter 6, net zero emissions (NZE) attempt to stabilise the load of contaminants in KN sinks. The key feature is the requirement that discharges into KN sinks are countered by their reabsorption. Three questions with this feature can be stated as: *When Where* and *How*? With reference to the question of *When*, most countries appear to set the timing to reach NZE as between 2050 and 2060. They agree that emissions must be reduced in the interim period and seem to suggest a transition away from fossil fuels towards renewable energy sources. The location of reabsorption is addressed by the question *Where*? As indicated in Chapter 6, the receptacle of the discharge and the venues of reabsorption are different and possibly dispersed. Hence, the issue of whether the discharge would be sufficiently offset by reabsorption remains. The question of *How* is related to that of *Where*. As per our brief review in Chapter 6, the reabsorption of discharges is dominated by nature-based methods and involves expansion of KN sinks. Following our analysis in Chapter 18, it is likely that KN stock sizes in most economies are on the decline. Hence over optimism regarding the effectiveness of NZE is likely to be misplaced.

Commodity Prices, Interest Rates and Wages

Price policies span a wide range of markets. A standard principle in social cost-benefit analysis is that the adoption of shadow prices, namely prices that manifest true social opportunity costs, will enhance social welfare. As discussed in the literature on cost-benefit analysis (for example, Dasgupta and Pearce 1972, Sugden and Williams 1978, Sinden and Thampapillai 1995) shadow prices are synonymous with prices that would prevail in the context of perfect competition. However, the discussion in Chapter 9 reveals that the conditions for perfect competition fall short of assuring the sustainability of income, and we will consider this subject under 'Policies aimed at minimising damages and restoring KN' below. Nevertheless, correcting the prevailing market prices to shadow prices can reduce KN damages by reducing excessive consumption and production and by achieving better allocation of resources.

The correction of commodity prices for market imperfections and KN externalities could prompt changes in market choices. The environmental taxes considered above contribute to this correction. For example, revising the price of a tonne of coal upwards from $90 to $200 – for the sake of argument – could render solar panels and other less carbon-intensive energy technologies clearly competitive in the energy market. A useful illustration here is the events that unfolded with the onset of the quadrupling of oil prices by OPEC in 1973. Almost every oil-importing country in the world embarked on the quest for alternative energy sources and technologies. However, the entry of North Sea oil into the oil market nearly a decade later and the resultant fall in oil prices led to such quests being either shelved or terminated. More recently, the aggressive development of shale oil and shale gas deposits – involving huge (uninternalised) KN costs – has further negated the pressure to promote alternative energy sources. The broader appreciation of the association between fossil fuel utilisation and climate change has somewhat tamed this aggression.

Another approach to proper costing of energy may be to simply remove subsidies – hidden and not-so-hidden – to firms extracting and burning fossil fuels. As a consequence, the prices of their products would increase, thus levelling the playing field with firms using renewable energy as their fuel source.

Interest rates and exchange rates often move together in the same direction. The determination of interest rates usually falls within the domain of the Central Bank (CB). During periods of recession, the CB will increase money supply by buying up bonds and securities and thereby reducing interest rates. During periods of expansion and inflation, the CB will do the reverse. The subject of cross-compliance warrants consideration in this context. That is, when fiscal instruments such as environmental taxes are introduced to curb KN damages, the easing of monetary policy by reducing interest rates can counter the outcomes of fiscal policy. (Recall also the discussion in Chapter 13).

KN capacity constraints can restrict growth. Attempts to stimulate economies in the context of such constraints can only prompt inflation. Then subjecting the economy to loose monetary policy without any reference to KN capacity constraints will be ineffective. In essence, the management of KN must fall within the domain of monetary policy. Towards this end, two sets of institutional changes can be considered. Note that the trading banks, which perform most of the investment lending, have a borrowing window with the CB. First, the trading banks could introduce strict criteria pertaining to KN stewardship in their checklist of lending criteria. For example, home-buyers could qualify for a home loan only if the home they purchase satisfies some efficiency criteria with reference to KN, such as rainwater tanks for water conservation and/or solar panels and electronic lighting ballasts for energy conservation. Further, the trading banks could also offer discounted lending to clients who demonstrate clear acts of KN stewardship in their activity portfolio. Second, the CB could institute differential lending windows to the trading banks instead of a single lending window. That is, it could open up a discount-lending window to trading banks that include KN stewardship in their practices.

Monetary policy in developing countries has an added dimension. The poor are not credit-worthy in the traditional sense. They lack collaterals and respectable guarantors and thus fall outside the realm of institutional credit. The poor are also agents of KN damages. Hence, the efforts to institute micro credits as discussed in the development literature also fall within the domain of KN conservation policy.

As indicated in Chapters 12 and 13, the subject of wage policy for compliance with the recognition of KN is in order. The literature on shadow prices deals dominantly with methods of pricing unemployed labour within the calculus of cost-benefit analysis. The rationale here is that the pricing of unemployed labour at opportunity costs, which amounts to near zero in developing countries, would lead to the adoption of labour-intensive programmes. As a result, both poverty and unemployment would be reduced. Where poverty and unemployment are also agents of KN damages, employment strategies stemming from such shadow pricing methods are pertinent.

Consideration must also be given to the setting of wages far above opportunity costs – especially among chief executive officers (CEOs) in the managerial sector. For example, as in most OECD economies, Australia has been experiencing a growing income divide. Atkinson and Leigh (2007) reported that, in 2002:

- The richest 10 per cent held nearly 31 per cent of national income
- Annual wages of CEOs on average exceeded 3 to 4 million Australian dollars
- The richest 200 persons held nearly 2 per cent of national income in 2001

As of 2023/2024, the picture appears not to have changed significantly. The highest-paid CEO in Australia earned around 30 million dollars in 2023. Similar, if not more pronounced, trends are apparent in many other industrialised counties, most notably the United States of America, where the gap between top earners and the rest of the workforce continues to grow.

Major sources of such wage inequality lie in the pervasion of asymmetric information and the quest for higher wages by principal agents. The implication of high wages for KN damages is contained in the following narrative from Thampapillai (2011) with reference to the global financial crisis:

> As many commentators concede, the main driver of the financial crisis was the prevalence of gross irregularities within an unregulated financial sector. Foremost of these irregularities is perhaps the excessive wages commandeered within the financial sector — driven primarily by the pervasion of asymmetric information and the formulation and delivery of risky financial products — which in turn sustain high wages. The emergence of an elite high-income group delivered clear demand-price signals for the market entry of goods and services that are energy intensive and environmentally damaging.

We referred to this narrative in Chapter 13, and reiterate it here to emphasise the role of wage inequality at higher end triggering KN damages. Clearly, a wage policy based on true opportunity costs (instead of contrived opportunity costs) could curb extravagant patterns of consumption and in turn help reduce KN damages.

Product Differentiation

Recall our discussion in Chapter 7 on the differentiation of consumption goods in terms of environment friendly and unfriendly attributes. Reinhardt (1999) explains that the idea behind product differentiation is to create products or processes that offer larger KN benefits or smaller KN costs than those of their competitors. The product or process that is designed this way may cost more to produce but, when the benefits are large enough, a higher price and/or a higher market share would recover the higher costs. Reinhardt gives the example of the Swiss manufacturer of textile dyes, Ciba, which was able to offer a differentiated product – a special dye that needed less salt and less waste treatment by textile-makers. This product was readily accepted in the market because it lowered operating costs for the customers. According to Reinhardt (1999), product differentiation will work if the business manager is able to:

i Identify customers who are willing to pay more for the differentiated product
ii Clearly explain the savings that customers will reap by incurring a higher initial outlay
iii Protect the business from rivals who would attempt to imitate the process

However, the question is: Would firms and their managers be sufficiently stewardly, even if cost savings and higher profits were absent? Would they accept a reduction in profits in order to safeguard KN? The difficulty here is that low-cost rivals with non-stewardly production processes can damage KN that is common to everyone. The discussion on the endowment effect in Chapter 7 suggests that lowering prices and reducing the profit margin initially could lead to a greater market share in the longer run. Nevertheless, the adoption of KN stewardship among both the consuming public and firms that produce various goods and services is a matter of attitudes that stem mostly from education. We will consider this next.

Education Attitudes and Lifestyles

The individual psyche ultimately governs actions. The psyche is the product of a system of beliefs. These beliefs in turn are a product of what one learns. A recent anecdote from Singapore illustrates this point. A busy suburban mother was wheeling her shopping trolley in a supermarket with her primary-school daughter tagging along. The child was vocal and visibly unhappy with her mother's shopping behaviour. The mother was hurriedly throwing items into the trolley. The child wanted the mother to read the labels and ascertain whether the items were nature-friendly. Either Singapore's Ministry of Education or a well-meaning teacher had taken on the task of instilling KN stewardship among children. It is imperative that environmental science and management are introduced into the core (mandatory) curriculum at all levels of education – primary, secondary and tertiary.

In this text, we have highlighted the lack of symmetry between the equilibria in various markets at the microeconomic level and the overall macroeconomic equilibrium in the economy. This imbalance, as we argued in previous chapters, emerges primarily because not all market externalities are fully internalised resulting in residual externalities, which accumulate over time. In this regard, indicators such as GDP, GNP and GVA do not readily capture the adverse effects of the residuals and their accumulation on economic performance. Hence, as suggested in our text, the macroeconomic equilibrium, which serves as the basis for macroeconomic policy analysis, requires revision. Recall that the revised equilibrium is $[Y = GVA - D_{KN}]$. Further, given that the basic benchmark for efficient markets is premised on perfect competition (PC), we also proposed in Chapter 9 that the PC benchmark also requires revision to include sustainability as an important condition besides the five standard conditions of anonymity, perfect information, homogeneity, perfect mobility and full employment. Most standard introductory textbooks in economics (for example Mankiw 2004, 2014; Frank et al. 2015; Hirshleifer et al. 2005; Parkin 2016) do not deal with the revisions that we proposed in this text. Without these revisions the content of basic texts in economics such as those indicated above potentially present a mixture of information asymmetry and moral hazard and misguides perception and policy formulation. The issue is serious because these basic texts represent a cornerstone in a student's learning of economics. In a vein similar to that of the endowment effect exposited by behavioural economists (Knetsch 1989; Kahneman et al. 1990), it would be difficult to undo the learning accomplished from basic texts. Some students especially those pursuing a study in commerce do not explore economics beyond the foundation module. Even with students pursuing a major in economics, authors and instructors often mistakenly assume that the issues we raise here could be gleaned from later year courses. However, sustainability must be integral to any interpretation of economics as the discipline that is dedicated to the optimal allocation of scarce resources to meet human needs. Furthermore, the topic is covered in only some courses on environmental economics and it is possible for students to graduate without ever having followed such subjects. Therefore, we assert that content on sustainability must make a dedicated appearance in basic texts in economics and as a result, the many authors of such texts would have to also alter their approach to expositing the principles of economics. For example, recall the welfare maximising outcome that 'smaller' could be better than 'larger' when sustainability is added to the conditions of perfect competition. Such a perspective puts traditional economics upside down (or to be more accurate, places the feet of economics squarely on the ground, rather than putting economics on its head).

We wish to reiterate that the appreciation of the scientifically driven linkages between KN and society must become an important component of the curriculum across several disciplines. The exclusion of such knowledge can lead to the emergence of ineffective policies, and mistaken premises and policy tools. An appreciation of the foundational role of KN in supporting the economy would lead policymakers to seek very different types of policies and practices compared to those in effect now. In other words, the choice of appropriate actions to occupy the policy portfolio will be contingent on the training of those who will formulate policy. As explained in Chapter 1, the training of economists and business managers is inadequate, given that environmental economics remains marginalised as an optional unit of study within the economics curriculum.

Policymakers, armed with an in-depth appreciation of the linkages between KN and the economy, would also consider lifestyle changes in their policy mix. Such changes are far more cost-effective compared to many technical options. In Chapter 9 we examined the importance of moderation. High-income economies need to cut back on economic growth and instead focus more on what actually helps generate further improvements in quality of life – much of which may come with increased spare time, a sense of community, and a healthy natural environment – instead of amassing material wealth and accumulating wastes. Low-income countries need to set modest aspirations for improvement in their standard of living. In some cases, this may mean further growth of their economies, but increasingly issues of income disparities, educational achievement, as well as healthy communities and KN come to the fore as essential elements of improved standards of living.

Consider, as an illustration, some of the wasteful and extravagant consumption behaviour in high-income countries. Most people in high-income countries drive to work – they even drive to their nearby suburban shop – mow their lawns and trim their edges with power tools, brush their teeth with electric toothbrushes, and carve their meat with electric carving knives. By the same token, low-income countries seem convinced that development means multi-storey, billion-dollar hotel complexes along their coastlines, six-lane freeways, nuclear power stations, their own airlines, and virtually everything else that the high-income countries have acquired.

While the rich countries need to moderate their lifestyles, the poorer countries need to recognise the price they pay for trying to achieve the lifestyles of the rich. That price includes not just damages to KN, but also to social cohesion, and personal and mental well-being. Settling for less could include power generation through windmills and solar panels and the construction of buildings with locally available and less energy-intensive materials. We often think of leap-frogging past developments as advantages for developing countries – for example, by skipping the development of wired communication with investments in wireless technology. Yet, we do not think of learning from the adverse environmental and social lessons from industrialised countries as motivators for environmental and social leap-frogging.

In the meantime, mainstream economists need to change their attitudes, too. For example, when an economy needs to be lifted out of a recession, economists need to shift their attention from the usual mechanisms of pump-priming the economy and consider incentives for investments in technologies that would render KN more effective, or go so far as to invest in KN directly, to build the assets on which societies can be maintained and thrive.

Population pressures

Increasing population size will inevitably lead to increasing pressure on KN stocks. Therefore, from the environmental policy perspective, keeping population size stable, while keeping per capita material and energy use in check as well, is essential for maintaining a steady stock of KN.

During the period 1960–1990, both environmental policy and growth policy shared a common ground in terms of population size. The focus was on reducing the rate of population growth and keeping the size of the population stable. The rationale from the growth perspective was that the rate of population growth must not exceed the rate of employment growth. Recall our discussion of this issue with the Swan–Solow model in Chapter 14. The growth-development literature identified population growth in developing countries as a result of economic decisions by parents. Having additional children in the family, beyond the first two or three who satisfy the intrinsic need for family formation, was to insure family income and aged care. While developed countries tried to practise zero population growth during these three decades, developing countries strove to reduce their rates of population growth in order to reduce the pressure of unemployment. For example, China practised the one-child policy and India even ventured into forced sterilisation. Several incentives to reduce the demand for additional children were also enacted. These included increasing wages and employment potential to mothers outside their homes, instituting firm child-labour laws and raising the costs of having additional children. The measures introduced by Singapore during the 1970s to curb the demand for additional children are interesting:

- Although housing was scarce, houses were assigned without reference to family size.
- Maternity leave was granted for a maximum of two children.
- Income tax relief was given for a maximum of three children.

However, after a lapse of nearly 30–40 years, it appears that environmental policy and growth policy no longer share a common ground. This is because of ageing populations and the reduction in the proportion of people of working age relative to total population size – especially in high-income countries such as Germany, Singapore, South Korea and Taiwan. Even a populous country like China has begun to demonstrate concerns with the issue of ageing and started to relax its one-child policy. Sentiments in favour of increasing population size in Australia was explicit in the words of a former treasurer, the Hon. Peter Costello, on the subject of having children: 'One for Mum, One for Dad and One for the Country'.

The issue of population size has another dimension beyond that of ageing. The disproportionate growth of population in urban areas exerts pressure on KN systems that support these urban areas. The mandate for the policy on population size is clear. The size needs to be kept stable while the working age demography is also maintained. The means of achieving this, however, are not straightforward.

Recent attempts include an extension of the working (retirement) age and, in some instances, the abolition of the retirement age. Further, such attempts need to be supplemented by the promotion of healthy lifestyles.

Property rights and land reform

Following the discussion in Chapters 3 and 6, the absence of tenure or limited tenure can often act as a disincentive to the adoption of KN stewardship. Consider issues pertaining to the littering of beaches and oceans, and the dumping of wastes into waterways. The reason for such unlawful behaviour rests at least in part on the fact that beaches and waterways are KN resources with common free access and unspecified ownership. Hardin's (1968) tragedy of the commons is the result of each individual with free access maximising his or her utility from such resources until the resource is depleted. The alternative is to confer collective ownership to the users of these resources. As indicated in Chapter 6, when KN resources are collectively owned their owners tend to refrain from indiscriminate exploitation. The evidence manifests in forests and agricultural lands that have survived centuries of utilisation, despite their common ownership. This is because the rights of ownership are well defined and these rights are contingent on clear rules of resource utilisation. It is not possible to confer the rights of ownership on KN assets such as beaches and rivers to selected individuals or groups, because this would create other adverse effects – specifically the aggravation of income and wealth inequalities.

Yet the conferring of ownership is possible in some cases. Recall the example of farmland in Australia, where most properties are leaseholds with the State. Surveys of farms in Australia and elsewhere have shown that the adoption of soil conservation and related methods of land care are less evident in leaseholds compared to freeholds. Converting leasehold titles to private ownership on farm properties is less arduous compared to the issue of privatising beaches and rivers.

In developing countries, most of the problems associated with common-access resources, such as deforestation, stem from the landlessness of labour. In many developing countries, a small minority of landowners employ a large number of tenant farmers to cultivate the land. For example, in Latin America, some 1.3 per cent of landowners hold nearly 72 per cent of the land under cultivation (Todaro and Smith 2006). Furthermore, in most cases, the rules of tenancy are so poorly defined that the eviction of the tenants happens quite frequently – often when the landowners find more lucrative uses for their land than tenant farming. Hence, land reform is inevitably an instrument of environmental policy. The conservation of forests and hill slopes in both South Korea and Taiwan stems partly from the fact that these countries achieved land reform early in their history.

Improving the productivity of KN

Several examples illustrate the importance of efforts to enhance the productivity of KN. Ambani (2011) reports on the Solar Bottle Bulb – a simple innovation developed by students at the Massachusetts Institute of Technology, costing barely $2 per bulb. It is simply a plastic bottle filled with bleach and water that is inserted through the metal roof. Although the bulb does not work at night, it provides much-wanted indoor illumination during the day in crowded settlements in Manila (Philippines) and elsewhere.

Lovins et al. (1999) give the example of Interface Corporation, which is a worldwide carpet-maker. One of Interface's engineers was able to make some very simple design changes to the length and width of pipes that were used for pumping fluids in the carpet factory. As a result, it was possible to replace a 95-horsepower pump with a 7-horsepower pump. This gave the carpet factory significant savings in energy costs and widened the profit margin. Similarly, Lovins et al. (1999) argue that the United States industry would save over US$100 billion, if all firms shifted from ordinary lighting ballasts (the controls for normal fluorescent lamps) to electronic ballasts that automatically change the intensity of illumination to match available daylight. Other examples include achieving cost savings through the efficient use of paper; for example, double-sided printing reduced AT&T's paper costs by 15 per cent, while Johnson & Johnson managed to cut expenses by US$2.8 million by managing their packaging more effectively. Lovins et al. (1999) argue that these savings are equivalent to saving 330 acres of forest each year. The list of opportunities to retain, if not increase, standards of living while reducing one's environmental footprint is long. Yet, there is no real concerted effort to aggressively exploit these opportunities for the betterment of the human condition and the environment at the same time.

The greenhouse gas concerns have already ushered in several alternative energy initiatives. These initiatives range between energy-saving devices and the development of alternative energy sources to replace fossil fuels. These include wind power options and electric vehicles (EVs) that could utilise recharging stations on the mobile phone model. However, recall our discussion in Chapter 6 about challenges surrounding EVs. Most important of these are perhaps the reliance on batteries and their disposal, as well as the need for mining and extracting lithium. And of course, advances in technology could reduce such reliance.

Not all the technology-induced challenges to KN quality need a technological solution. While EVs can serve as a meaningful alternative to vehicles using internal combustion engines, they continue to perpetuate current mobility choices – with all their impacts on land use and wildlife, for example. A broader systems view to solving the problem will be required, such as changing to zoning regulations so that small shops can thrive in the vicinity to consumers, thus eliminating the longer-distance shopping trips that are so typical in many urban and suburban areas of wealthy countries. Or different agricultural policies that help promote local food production, and thus reduce the reliance on long-distance hauls of produce, which, whether done with EV or conventional trucks, is still more environmentally harmful and socially undesirable than supporting local farmers.

Such measures, nevertheless, while aiming to reduce KN damages, fall short of reducing cumulative KN damages.

Policies Aimed at Minimising KN Damages towards Zero and Restoring KN

Let us recall our discussion in Chapter 9 on the joint consideration of the sustainability of KN with the conditions of perfect competition. Such consideration changes the choice criterion from mere welfare maximisation to welfare maximisation over an infinite length of time. This means that the configurations of markets should be such that consumption and production can continue indefinitely. The policies considered in this section strive to meet this end.

Closed-Loop Production Systems

The closed-loop manufacturing model has already permeated business and is reported to have generated for the United States manufacturing industry revenues amounting to US$53 billion way back in 1996 alone. The Ellen MacArthur Foundation (2017) provides several examples and case studies in this domain area. Sometimes also referred to as the circular economy, the closed loop system is one in which waste is either recycled as an input into another process or returned to nature as harmless material. We need to strive for the prevention, as far as possible, of pollution loads from entering environmental sinks such that the sinks have sufficient time to heal and recommence the provision of service flows. The literature on industrial ecology provides several examples of such systems (Zhu and Ruth 2013, 2014). For example, the Dutch flower industry considered above is a near closed-loop system. So is the example of the industrial park in the Kalundborg District of Denmark (discussed in Chapter 8), where several industries had formed a symbiotic relationship between themselves to reuse waste generated. Several examples in the literature illustrate this system. We consider here three examples drawn from waste management, energy conservation and management, and agriculture.

Waste Management

Sanitation and hygiene are not the only issues that surround waste treatment and management. For example, consider sewage treatment, which eventually relies on a variety of environmental sinks – oceans, lakes, rivers and subterranean ecosystems. Efficient methods of sewage treatment have far-reaching implications for sustainability. In this context, Singapore's Public Utilities Board (PUB) offers a framework that other cities can and perhaps must emulate. In this city-state, every single dwelling is connected to a sewerage system where the treatment is so advanced that the recovered water is reusable for both industrial and potable purposes. However, besides the issue of water conservation, it is equally important to note that the extent of treatment renders inert the residues that are deposited into the ocean sink. Most coastal cities in the world deposit partially sewage into ocean sinks on the grossly mistaken assumption that the ocean is a KN sink of infinite size. For example, Sydney (Australia) pumps out each day at least 12 million litres of partially treated sewage into the Pacific Ocean. Imagine the cumulative load of pathogenic material that will have accumulated if one tallies all other coastal pumping stations in Australia and the number of years of this activity. Sceptics would of course argue that the ocean is not a static body of water and that wave actions, oceanic movements and biological activity will render the deposits harmless. This may be true if the loads of deposition are small and not continuous over time. Scientific evidence now indicates that the quality of the Pacific Ocean on the Eastern Seaboard of Australia may be seriously compromised. It is this type of practice (among others) that had rendered nearly half the Baltic Sea (below a certain depth) lifeless. It is plausible to argue that the sink capacity of the oceans spanning every continent is compromised, owing to improper methods of sewage treatment. If all cities in the world were to replicate the Singapore model, then the implications for global warming would be immensely significant. This is because oceans are the world's largest naturally occurring carbon sink, due mainly to the presence of phytoplankton and their photosynthetic activity. Waste deposition into the oceans destroys these plankton. The sink capacity of the oceans, when restored through adequate waste management, could in turn help restore balance to the carbon cycle. However, the ruling on 21 May 2024, of the International Tribunal for the Law Seas declares that GHG emissions do represent marine pollution. This ruling, which is based on scientific evidence, implies that oceans too have finite capacity to deal with GHGs.

Energy Conservation and Management

Compared to 20 years ago, every dwelling and building in almost every city is now equipped with an air conditioner (in tropical countries) or a heat pump (in temperate countries) or a reverse-cycle unit (mainly in temperate countries). Air conditioners generally pump out hot air – depending on the indoor temperature setting – and thus raise the outdoor ambient temperature. Heat pumps in winter gush out colder air and thereby lower the external ambient temperature. These technologies, in short, create desirable indoor temperatures at the expense of temperatures outdoors. The sceptics' response would be that the change in temperature prompted by each unit is minuscule relative to the volume of the earth's troposphere. On top of that, the generation of electricity for these pieces of equipment, mainly from fossil fuels, adds to the heat-trapping processes associated with climate change. If one were to take a tally of the number of dwellings across the globe and the time continuum, then the cumulative effect is quite likely to be significant. Closing the loop (at least partially) on this type of system would be to search for dwelling/building designs that reduce the demand for heating and cooling and at the same time enable the capture of the heat/cold emission for reuse within the dwelling/building. For example, in a shopping centre complex, the design aspects could be construed such that dry-cleaning establishments receive the hot air exhausts (in tropical countries) and cold rooms for refrigeration receive the cold air exhausts (in temperate countries). In both cases, the partial loop closure will reduce the energy demand imposed on the grid.

Agriculture

The above two examples of closed-loop systems are perhaps technical as well as energy-intensive. Khatri (2009) offers a simple example of Montfort Boys Town, an institution, which helps disadvantaged teenagers in Fiji at a location not far from the capital Suva. The project involved transferring waste from a nearby brewery, mainly spent grain, to the Boys Town, where several types of zero waste bio-systems management practices are adopted and taught to the students. These included an animal husbandry unit to prepare animal feed from the spent grain. The animal waste is then used for biogas production as well as barley crop production. The barley grains are of course transferred back to the brewery.

Natural Sequence Farming (NSF) was introduced by an Australian farmer – Peter Andrews. As explained in Andrews (2006, 2008), NSF is a collection of structural and non-structural methods that aim to minimise soil erosion and increase groundwater recharge. The methods also involve reintroducing some native species of flora (mistakenly deemed as weeds) so that soils can recoup their original properties in terms of organic matter.

Ellen MacArthur Foundation (2017) provides the example of a Canadian firm Ostara that demonstrates circularity involving three entities, namely the urban sector, a waste recovery facility and agriculture. The aim is to deconstruct the prevailing linear model in agriculture. The premise adopted here is that nutrients extracted from soils through agriculture are returned at least partially to urban waste repositories. The circular (closed-loop) system involves connecting the urban waste repositories to nutrient management facilities that recover phosphorous and other important nutrients, which in turn are then returned to agriculture.

Bio-Mimicry

Innovative methods of commodity development involve attempts to replicate nature's methods of dealing with waste. Benyus (1997) provides several examples under the heading of 'bio-mimicry', which is defined as 'a new science that studies nature's models and then imitates or takes inspiration from these designs and processes to solve human problems'. Examples of bio-mimicry abound, the web that spiders weave results from the residues of their diet, and Benyus illustrates how similarly strong fabric can be made without powerful chemicals such as sulphuric acid. Abalone make their shells without the heating and furnaces that accompany most ceramic factories, yet the properties of these shells suggest that the processes by which they are crafted can serve as close substitutes for traditional processes. Several other examples of bio-mimicry can be found in Vanderbilt (2012). These include establishing a greenhouse in the Qatari desert by mimicking the evaporation tricks of a camel's nose to induce condensation; and establishing devices to trap heat from the study of birdwing butterflies.

Edward Linacre from the Swinburne University of Technology in Melbourne (Australia) won the James Dyson award in 2011 for developing a low-cost technology to trap moisture from thin air. The device, labelled 'Airdrop', mimics in part water harvesting performed by desert plants such as rhubarb. The device draws air from the atmosphere below the ground using a solar-powered pump and the resulting condensation within a network of pipes permits water to be collected and then used for subsurface irrigation. Although not commercialised as yet, innovations such as these suggest that desertification in arid areas can be reversed and the capacity of KN expanded. Consider the costly conflict between the states of Karnataka and Tami Nadu in India when the Supreme Court deemed that the waters of the river Kaveri that originates in Karnataka must be shared with Tamil Nadu. Tensions pertaining to water sharing in this and many other regions could be minimised if cost-effective technologies such as airdrop irrigation were commercialised.

Changing the Fundamental Nature of the Business Model

The traditional business model is one that involves the production and sale of goods. Profits from these activities tend to be privatised while the social and environmental costs are spread across society. One way to shift from this tradition is to embrace a service-oriented enterprise so that environmental resources can be conserved through recycling and resource conservation. The underlying philosophy is encapsulated in the following phrases: *Do not sell light bulbs, but rather sell illumination* or *Do not sell heaters, but rather sell heat*. Lovins et al. (1999) give examples of how Interface (the carpet-makers) and Schindler (makers of elevators) have shifted to a leasing model so that considerable savings are made through recycling the original. For example, Interface does not sell carpets, but rather leases floor covering and manages the maintenance of it after it has been leased. When the floor covering needs replacement, Interface can recycle the original and thus avoid adding it to the materials that accumulate in landfills. In this business model, the firm is responsible for the product that it makes throughout the product's lifecycle. Xerox Corporation, which had a near monopoly with its photocopiers, was able to fight back competition with new arrivals by adopting this service orientation and putting into practice the maintenance and recycling of photocopiers. Reinhardt (1999) states that Xerox saved nearly US$50 million in its first year of restructured operation due to the resultant drop in raw material purchases to make new equipment.

Restore and Expand KN Assets

In Chapter 1, we presented a sample of damages inflicted on KN. These are just a few illustrations from a myriad of damages across the globe. Dominant among these is climate change. The damages represent the detrimental effects of sustained interference with KN endowments over many centuries. The primary drivers of such interference are economic growth, conquest and conflict. These drivers are interrelated. For example, colonisers having conquered alien lands, cleared mountain ranges of tall-tree forest canopies and replaced them with shrub like commercial crops to enhance trade and income. As indicated earlier, changes in canopy density can trigger hydrologic disturbances. We consider the issues of conquest and conflict separately as an overarching policy item below. For the moment we deal with economic growth as a driver of KN damage.

As explained earlier in Chapters 10–15, economic growth is mistakenly deemed an essential prerequisite for employment creation and income. This is primarily because of the longstanding notion that the rate of increase of national product must exceed that of population growth to ensure that per capita income does not fall. However, as we have explained in preceding chapters, economic growth inevitably results in KN depletion. Hence, revitalisation of KN endowments becomes imperative. That is, the need to make explicit KN investments – we referred to this as (I_{KN}) in chapter 11. The challenges associated with an economic growth mentality become even more questionable when growth is perceived – or sold by investment and policymakers to the general public – as a panacea for environmental problems. In short, the notion that economic growth can solve the problems caused by economic growth seems nonsensical at its core.

Policy formulation needs to be guided by scientific information and research. Therefore, expansion of research funding for environmental science becomes a priority area. The current focus appears to be on restoring the Carbon Cycle by reducing emissions of GHGs. Yet, a broader approach spanning several KN investments would facilitate reduction in GHG loads. The following two examples of policy initiatives with reference to forests and rivers illustrate this observation.

- Reforestation will always be required as long as economic growth persists because forests are regularly cleared to support the building and other industries. Forest clearance also makes way for other economic activities such as mining and agriculture. Several economists refer to Australia's display of 30 years of sustained GDP growth between 1990 and 2019 as a record performance. But they fail to realise that during this period, Australia lost nearly 15 million hectares of forest. Ironically, Australia lost nearly the same amount of forest within a few months due to the wild summer fires of 2019–2020. Deforestation that accompanied economic growth could have been a causal factor underlying these catastrophic fires. Loss of forests due to wildfires is universal across all continents. In this regard, two observations are warranted. First, Qiu et al. (2019) and Bystriakova et al. (2003) demonstrate that bamboo is a potential substitute for hardwood. Hence, the establishment of bamboo plantations on marginal lands, as suggested by Buckingham (2014), could reduce the demand for hardwood timber and hence dilute the need for deforestation. Second, it is imperative that a prohibition order be issued on any clearance of old growth forests primarily for reasons of preserving canopy density and secondarily to protect biodiversity so important for regeneration in the long run. Recall our discussion in Chapter 5 (Box 5.2) of how old growth forests are harvested for commercial interests. So, preservation of existing forests is in order and rapid reforestation – which takes time to take hold – remains essential. As illustrated in Thampapillai (2020), rapid planting of seedlings using drone technology has been trialled in Australia.

Revitalising the forest system will in turn help restore the hydrologic system of which forests are an integral part.
- Rivers and streams require revitalisation because they are essential components of the hydrologic system that have been subject to extreme stress. The main driver causing stress to this system is the abstraction of water for irrigation, mining, commercial purposes and domestic use. In several instances, the abstractions have been too severe to the extent diverting the natural course of rivers and causing the collapse of supporting ecosystems. A classic example is the death of the Aral Sea, owing to the diversion in the 1960s of the Syr Darya and the Amu Darya rivers by the former Soviet Union. These rivers were diverted away from the Sea and its basin to support agricultural enterprises, especially cotton. As a result, the Aral Sea, which was once the fourth largest lake in the world, became a dry bed. Efforts to restore flows into the Northern part of the Sea began only around the early 2000s by Kazakhstan in collaboration with the World Bank. These efforts now show promising results. In such contexts, adopting the principles of socio-hydrology, pioneered by Professor Sivapalan from the University of Illinois, are pertinent. Sivapalan et al. (2011) present persuasive case for managing the needs of KN alongside those of humans as per the principles of socio-hydrology. This means that several investments need to be made on the river systems. For example, large-scale abstractions that led to the creation of storage infrastructures would have to be demolished and the impounded water returned to natural river flows. Further, riverbanks that were cleared of mangroves and other native vegetation to foster various economic activities need to be reintroduced. Proactive measures to clean up polluted and weed infested river systems such as those cited by Bush (2024) must be adopted. Widespread adoption of Israel's innovation in drip irrigation would reduce the need for largescale abstractions.

A quick note about mining is in order. Minerals beneath the ground are important source materials for the functioning of any economy. But, as Ross et al. (2018) illustrate, the preservation of mining areas and leaving them without being mined, stabilises the earth's surface, and also prevents the loss of sulphur – a key source nutrient for all life forms. It is not just the relatively small areas where mining takes place that should be of concern, it is the long shadow that mining operations cast across the earth – mine tailings that leach into rivers, streams and ground water bodies, and thus affect large swaths of land well beyond the places where the soil is disturbed, and manufacturing processes that require immense amounts of energy, which, in turn, is extracted from mines as well as oil and gas wells. Hence it is imperative that mining is streamlined and performed under careful regulatory control. Mining can reduce forest cover and also involve abstraction of water. Further details about the effects of mining on KN with reference to Australia can be found in Thampapillai (2020). And as Trenberth (2005) explains, the carbon and hydrological cycles, upon which mining has its impact, are interrelated. A disturbed hydrologic cycle is invariably one where prolonged periods of drought, are followed by: intense precipitation, floods and storms. Recent history provides evidence of such events. That is, stabilising the hydrologic system does help to stabilise the carbon cycle.

Overarching Policies – Social Capital (KS) – A Precondition

Implementing policies on the basis of the foregoing narrative could bear limited results and may even be futile, if society's landscape is riddled with conflict, violence and war. Attempts to preserve and nurture KN must be also accompanied by the prevalence of a high degree of social capital (KS). Putnam (2001) defines KS as generally a collection of cohesive social networks and organisations that achieve social stability. The presence of a high degree of KS in any society is manifested in the display of a diverse array of attributes consisting of items such as: happiness, peace, law and order, absence of corruption and good health. Therefore, a society blessed with a high degree of these attributes serves as a platform, for the smooth functioning of any economy including the pursuit of quality standards for KN. The same applies to the global community and the international economy.

Reality though is one where conflict and violence have pervaded every space of the globe ranging from within households to territories spanning nations. Notable international conflicts of 2024 include: Russia's invasion of Ukraine and the resulting war between the two states; Israel's war against Hamas and complete disarray in Palestine; China's territorial claims in the South China Sea and its forthright rejection by especially Vietnam and the Philippines; Rebellion in Myanmar by tribal and ethnic groups; Gang Violence in Haiti; and Civil War in Sudan. The analysis of such conflicts is beyond the scope of this book. Yet, we wish to make the following pertinent observations.

- The relationship between KN and KS is dual. The breakdown of either results in a collapse of the other. For example, conflict and war unleash, apart from human misery, the collapse of several vital ecosystems. Recall our reference to Bergman (2023) in Chapter 10. We reproduce the statement from the editorial here:

  > Artillery fire wipes out forests. Landmines maim animals and humans or wash up after climate-fuelled flooding. Toxic chemicals lace the soils. If all the world's military forces were a country, they would pump out enough greenhouse gases to be the fourth-highest emitter.

  Therefore, nurturing and preserving KS is tantamount to safeguarding KN.
- An important driver of KS damage is corruption, which underlies the spread of arms and weapons that perpetuate conflicts and violence as well as the permeation of illegal drugs that lead to substance abuse in society. Anecdotal evidence suggests that arms dealers and traders fuel conflicts by selling as well as training various parties to the conflict. Hence, strict market regulation of arms and weapons as well as prohibition of markets for harmful substances are in order. Both the collapse of KN as well that of KS result in people being rendered homeless

- and refugees. In these contexts, perverse markets of 'people smuggling' that extort funds from innocent victims have emerged. Policies to curb corruption are dealt within the literature of public policy and administration; for example, see Bali et al. (2021), Graycar (2020), and Howlett and Ramesh (2016).
- Although much easier said than done, dialogue and negotiation must always take precedence over the employment of arms and weapons. Recall our reference to the concept of cross-compliance earlier in this Chapter. That is, implementation of one policy must not negate the effectiveness of another. Observing cross compliance between KN and KS is of paramount importance. Hence, it would be superfluous for politicians to speak of mitigating climate change and embarking on a net-zero journey, when they are unable to give priority to negotiant and dialogue.
- The late Professor Lynn Stout (2011) in her book on Social Conscience – an important attribute of KS – illustrates how the media often overlooks acts of stewardship and kinship by the community during times of crises. She gives the example of Hurricane Katrina – a KN disaster – that caused widespread destruction and havoc in the city of New Orleans in 2005. Whilst the media reported the damage and scenes of looting, it failed to show how members of the community rallied around to help each other. Stout's (2011) thesis rests on the need for cultivating social conscience through various avenues – something, which must find a place in the formulation of economic policies.

Concluding Remarks

The choice of an appropriate mix of actions within the policy portfolio must inevitably follow the dedicated path to restore KN endowments that have been damaged by years of neglect, false promises and complacency. As indicated, safeguarding KS is at least as important. While a zero-emission regime may prove difficult in the short run, it may be feasible in the long run. In his report on KN repatriation, Stern (2006) gives only 50 years as the lower end of the time spectrum within which solutions must be found. Although this estimate is probabilistic, it is prudent to be aggressive in the search for appropriate measures such as some of those mentioned above. Further, Stern's estimate may not have anticipated the various wars and conflicts that have emerged since the submission of his report. Some commonsense approaches, such as globalising the Singapore–PUB model of sewage treatment are not readily apparent to many but can be put in place perhaps much sooner than global Carbon trading markets.

Equally important is the role of Environmental Economics in guiding policy formulation. As illustrated in Chapter 8, when KN is explicitly recognised in the production function, the basic theory of the firm that guides business decisions becomes altered to such an extent that voluntary stewardship by firms with respect to KN may be feasible. However, the theory of production involving KN is nowhere to be found within the Economics Curriculum of business schools.

Finally, in most instances, the ownership of KN stocks does not rest with individual nations, but with the global community. Hence, many of the technologies and innovations to expand the size of KN must be pursued in a spirit of global partnership and cooperation and territorial conflicts that destroy both KS and KN have to be made redundant.

Review Questions

1 Environmental policies were classified into three broad categories in this chapter, namely (i) those that reduce KN damages; (ii) those that minimize KN damages towards zero KN damages; (iii) overarching policies. Which of the policies considered in the first category have the clear potential to move to the second category?
2 Review the types of lifestyle changes that need to be considered in both developed and developing countries to achieve the sustainability of global ecosystems.
3 Provide examples of 'closed-loop' production systems (beyond those provided in the text) that could be easily implemented.
4 Provide illustrations of how global partnerships could work in implementing 'closed-loop' production systems.

References

Ambani, P., 'An innovative and cheap 'solar bottle bulb' solution lights homes in Manila', *Ecopreneurist*, September 2011 (http://ecopreneurist.com/2011/09/14/an-innovative-and-cheap-solar-light-bulb-lights-homes-in-manila/)

Andrews, P., *Back from the Brink: How Australia's Landscape can be Saved*, ABC Books, Sydney, 2006.

Andrews, P., *Beyond the Brink: A Radical Vision for Australia's Landscape*, ABC Books, Sydney, 2008.

Atkinson, A. B. and Leigh, A., 'The distribution of top incomes in Australia', *Economic Record*, 83(262): 247–261, 2007.

Bali, A. S., Howlett, M., Lewis, J. M. and Ramesh, M., 'Procedural policy tools in theory and practice', *Policy and Society*, 40(3): 295–311, 2021.

Bergman, J., Editorial, *The Conversation – Australian Edition*, 15 November 2023.

Benyus, J., *Biomimicry*, Harper Collins, New York, 1997.

Buckingham, K., 'Bamboo: The Secret Weapon in Forest and Landscape Restoration?', *Insights*, World Resources Institute, 28 February 2014. (https://www.wri.org/insights/bamboo-secret-weapon-forest-and-landscape-restoration)

Bystriakova, N., Kapos, V., Stapleton, C. and Lysenko, I., *Bamboo Biodiversity: Information for Planning Conservation and Management in the Asia-Pacific region*, UNEP-WCMP, UK 2003

Bush, J., 'Making merry: How we brought Melbourne's Merri Creek back from pollution, neglect and weeds', *The Conversation*, 7 May 2024.

Dasgupta, A. K. and Pearce, D. W., *Cost-Benefit Analysis: Theory and Practice*, Macmillan, London, 1972.

Dasgupta, P. and Heal, G. M., 'The optimal depletion of exhaustible resources', *The Review of Economic Studies*, 41(1): 3–28, 1974.

Ellen MacArthur Foundation, *Towards the Circular Economy: Economic and Business Rationale for an Accelerated Transition*, Cowes, United Kingdom, 2017.

Frank, R. H., Bernanke, B. S. and Lui, H.-K., *Principles of Economics* (Asia Global Edition), McGraw Hill Education, New York, 2015.

Graycar, A. (Ed), *Handbook on Corruption, Ethics and Integrity in Public Administration: Elgar Handbooks in Public Administration and Management*, Edward Elgar, London, 2020.

Hardin, G., 'The tragedy of the commons', *Science*, 162(3859): 1243–1248, 1968.

Hirshleifer, J., Glazer, A. and Hirshleifer, D. *Price Theory and Applications – Decisions, Markets and Information*, Cambridge University Press, New York, 2005.

Howlett, M. and Ramesh, M., 'Achilles' heels of governance: Critical capacity deficits and their role in governance failures', *Regulation and Governance*, 10(4): 301–313, 2016.

Kahneman, D., Knetsch, J. L. and Thaler, R. H., 'Experimental tests of the endowment effect and the Coase Theorem', *The Journal of Political Economy*, 98(6): 1325–1348, 1990.

Khatri, K. 'Integrated biosystems -Montfort Boys Town, Suva, Fiji Islands', in Nair, J., Furedy, C., Hoysala, C. and Doelle, H. (Eds), *Technologies and Management for Sustainable Biosystems*, Nova Science Publishers Inc., 2009.

Knetsch, J. L., 'The endowment effect and evidence of nonreversible indifference curves', *American Economic Review*, 79(5): 1277–1284, 1989.

Lovins, A. B., Lovins, L. H. and Hawken, P., 'A road map to natural capitalism', *Harvard Business Review*, 77(3): 145–58, 1999.

Mankiw, N. G., *Principles of Macroeconomics*, 3rd edn, Thomson South-Western, Mason, OH, 2004.

Mankiw, N. G., *Principles of Economics*, 7th edn, Cengage Learning, Stamford CT, 2014.

Parkin, M., *Economics*, 12th edn, Pearson Education Limited, Harlow, 2016.

Pearse, R. and Böhm, S., 'Ten reasons why carbon markets will not bring about radical emissions reduction', *Carbon Management*, 5(4): 325–337, 2014. https://doi.org/10.1080/17583004.2014.990679

Porter, M. E. and van der Linde, C., 'Toward a new conception of the environment-competitiveness relationship', *The Journal of Economic Perspectives*, 9(4): 97–118, 1995.

Putnam, R. D., *Bowling Alone: The Collapse and Revival of American Community*, Simon and Schuster, New York, 2001.

Qiu, H., Xu, J., He, Z., Long, L. and Yue, X., "Bamboo as an emerging source of raw material for household and building products," *BioRes,* 14(2): 2465–2467, 2019.

Reinhardt, F. L., 'Bringing the environment down to earth', *Harvard Business Review*, 77(4): 149–158, 1999.

Ross, M. R. V., Nippgen, F., Hassett, B. A., McGlynn, B. L. and Bernhardt, E. S., "Pyrite oxidation drives exceptionally high weathering rates and geologic CO_2 release in mountaintop-mined landscapes", *Global Biogeochemical Cycles*, 2018 (https:doi.org/10.1029/2017GB005798)

Ruth, M., Zhu, J., Lee, N. S. and Mirzaee, S., 'Creating and evaluating co-benefits and co-costs of environmental policies and investments in urban areas', in J. Woltjer, E. Alexander, A. Hull and M. Ruth (eds), *Place-based Evaluation for Integrated Land-Use Management*, Ashgate, London, England, 27–56, 2015.

Sinden, J. A. and Thampapillai, D. J., *Benefit-Cost Analysis*, Longman, Melbourne, 1995.

Sivapalan, M., Savenije, H. H. G. and Blöschl, G., "Socio-hydrology: A new science of people and water", *Hydrological Processes*, Invited Commentary, 2011.

Stern, N., *Stern Review: The Economics of Climate Change,* Cambridge University Press, Cambridge, United Kingdom, 2006.

Stout, L. A., *Cultivating Social Conscience: How Good Laws Make Good People*, Princeton University Press, Princeton, 2011.

Sugden, R. and Williams, A., *The Principles of Practical Cost-Benefit Analysis*, Oxford University Press, New York, 1978.

Thampapillai, D. J., 'Economic fixes should not worsen environmental crisis', *YaleGlobal Online*, October 2011.

Thampapillai, D. J., Wu, X. and Tan, S. L., 'Fiscal balance: Environmental taxes and investments', *Journal of Natural Resources Policy Research*, 2(2): 137–147, 2010.

Thampapillai, D. J., Bali, A., Lodhi, I. A. and Thampapillai, Dilan J., *Covid 19 and its Contagion,* https://scholarbank.nus.edu.sg/handle/10635/201009 (posted 30 September 2021).

Thampapillai, V., 'Limits to government water buy-backs for environmental flows in the Murray-Darling Basin (part 1)', *Environmental Policy and Law*, 39(4–5): 247–265, 2009.

Todaro, M. P. and Smith, S. C., *Economic Development*, 9th edn, Addison Wesley, New York, 2006.

Trenberth, K. E., "The impact of climate change and variability on heavy precipitation, floods, and droughts", in Anderson, M. G. (ed), *Encyclopaedia of Hydrological Sciences*, John Wiley & Sons, Ltd. UK, 2005.

Vanderbilt, T., "How biomimicry is inspiring human innovation", *Smithsonian Magazine*, September 2012; (Accessed at: https://www.smithsonianmag.com/science-nature/how-biomimicry-is-inspiring-human-innovation-17924040/)

Zhu, J. and Ruth, M., 'Exploring the resilience of industrial ecosystems', *Journal of Environmental Management*, 122: 65–75, 2013.

Zhu, J. and Ruth, M., 'The development of regional collaboration for resource efficiency: A network perspective on industrial symbiosis', *Computers, Environment and Urban Systems*, 44: 37–46, 2014.

20 An Environmental Economics for Sustainability

The Premises for Sustainability

This book was largely motivated by the recognition that any form of economic analysis must be cognisant of the fact that neither production, consumption nor the exchange of goods and services can occur without drawing on KN. Extraction of mineral resources, the use of nutrients in soils, conversion of energy from either fossil deposits or solar radiation, for example, all have their impacts on KN. It is important to note that the use of KN resources for making infrastructures, food and a host of other services to enhance society's welfare, does tend to concurrently draw down the KN assets on which that wealth-generation depends. Even resources that are in principle renewable can turn into non-renewable ones, for example when fish catch exceeds the populations' regenerative capacities, or when forests are cut at rates higher than their re-growth. The use of non-renewable resources also can push renewable resources past the point at which they can recover, for example when the burning of fossil fuels results in sulfur emissions that acidify lakes and soils, or when the accumulation of greenhouse gases from land conversion and industrial activity alters the climate and triggers increasingly frequent severe weather events that adversely impact the structure and function of ecosystems, let alone of economies.

Overutilisation of renewable resources and impairment of their regenerative capacities as a result of non-renewable resource consumption both undermine the sustainability of renewables. These fundamental biophysical constraints on economic activity have direct consequences not only for policies that seek to foster economic growth (EG) and environmental quality (EQ) but also for issues of intergenerational concerns (IGC). We ask: Under what circumstances would the erosion of KN justify the accumulation of human-made capital (KM)? How much of KN would be required to maintain KM? Is there a realm within which the economy can function without losses to KN and KM, and if so, what is that realm? Since answers to these questions require us to take a long view on economic performance, they inevitably affect judgments about the needs and wants of future generations, the technologies available to them, and the conditions of KN needed to support them all.

Aside from biophysical limits on the rates at which wastes can be assimilated by ecosystems or at which renewable resources can re-generate, there are other constraints on economies that tend to be outside the purview of traditional economics. Most notable among these are the laws of thermodynamics that constrain the use of materials and conversion of energy. The first law states that matter can neither be created nor destroyed, which implies that the wastes we generate are always with us in one form or another. Reusing, recycling or processing them will require energy. As the second law of thermodynamics states, that conversion of energy is always associated with a loss of quality. The concentrated energy in the fuel used to drive a car, for example, ultimately ends up as low-quality energy that is dissipated into KN sinks in the form of friction and heat, or via tailpipe emissions. None of them can be brought back to fuel the car, and even if technical means were available to do so, they would require on the net a greater degradation of energy than what would be recovered. As we pointed out in Chapter 2, the combination of these two laws of thermodynamics – not just the first law alone – are at the heart of what causes absolute scarcity.

With an economy that must do with finite material endowments on Earth and that is constrained by the energy flow onto it and back into space, the potential for absolute scarcity is real. Only an economy that makes use of materials and energy in ways that leave KN intact can flourish in the long run. Any depletion of fossil fuels must therefore be accompanied by their replacement with renewables at rates that compensate for the depletion of these fossil fuels *and* make energy available to reuse and recycle the materials that are dissipated in the processes that extract and use fossil fuels and that make and maintain renewable energy systems. Because all processes occur with a net increase in entropy, typically less rather than more consumption is better (assuming that basic needs are met). This is in stark contrast to a key premise of consumer theory, which presumes that more is better. In short, the standard distinction between a *good* and a *bad* becomes blurred when we recognise the pervasive nature of the laws of thermodynamics. And as a consequence, the standard depiction in economics of 'optimal' consumer choice must be adjusted for the biophysical reality within which that choice plays out.

Building KN into Economic Models

Economics is conventionally defined as the discipline that deals with the optimal allocation of scarce resources to meet human needs. To do so in a meaningful way requires that KN must explicitly enter the analysis as capital, alongside other contributors to economic wealth, such as L and KM. Since both the sink and source functions of KN are limited, sustainability cannot be achieved without proper treatment of KN. As a consequence of this observation, we have expanded in this text the standard economic models of production, consumption and exchange to explicitly account for KN. We have done this largely in a two-step process, first by understanding the purpose, nature and limitations of the standard models, and then by comparing the results from these models in a second step to the results once KN has been introduced into them.

As much as possible, we retained the original models on which much of modern economics rests. We did not question, for example, conventional notions of optimality when we derived equilibrium conditions for the interactions of demand and supply, or the underlying assumptions that decision makers – households and firms, for example – operate rationally, let alone have all the relevant information at their disposal to assess at any given point in time the ramifications of actions that may extend over long periods of time and across large geographies (Ruth 2018).

We also stuck to the notion that human behaviour is guided by incentives and that proper design of incentive systems can steer behaviours towards desirable outcomes. Well-functioning markets may give the signals on which to act – rising prices would suggest that the scarcity of a resource is increasing and thus stimulate a decline in demand and a search for alternatives. Where unintended consequences of production and consumption are encountered, namely externalities may, in principle, be internalised to adjust prices and the signals they give to change behaviours and technologies. Establishment of property rights and creating markets for the aspects of KN in question may help in that regard. But what if the world is more complex than presumed here? What if prices are not only the product of rational economic decision-making and instead reflect deep-seated social tensions and long-overlooked environmental constraints? For example, in societies in which women and children hold limited rights for self-expression and self-determination, the wages paid to them for their labour will be depressed. In places where standards for preserving KN are low, resource extraction and pollution may cause harms that remain unaccounted for in economic decision-making. The prices of goods and services in conditions of social and KN exploitation are then not worth much with respect to their ability to guide economic decisions towards optimal outcomes (Røpke 1999). More likely, they will entrench unsustainable practices.

Within these conceptual confines we identified the conditions for economic decisions and for policy to help promote sustainable uses of KN. Unlimited economic growth – the steady state of traditional economics – violates sustainable uses of KN. The steady state of economics is in fact an unsteady state when KN is considered because the growth trajectory coincides with a KN depletion trajectory (Thampapillai and Ruth 2019). These differences permeate all of economics. For example, when KN is introduced as an additional input in a standard production function alongside KM and L, we find that there are some important changes. First, it is not so straightforward to pursue an expansion path when the budgetary conditions improve. This is because higher levels of KM and L intensities are associated with increased fragility of the KN endowments as we illustrated in Chapter 8. Recall the conceptualisation regarding the shrinking of KN's domain of assimilative capacity and the increase in the potential gradient of KN degradation, as increases in Y are funnelled by increasing factor intensities of L and KM. Second, and closely related, the benefits from economies of scale that are observed with higher production quantities may be overstated. In short, there are fewer degrees of freedom for the optimal ways to allocate scarce resources to meet human needs than traditional economics leads us to belief. As a consequence of the additional constraints, the size of the economy, as measured by its throughput of materials and energy, must be smaller than traditional economic models suggest.

To close that gap requires policies cognisant of the fundamental roles that KN and the laws of physics play in wealth generation.

The Need for a New Kind of Policy

One of the many challenges associated with the use of KN lies in the long lead times that often exist in identifying the solutions to problems pertaining to KN, implementing solutions and actually reversing the KN damage that has been done in the meanwhile. In the case of global warming, for example, the problem itself was largely caused by the rapidly rising use of fossil fuels during and following the industrial revolution of the mid-to-late-1700s. The connection between burning of fossil fuels and what is called the greenhouse effect – the trapping of long-wave radiation as a result of accumulating greenhouse gases in the atmosphere – was not postulated until the end of the 1800s by the Swedish scientist Svante Arrhenius. It took then almost another hundred years for international negotiations to begin calling for the reduction in emissions of greenhouse gases, which, by the way, would not mean a reduction in atmospheric concentrations for more than at least yet another hundred years to come, given the long mean residence time of greenhouse gases in the atmosphere. And despite these calls for action, emissions have been rising, and continue to be on an upward trend for the foreseeable future notwithstanding the pledges to net zero targets. This is in part because of the considerable inertia of the economic systems that are built to use fossil fuels and designed to grow, and in part because of a lack of forceful actions by policymakers, many of whom still question the fundamental science that has been established so long ago.

Meanwhile, economists point at the role of markets in allocating goods and services among producers and consumers, and the lack of markets for many of the elements that make up KN. Their logical conclusion, then, is to establish markets for those elements of KN. But even if property rights markets could be defined for KN, this does not guarantee use of KN in sustainable ways.

A commonly acknowledged reason for the misallocation of KN are standard market failures. Beyond that, almost all KN goods and services lack the essential features required for a system of private property rights to function effectively. Recall from Chapter 3 that these are enforceability, exclusivity and transferability. Enforceability means that rights of ownership can be enforced by law, while exclusivity refers to the exclusiveness of the ownership rights once they have been enforced. Transferability refers to the ability to transfer the ownership rights at a price. These three conditions of property rights are satisfied only when the goods concerned can be exchanged through the market system. That is, a system of private property rights and the market mechanism work together in unison when goods are exchanged in the market.

Other challenges associated with the use of KN must do, for example, with the complexity of economy-KN interactions. In many instances, observed changes in KN in response to economic actions are non-linear and time-lagged. Doubling in the release of pollutants from a production process into a nearby lake, for example, may not show any noticeable impact right away, if those emissions have been small to start with. Another doubling, however, may suddenly flip the system to one of rapid species loss. Considerable surprises can come with such non-linear and time lagged relationships between actions and their effects. And not only the economy but also KN consists of myriad interdependent subsystems that are connected with each other in many intricate, non-linear and time-lagged ways. Hence, deep uncertainties prevail about cause-effect relationships. Making decisions that ensure sustainable development will therefore be easier if the natural system is not pushed to its limits. Sometimes, this is called the precautionary principle – a call for caution in the light of incomplete scientific knowledge about the impacts of decisions on outcomes.

Other challenges surrounding the appropriate levels at which KN is used are related to practical difficulties in implementing actions designed to achieve an optimal level of pollution control. To begin with, it is difficult to bestow the ownership of an endowment such as a river on a group of individuals or a firm. This difficulty is compounded when there are several stakeholders along the banks of the river. Further, even if the ownership rights were vested with one group among those involved in a conflict, it is not certain that an optimal solution may be reached by negotiations involving compensation. This is because of the time lags that are usually observed with compensation and the complexities that surround issues pertaining to KN. For example, consider water quality and associated KN attributes, as well as cause-effect relationships between water quality and harm to humans. Quantifying any of these complexities is ripe with uncertainties and open to interpretation and conflict, which can drag out the implementation of pollution control, all the while, undermining KN.

Similarly, taxes, like standards, suffer from operational difficulties. Both require well-advanced methods of monitoring, measurement and enforcement. Further, their implementation requires the existence of well-established institutions. Almost all countries now have a national agency dedicated to KN issues, but considerable variation exists among them in terms of their mandates and capabilities. In most cases, the countries with agencies for KN have also formalised the process by which experts and the lay public can make their voices heard before decisions are made. The rules, regulations and policies that may ensue, often try to strike a balance among the perspectives and goals of those who engaged in the decision-making process. Given the invariable contradictory agendas of the parties involved, it is no surprise that the outcomes may not necessarily protect KN at the levels needed.

As we have shown in Chapter 12, a mandate to achieve sustainability calls for a serious departure also from the standard conceptualisation of the steady state. Standard macro-economic models are based on the premise that, in the short run, economies must be stabilised in the light of wayward oscillations of national product, prices (inflation) and employment. Such stabilisation is expected to yield a smooth transition across time periods in such a way that the long run is represented by a steady upward trajectory of output and employment with inflation under control. However, because KN is finite, perpetual growth of national product is impossible.

The application of the standard macroeconomic framework overstates targets and performance. Policymakers tend to pursue mistaken income targets. And since such targets exceed true productive capacity, higher-than-usual inflation outcomes would be inevitable. Further, if prices and wages are used as policy tools, and if the role of KN for economic prosperity were disregarded, prices would not be adjusted fast enough to assure sustainability. There is a need to adjust prices at a rate faster than that implied by the standard framework. Our simple expansion of the standard Keynesian model of employment, prices and income, for example, shows that during times of unemployment and a general economic recession, investment in KN can help stabilise the economy.

Throughout this volume we have shown how standard micro- and macro-economic models can be amended or revised to account for the fundamental reality that the economy is a part of and dependent on, and not separate from, the larger ecosystem. The theoretical insights generated by our analyses point towards uses of KN as sources and sinks in ways that *directly* benefit both the economy and KN, and *indirectly* helps sustain economic activity in the long run by benefitting KN. At the same time, in this last chapter we have also raised concern about the fundamental constraints that limit the abilities of markets (via the price mechanisms) and other institutions (such as government agencies assigning property rights, implementing standards or taxes) to achieve sustainability. The ensuing challenges are unlikely to be solved by the markets or institutions that are in fact at the core of the problem. Instead achieving sustainable uses of KN, and sustainability of the economic enterprise more generally, will require a change in more fundamental parameters that guide market behaviour and policymaking – the development of an environmental ethic; harmonious relations among the people who plan, organise, produce and consume; the ability to make decisions within the confines of nature's abilities to provide materials and energy and absorb wastes; and the willingness to make these decisions in ways that leave room for the fact that the effects of human action will have surprising and sometimes novel outcomes that may require continuous adaptation. Treatment of these fundamental determinants for sustainability is outside the purview of this text. That said, economic models presented here, and the insights they generate, must be part of the decision-making process to help ensure that the opportunity costs of alternative uses of KN are likely minimised in both the short and the long run.

References

Røpke, I., 'Prices are not worth much', *Ecological Economics*, 29(1): 45–46, 1999.

Ruth, M. 'Regional science in a resource-constrained world', *The Annals of Regional Science*, September, 2018. (https://doi.org/10.1007/s00168-018-0879-0).

Thampapillai, D. J. and Ruth, M., 'Contemporary economics and contraindications for climate maladies: Lessons from environmental macroeconomics', in M. Ruth (ed), *Research Agenda for Environmental Economics* (pp. 106–121), Edward Elgar, London, 2019.

Index

Note: **Bold** page numbers refer to tables; *italic* page numbers refer to figures and page numbers followed by "n" denote endnotes.

abatement *87*, 87–88, 92, 95, **221**
Aboriginal and Torres Strait 154
accumulation 9, 12, 230–231, 235–237, *236*, *239*, 239–241; valuation and 309–312
Acworth, R.I. 289
aggregate demand (AD) 17, 159, 165–168, *166*, 177; AD–AS framework 168–174, *169–172*, **214**, **221**, 224–225; and environmental macroeconomics 194, 196, 197–199, 203, 212, 216–218, 222; and environmental policies 319; and valuation 277
aggregate supply (AS) 38, 159, 167, *167–168*, 177; and environmental macroeconomics 196, 212, 215–216, 222; and environmental policies 319
agriculture 10, 11, 25, 82, 86, 136, 188, 290, 331, 333, 335
air quality 132, 278; NEPM ambient air quality standards 89, **90**
alternative efficiency indicator for utilised KN **304**
Ambani, P. 329
Anderson, J.R. 65
Aron, P.G. 80
Asian financial crisis *170–172*
assimilative capacity (AC) 4, *123*, 123–128, 130–131, 133, 135; and environmental capital 184; and public goods 55–58, *58*; and revised economic system 21–22
Atkinson, A.B. 324
Australia 46, 66, 82, 88, 93, 136; consumption tax in *170–172*; and important concepts in macroeconomics 166, 170–172; Liverpool Plains *289*, 289–290, 320; macroeconomic indicators and estimates for **214**; Murray River 53, 137; National Environmental Protection Measure (NEPM) 88–89; standard macroeconomic framework for *219*
average variable costs (AVC) 37, 38

Bali, A. 338
Bandara, R. 273
Bell, F.C. 78
Bennett, J. 275, 308
Benyus, J. 333
Bergman, J. 174, 337
Berlemann, M. 309
Bernanke, B. 7, 159, 167, 242
BHP-Billiton 289, 320
Böhm, S. 322
Bolat, A. 208
Boulding, K. 144
British Thermal Units (BTU) 20
Brundtland Commission 4, 12
Buckingham, K. 335
Bueren, M. van 275
Bunce, A.C. 5
Bureau of the Census 262
Bush, J. 308, 336
Bystriakova, N. 335

cap-and-trade system 322
capital 4–5; *see also* environmental capital; investment; manufactured capital
cause-effect relationships 10, 346
Central Bank (CB) 323, 324

350 *Index*

Central Monetary Authorities (CMA) 166, 173
chief executive officers (CEOs) 324
China 58, 63, 166, 181, *182*, 206–208; emissions trading in 94; and environmental policies 328; international trade 258, 262, 265, 267; and valuation 289
China National Renewable Energy Center (CNREC) 94
Chindarkar, N. 178
Cicchetti, C.J. 281, 286
Clawson, M. 276
Cleveland, C.J. 230
climate change 10–12
Coase, Ronald 69; *see also* Coase theorem
Coase theorem *69*, 83
Cobb–Douglas (CD function) 213, 223, 249–251
Cobb, J.B. 24, 191, 205, 230
Compensation of Employees (CE) 163
competition 334; imperfect 13, 44–45, *45*, 140; *see also* perfect competition (PC)
Conference of Parties (COP) 86
consumer demand 102–104, 116–117, 119, 259; and elasticity of demand 113–116; and the market demand curve 112–113; and standard theory 104–107
Convention on International Trade of Endangered Species (CITES) 254–255
costs 37–44, 119, 127, *134*, 135–136; capital 181–184, 190, 191; and consumer demand 103–104, 108, 116, 117; cost–benefit analysis 155, 275, 287, 322–324; cost savings 70–72, 96, 117, *285*, 285–286, 325, 330; economics of resources 61–62, 64, 76, 78, 83; environmental 67–68, *68*, 78, 262, 324, 334; environmental macroeconomics 210, 220, 224; and environmental policies 319–321, 327, 328, 330; and equilibrium income 203; and important concepts in macroeconomics 169, 173; and international trade 260–261, 263, 266; marginal private cost (MPC) 53; marginal social cost (MSC) 53, 76–77, 83; marginal user 73, 75, 83; and market organisation 139, 141, 147; and policy analysis 204–205, 208; and public goods 49–51, 54–57; and renewable services 87–88, 92, 94, 98–99; and the South Korean economy 243; and sustainability 347; theory of 132; user costs 5, 80–81, 85, 180; and valuation 272, 280–283, 297, 299, 305, 312; *see also* travel-cost method (TCM)
Cropper, M.L. 177

Daily, G. 99
Daly, H.E. 23–25, 191, 205, 230, 232, 248
Dasgupta, P 141, 144
decision-making 6, 10, 12, 64, 122, 129, 133, 136; consumer demand and 103–104; sustainability and 344, 346–347; valuation and 281–282, 288, 291
defecation *see* open defecation (OD)
deforestation 188
demand 18, 33, *35*, *40*, 42–44, 47–48; and choice modelling 275; and the contingent valuation method 273–275; and dose–response methods 279, *280*; and economics of resources 73–76, 78; and environmental policies 328, 332; and game theory methods 277–278, *278*; and hedonic prices 278–279; and international trade 260–265; and market organisation 142–143; and public goods 53–54, 56; and sustainability 344; and travel-cost method *276*, 276–277; and valuation 286; *see also* aggregate demand (AD); consumer demand; demand curve
demand curve 35–37, 39–41, 44, 52, 70; and consumer demand 104, *106*, 106–107, 110, *111*; and environmental effects 112–113; and valuation 272, 275–277
depreciation 164, 249, 250–251, 307, 309–310; mine depreciation *183*, 183–184, 208; *see also* depreciation of KN
depreciation of KN 130–131, 164, 174, **178**, 178–182, 190–192; comparison of income paths *245*; conceptual framework for 184–191, *185*, *190*; and dose–response methods 279; and efficiency of utilisation with respect to KN **304**; and environmental macroeconomics 196, 220–224, 228; and environmental policies 317, 320, 325; equilibrium income with 199–204, *202*, *204*; equilibrium income without 197–199; and the Harrod–Domar growth model 235, 238; and the perpetual inventory method 311, **312**; and policy analysis 205–206, 208–210; and relative scarcity of KN **300**, 302; and the South Korean economy 243, **244**, *245*, 245–246; and the steady state 231–233; time

Index   351

paths of *182*; and use value of KN in macroeconomics 296–297
Devonshire, P. G. 285
Dey-Chowdhury, S. 309
Dhareshwar, A. 309
Direct Air Capture (DAC) 98
Domar, E.D. 6, 233
dose–response methods 279–280, *280*
Du Preez, M. 276
Dyson, J. 333

earthquakes 5, 10, 62–63, 259
economics *see* economic system; neoclassical economics
economic system 16, 28, 181; adaptations of 25–28; and the entropy law 23–25; and materials balance *19*, 19–23; revised for leakages and injections *18*; revised for materials balance and entropy *26*; standard version 16–18, *17*
efficiencies 80, 120, **304**; first law 20–21
elasticity 103, 107, 116; of demand and revenue *114*; of demand and the environment 113–116, *114*; of environmental friendliness 113, 116–117
electricity and power generation 86
electric vehicles (EVs) 97–98
Ellen MacArthur Foundation 331, 333
Ellison, K. 99
Emission Receptor-Sink 97
emissions trading 8, 94, *95*, 96
emissions trading schemes (ETSs) 94
Emmanuel, K. 62
endowment effect 104, 107–110, 116, 325–326
endowments 4–5, 25–28, **178**, 178–179, 271–275, 290–291, 307–308; factor endowment 256–259, *259*; and indifference curves *109*; *see also* endowment effect
entropy 25–27; desired output and *125*, 125–126; and diminishing marginal returns *126*; the economic system revised for *26*; the economy and 23–25; transfer of *126*, 126–128, *128*
environmental capital (KN) 3, 5–11, 119–120, 135–138, 176–178, 191–192; and analysis of costs 132–135, *134*; assets 334–337; and the basis for valuation 271–272, *272*; and choice modelling 275; and Cobb–Douglas production function 250–251; and commodity prices 322–324; conceptual basis for estimation of *222*; conceptual framework for 184–191, *185*, *190*; and contingent valuation method 273–275; depreciation function 184, *185*, 186, *186*, *190*, 191; depreciation of 179–180; dose–response methods 279–280, *280*; and education attitudes and lifestyles 325–327; efficiency of utilisation with respect to 303–304, *303–304*, **304**; empirical evidence for 191–192; empirical illustration for **312**, 312–313; and the environmental macroeconomic framework 220–224, 228–230; and environmental policies 317–319; and equilibrium income 201–203; game theory methods 277–278, *278*; H-D model revised with 234–235; and hedonic prices 278–279; and important concepts in macroeconomics 159, 163, 164, 167–168, 174; improving productivity of 329–330; and interest rates 323; and international trade 254–255, 257–261, 263–268; isoquants and 128–132, *129–130*; KN investments **178**, 178–179, 210; and limits to growth 6–7; and the market model 33, 46–47; and market organisation 140–143, 148, 155–156; and methods based on opportunity costs 280–290, *284–285*; microeconomic basis for 290–291; perfect information and full employment 153–155; and the perpetual inventory method 309–312; and policy 205–206, **318**, 331–337; and population pressures 328–329; and product differentiation 325; production and assimilative capacity and *123*, 123–128, *125–126*, *128*; production function and *121*, 121–123; and property rights and land reform 329; and the Ramsey model of utility estimation 292–293, **292–293**; and regulation and tradable permits 321–322; relative scarcity of 300, **300**, 301, *301*, 302; and renewable services 98–99; resilience of 33, 99, 119, 124, 125, 130, 131, 140, 155, 163, 176, 184, 186, 187, 191, 194, 220, 221–223, 234, 255, 297, 307–313, 317; restoration of *207*; and revised economic system 16–17, 19–23, 25–28; and the South Korean economy 243–248; and the standard framework 212–213; and the steady state 231–233; stocks and flows of 307–309; and substitution 131–132; and sustainability 342–347; and

sustainable income 180–184, *182–183*; and the Swan–Solow model 237–242; and taxes and subsidies 319–321; and travel-cost method *276*, 276–277; use value of 296–299; utilisation of *246, 247*; and valuation 304–305, 313; and wages 324–325
environmental friendliness 104, 116–117; and consumer goods 13, 56, 107–108, 111–112, 325; elasticity of 113, 116–117; and house designs 99; index of 116
Environmental Kuznets Curve (EKC) 8–9
environmental macroeconomic indicators and estimates 221
environmental macroeconomics 194–197, *195, 209*, 210, 212, 217, 220, 224–225, 228–231; and endogenous growth models 240–243; and equilibrium income 197–204, *198, 202, 204*; framework for 220–228, **221**, *222, 225*; and the Harrod–Domar growth model 233–235; and policy analyses 204–209, *206, 207*; and South Korean economy 243–248, **244**, *245–247*; and the standard framework 212–220, **214**, *219*; steady state in *195*, 231–233; and the Swan–Solow model 235–240, *236, 239*
environmental policies 325, 328–329, 331–334
equilibrium 16–18, 40–42, 143; in the adapted economic system 27–28; and environmental macroeconomics 217–218, 231–233, 237–238; equilibrium income 197–204, *198, 202, 204*, 208; macroeconomic 168–169, *169*, 177, 231, 326; between marginal costs of abatement and pollution *87*; market *40*, 40–42, 139–143, 168, 260, 319; steady-state 235, 237–238, 240, *246*, 249, 251
exports 18, 24–25, 163–164, 254–255, 257–264, 281–282; market for *260*
externalities 52–54, 61–62, 80–81, 177–178, 241–243; and environmental policies 319, 326; illustration of *54*; open defecation and 57–58; public goods and 54–57, *55*

factor endowment *see under* endowments
Farr, M. 276
financial crisis *see* Asian financial crisis; global financial crisis
Fisher, I. 5, 8, 176, 177, 230, 281, 307
food and beverage industries 85

foreign exchange 263–265, *264*
forests: old-growth *see also* reforestation
Forrester, J.W. 6
fragility *125*, 125–127, 135–137; and sustainability 344
Frank, R. H. 7, 50, 156, 159, 167, 242
Franklin, R.S. 8
free trade agreements (FTAs) 265
Fryklbom, P. 274

game theory 277–278, *278*, 290; Ramsey model 277, 292–293, **292–293**
Georgescu-Roegen, N. 23, 230
global financial crisis (GFC) 91, 155, *170–172*, 172–173, 213, 226–227; and environmental policies 324
Global Forest Watch 65
globalisation 254–255, 267–268, 338; and comparative advantage 256; and foreign exchange 263–265, *264*; and the market model *260–261*, 260–263; and the need for a global paradigm 265–267; and the revised framework 258–259, *259*; and the standard framework 256–258, *259*
goods 3–4, 25–26, 43–46, 108–113, 143, 344–345; consumer 108, 258; continuum of *50*; environmental 42–51, 86; environmental macroeconomics 256, 259–261; and environmental policies 325; and important concepts in macroeconomics 165–167; and international trade 263–265; public 49–52, *50–51*, 54–57, *55*, 86
goods and services tax (GST) 172
Gowdy, J. 230
Graham, S. 308
Graycar, A. 338
Gray, L. C. 5
greenhouse emission 266
greenhouse gases (GHGs) 9, 10, 11, 59, 62, 63, 65, 80, 85, 86, 87, 93, 97, 98, 136, 149, 151, 154, 174, 205, 220, 227, 243, 255, 267, 312, 319, 330, 332, 335, 337, 342, 345
Greig, P. J. 285
gross domestic product (GDP) 17, 18, 163–165, 176, 181, 188, 189, 192, 196, 197, 206, 212, 213, 223, 326, 335
gross national product (GNP) 164, 176, 196, 326
Gross Operating Surplus (GOS) 163, 164, 298
growth 12–13; economic 162–165, 230–231, 233–235, 240–243, *243*;

employment 162, 328; growth models 231–235, 240–243, **244**, 248; growth policy 328; population 162, 235, 237, 328–329; rates of 247, *247*

Hahn, R.W. 96
Hamilton, C. 93
Hamilton, D.S. 290
Hanley, N. 275
Hansen, J. 208
Hardin, G. 329
Harrod-Domar growth model 232, 247
Harrod, R.F. 6, 233
Hartwick Rule 182, 208
Heal, G. M. 141, 144
Hecht, S. 65
Hester, G. L. 96
Hicks, J.R. 5, 317
Hiranuma, H. 63
Hirshleifer, J. 37, 105
Hosking, S.G. 276
Hotelling 5, 6, 180, 276
Howlett, M. 338
Hundloe, T. 93

imports 254–255, 257–264; market for *261*
income 7–8, 26, 103, 107, 159–168; equilibrium income 197–204, *198*, *202*, *204*, 208; income paths 209, *245*; sustainable 5, *198*, 199–201, *202*, *204*; *see also* income effect; national income
income effect 111–112, *112*
India 57, 65, 137, 181, 262; and environmental policies 328, 334; time paths for *182*
indifference 102–110, *109*, 256–257, 277–278, *278*, 292–293
Indonesia 65, 161, 171, 206–208, *207*, 254, 266
industrial revolution 10, 21, 64, 345
injections 18, *18*, 177
interdependence 53, 55
intergenerational concern (IGC) 72–78
Intergovernmental Panel on Climate Change (IPCC) 265
International Institute for Sustainable Development (IISD) 94
International Standards Office (ISO) 86, 91, 92, 261, 321
international trade 254–255, 258, 262, 267–268; and comparative advantage 256; and exchange rates 263–265, *264*; and the factor endowment 256–259, *259*; and global issues 265–267; and the market model *260–261*, 260–263
investment 162–164, 171–174, 199, 233–235, 254–255, 264; and depreciation **178**, 178–180; and environmental capital 176–178, 184–191; and environmental policies 319–320; reinvestment 182–184, *183*, 208–210, 320; and sustainable income 180–184; and valuation 309–310
irreversibility 281, 283–284, *284*
isoquants 13, 119–120, 127, 128–137, *129–130*
Israel 81–82, 169, 173, 336, 337

Jackson, D. 163
Japan 4–5, 62–63, 79, 90, 91, 263
Jevons, W.S. 5

Kariba lake shore 273
Katz, M.L. 50
Kay, J.J. 127, 242
Kenya 164, 265; and important concepts in macroeconomics 141
Keynesian framework 168, 347; and environmental macroeconomics 194, 199, 205–206, 208, 212, 215; equilibrium income within *198*; sustainable income within *202*, *204*
Keynes, J. 5, 168, 180, 212
Khatri, K. 333
KM *see* manufactured capital stocks (KM)
KNEF 111, 116
Knetsch, J.L. 274, 276
KN-Friendly (KNF) 13, 56, 99, 108, 110, 111, 113, 115–117, 152, 153
KN-sink 19
KN-source 19
KN-Unfriendly (KNUF) 13, 108, 110, 111, 115
Kohr, L. 144
Krutilla, J.V. 281, 286
Kyoto Protocol 8, 9, 254, 255

Larranaga, A. 108
leakages 18, *18*, 78, 177
Lebling, K. 97
Lecomber, R. 65
Leigh, A. 324
Levin, K. 97
lifecycle 116, 334
Linacre, E. 333
Linde, C. van der 321
Lipscy, P. 62

354 Index

Liverpool Plains *see under* Australia
local currency unit (LCU) 256, 263, 264
Lodhi, I. 290
Lovins, A.B. 330, 334

macroeconomics 12–13, 174; and AD–AS framework 168–174; and final expenditures 161; and goals 162–165; and inflation 165–168; and national product 159–160; and prices 160–161; steady state in *195*; *see also* environmental macroeconomics
Mahbubani, K. 173
Malthus 3
Malthusian limits 6
Mankiw, N.G. 4, 7, 22, 159, 241
manufactured capital stocks (KM) 6, 7, 121–126, 128–129, 131, 135, 344; in Australia **300**; in Australia and South Korea **214**, **221**, 243, **244**, *303*; and Cobb–Douglas production function 249, 251; and the environmental macroeconomic framework 220–224; and environmental macroeconomics 228–235; and equilibrium income 201–203; and fragility of KN *125*; and important concepts in macroeconomics 165, 166–168, 171, 174; and international trade 256–258, 264; KM – L 126, *130*, 130–132, 136–138; and long run analysis for environmental macroeconomics 248; and the standard framework 213; for steady-state equilibrium *246*; and the Swan–Solow model 237, 239–241; and valuation 279, 297–298, 303, 305, 308–311
marginal benefits of pollution control (MBP) *69*, 69–72
marginal cost (MC) 37, 38
marginal cost of controlling pollution (MCP) 67, 69–72, 87
marginal externality cost (MEC) 53, 56, 58, 59, 73–77, 79
marginal rate of substitution (MRS) 107, 109, 110, 129, 130
marginal returns 119–120, 125–127, *126*, 220, 235, 279
marginal utilities (MU) 109
market model 33, 47–48, 61, 66–69, 73, 256, 259; and the functions of the market 33–42; and market failure 42–43; and property rights 43–47; trade framework based on 260–263
market organisation 139–156, 231

Marshall, A. 5, 7, 8, 34, 176, 189, 230, 305
Martin, P. 187
materials balance *19*, 19–23, 25, *26*
McAllister, T. 154
Mcfarlane, S. 267
McInerney, J. 73–74, 76, 184
Meadows, D. H. 6
Messener, S. 230
microeconomics 12–13, 116–117, 119–120, 128–138, 231; and assimilative capacity 123–128; and conflicts between EG and EQ 66–72; and consumer demand 102–112; and elasticity of demand 113–116; and emissions trading 94–97; and intergenerational concern 72–78; and the market demand curve 112–113; and pollution control 85–95, 99–100; and the production function 121–123; and property rights 98–99; and resource management 61–66, 78–82; *see also* externalities; market model; market organisation; public goods
Mill, J.S. 230
mining 43, 85, 92, 99, 138, 182–184; earthquakes 62–63; and environmental policies 320; and international trade 261; KN investment and 178; Liverpool Plains and 289–290; and policy analysis 208; and valuation 277, 281, 283, 285, 289–290
Mishan, E.J. 139, 308
Mongolia 63, 208–209, *209*
monopolistic competition 151–152, *152*
monopoly 44, 140, 144–149, *146–147*, *149*, 334
Moran, S. 65
Morrison, M. 275
Murray-Darling Basin (MDB) 188
Murray River 137; *see also under* Australia

Nadkarni, M.V. 65
National Environment Protection Measure (NEPM) 89, **90**
national income 121–126, *125*, 127, *128*, 128–131, 344; and Cobb–Douglas production function 249, 251; and consumer demand 110; and dose–response methods 279; and economics of resources 64; and efficiency of utilisation with respect to KN 303, *304*; and environmental capital 176–182, 184–191; and the environmental macroeconomic

framework 220–224, **221**; and environmental macroeconomics 194; and equilibrium income 199–203; and important concepts in macroeconomics 161, 169; income paths *245*; national income accounting 13, 179, 230, 312; and the perpetual inventory method 311; and policy analysis 205, 208–210; and product differentiation 324; and revised economic system 17–18, 27–28; and the South Korean economy 243–245, **244**, 248; and the standard framework 213–216, **214**; and the steady state 231–235; and stocks and flows of KN 307; and the Swan–Solow model 237–243; time paths of *182, 206, 209*; and travel-cost method 276–277; and use value of KN in macroeconomics 296–299; and valuation **300**
national parks *51*, 51–52, 273, 280
National Pollution Inventory (NPI) 89
national product (NP), measurement 163–165
National Science Foundation 59
Natural Sequence Farming (NSF) 333
Nehru, V. 309
neoclassical economics 3, 5–8, 230, 231
NEPM ambient air quality standards *see under* air quality
net national product (NNP) 164, 176, 177, 179–181, 192, 196–201, 203, 204, 235
net taxes (T) 163, 212, 251n2
net zero emissions (NZE) 87, 97–98, 322
New South Wales (NSW) 88, 289–290
Nijkamp, P. 274
Nogrady, B. 90, 91
non-renewable resources 24, 27, 83, 99–100, 141; and conflicts between EG and EQ objectives 66–72; and emissions trading 94–97; and environmental capital 181, 184, 192; and environmental macroeconomics 205; and a framework for pollution control 87–88; and intergenerational concern 72–78; and objectives of resource management 61–66; and property rights 98–99; and resource management policies 78–82; and standards 88–92; and taxes and charges 92–94
North American Free Trade Agreement (NAFTA) 265
Numes, P. 274

Oates, W.E. 177
oil 4–5; North Sea oil *170*, 323; *see also* Organisation of Petroleum Exporting Countries (OPEC)
Okaru, Valentina O. 96
oligopoly 44, 45, 140, 149–151, 153, 155
open defecation (OD) 57–58, *58*, 70–71, 77
opportunity costs (OC): methods based on 280–288; and mining 289–290; and old growth forests 281–282
optimal control theory 73
Organisation of Petroleum Exporting Countries (OPEC) 150, 169, *170*, 323
outcomes, environmental *272*, 290–291
overcapitalisation 167, 239–240, 245
overconfidence 162
overexploitation 72, 178
overuse 26, 47, 60
overutilisation 61, 240, 245, 342

palm oil 266
Pearce, D.W. 276
Pearse, R. 322
Pérez-Peña, R. 63
perfect competition (PC) 13, 34, 41–42, 44, 48, 140–151, 155, 326; and environmental policies 317, 321, 326, 330; and international trade 267; and market organisation 155; and monopoly 144–149, *147*; and preference relations 143–144; and public goods 54; and sustainability 140–143, *142*
perfect competition and sustainability (PCS) 140–144, 147, 148, 150, 154–156
perpetual inventory method (PIM) 243, 308, 309–310, 312–313
Peters, G.P. 255
Pindyck, R.S. 7, 153
policies 317–319, **318**, 338; bio-mimicry 333–334; closed-loop production systems 331–333; and education and lifestyles 325–327; and population pressures 328–329; price policies 322–325; product differentiation 324; and the productivity of KN 329–330; and property rights 328–329; regulation 321–322; taxes and subsides 319–321; and the traditional business model 334
pollution 5, 9–10, 85–87, 99–100, 116, 344, 346; and abatement *87*; adaptation of trade in 96; and economics of resources 64, 67, 69–72, 78, 82; and emissions trading 94–97; and environmental capital 178, 181, 191;

356 *Index*

and environmental macroeconomics 205, 224, 243; framework for control of 87–88; and international trade 254; and property rights 98–99; and public goods 51–56; and revised economic system 20–21; and standards 88–92; and taxes and charges 92–94; and valuation 279–280, 286, 297
Porter, M.E. 321
price elasticity 113–116, *114*, 116
production 3–4, 12, 25–27, 132, 135–137, 342–343; and analysis of costs 132–135; and assimilative capacity 123–128; closed-loop production systems 116, 155, 203, 228, **318**, 331–333; and economics of resources 67–68; and international trade 256–257, 259–260, 262; and the market model 39–42; and market organisation 142–144; and public goods 54–57; theory of 121, 338; zones of *121*, 121–122; *see also* production function; production processes; productive capacity
production function 13, 119, 129–130, 132–138, 344; and assimilative capacity 123–124, 127; Cobb–Douglas 249–251; and environmental macroeconomics 201–203, 213, 215, 220–224, 232, 235, 237–240; and environmental policies 317, 338; and fragility of KN *125*; and international trade 258, 261; and KN 121–123, *123*; and valuation 278, 291; and zones of production *121*
production processes 7, 38, 115, 119, 136, 325, 345; thermodynamic constraints on 120
productive capacity 167–168, *168*, 171, 174, 191, 347; and environmental macroeconomics 196, 213, 226, 228
profit maximisation *67, 68*, 92
property rights 52, 54, 121, 344–345, 347; and economics of resources 72, 79; for environmental goods 46–47; and market organisation 139; policy and **318**, 329; price mechanism and 43–47; and renewable services 98–99
property value (PV) 143, 278, 279, 287, 291
public goods 49–61, 86, 147, 274
public space 57–58
Public Utilities Board (PUB) 331, 338
Putnam, R.D. 337
Qiu, H. 335
Quah, E. 139, 206, 308

quantity-price space 36
Quiggin, J. 93

Ramanathan, V. 59
Ramesh, M. 338
Reabsorption Sink 97, 98
recycling 22, 24, 47, 99, 203, 334, 343; water recycling 81–82, 123
reforestation (RF) 7, **178**, 206–208, *207*, 210, 265; and environmental policies 321; and valuation 282–283
Regional Greenhouse Gas Initiative (RGGI) 94
Reinhardt, F.L. 325, 334
reinvestment *see under* investment
renewable resources 24, 27, 28, 59, 99–100, 342–343; and conflicts between EG and EQ objectives 66–72; and emissions trading 94–97; and environmental capital 181; and environmental policies 319–320; and a framework for pollution control 87–88; and intergenerational concern 72–78; and market organisation 141; and property rights 98–99; and resource management 61–66, 78–79; and standards 88–92; and taxes and charges 92–94
renewable services 86, 99–100; and emissions trading 94–97; and a framework for pollution control 87–88; and standards 88–92; and taxes and charges 92–94
replacement costs 282–285
residual externality 58–59, 62, 96, 98, 177–179, 184–186, *186*, 187, 190, 197, 200, 201, 205, 231, 317, 319, 326
resilience, of environmental capital (KN) 33, 99, 119, 124, 125, 130, 131, 140, 155, 163, 176, 184, 186, 187, 191, 194, 220, 221–223, 234, 255, 297, 307–313, 317
Resource Rent Tax *see under* taxes
resource rent tax (RRT) 208, 209, 320
resources *see* non-renewable resources; renewable resources
revenue 38, 58, 66, 94, 99; and consumer demand 114, *114*; and environmental capital 182–184; and environmental macroeconomics 208–209; and environmental policies 319–320, 331; and international trade 256–258, 266; and market organisation 147; and valuation 281–282, 285, 289
Richardson, D. 80
Richter-Scale (RS) 62–63

Rosen, H.S. 50
Ross, M.R.V. 336
Rubinfeld, D.L. 7, 153
Ruth, M. 120, 230

Samuelson, P.A. 6, 109, 230
sand-mining 102
savings *see under* costs
Schickele, R. 5
Schneider, E.D. 127, 242
Schumacher, E. 144
Schumer, C. 97
Scott, A. 5
Semenova, G. 89
services *see* renewable services
sewerage treatment (ST) 206–208, *207*
Shenhua Watermark 289–290, 320
Sinden, J.A. 6, 139, 242, 276, 277, 279, 292, 308
Sivapalan, M. 336
social capital (KS) 16, 337–338
Solar Bottle Bulb 329
Solow, R. 6, 141
Southeast Asia 151, 154, 266; *see also specific countries*
South Korea 171, 243–248, 328–329; model estimates for **244**; valuation of KN in 299–304, *303–304*
standards 56, 72, 344, 346–347; for air quality **90**; and environmental policies 321, 327; and international trade 261; and market organisation 148; and renewable services 86, 88–92, 94, 100; taxes and *89*
Stavins, R. 177
steady state 13, 180, 344, 346; and environmental macroeconomics 194, *195*, 230, 231–233, 237, 245–246; and international trade 254; in standard macroeconomics *195*; and valuation 309
Stern, N. 220, 241, 243, 338
stocks *see* manufactured capital stocks (KM)
Stout, L.A. 338
substituting 24, 108, 131–132
substitution effect 111–112, *112*
subterranean resource systems 85–86
supply 17–18, 119, 132–135, 344; and economics of resources 73–76, 82; and environmental macroeconomics 224, 260; and environmental policies 323; and international trade 263; and the market model 34, 40–44, 47–48; and market organisation 139, 142, 143; market supply 37–40; and productive capacity *168*; and public goods 51, 52–54, 60; and valuation 272–273; *see also* aggregate supply (AS)
sustainability 3, 12–13, 131, 346–347; and consumer demand 102–104; and environmental macroeconomics 194, 205, 228; and environmental policies 317, 323, 326, 331, 339; and international trade 254–255, 267–268; and the market model 47–48; and market organisation 139, 144, 155–156; perfect competition and 140–143; the premises for 342–343; and revised economic system 16, 20, 28
sustainable development 140, 317–339
Swan–Solow (S-S) model 231–232, 235–241, *236*, *239*; and environmental policies 328; rate of economic growth in 250; and the South Korean economy 243, **244**, 245–247, *245–247*
Swan, T.W. 6, 235

Tan, S.L. 206
taxes 16, 131, 166, 168, 346–347; and economics of resources 72; and environmental macroeconomics 199, 210, **214**; and environmental policies **318**, 319–323; and market organisation 148; and public goods 53, 56; and renewable services 86, 88–93; resource rent tax 320; Resource Rent Tax 208–209; and standards *89*
Taylor, J.B. 7
textile and fashion industry 85
Thampapillai, D.J. 59, 123, 139, 173, 178, 181, 205, 206, 208, 242, 277, 283, 290, 292, 321, 324, 335, 336
Thampapillai, V. 322
Thangavelu, S. 206
thermodynamics 7, 13, 119–120, 127, 343; and environmental capital 176; and environmental macroeconomics 220; and important concepts in macroeconomics 163; and the market model 33, 38; and market organisation 139, 141; and renewable services 99, 102; and revised economic system 20, 22–23, 25–28; and valuation 305
time paths *182*, 205–206, *206*, *209*; and environmental policies 321; the South Korean economy 245; and the Swan-Solow model *236*, 237, *239*
Timms, W.A. 289
Tisdell, C. A. 277, 279

358 Index

Tokyo Electric Power Corporation (TEPCO) 90
Tompkins County Area Transit 36
trade: seemingly nonsensical 262–263; *see also* emissions trading; exports; imports; international trade
transport industry 85
travel-cost method (TCM) 70, *276*, 276–277, 290
Trenberth, K.E. 80, 336
Truth Telling 154, 155
Turner, K. 276

Uhlin, H.-E. 205
United Nations Commission on Trade and Development (UNCTAD) 265
United Nations Environment Program (UNEP) 265
United States 20, 91, 96, 99, 166; emissions trading in 94; and environmental capital 191; and environmental macroeconomics 205–206, 226–227; and environmental policies 324; and international trade 263; time paths for 206
United States Geological Survey (USGS) 63
unit price (PKN) 299, 302, 304; time trends of *301*; and utilised KN **304**; and the valuation of KN **300**
USA World Trade Statistics 262
utility functions 104–107, *105*

Valentine, S. 123
Valor, C. 108
valuation 5–6, 13, 43, 47, 49, 52, 109, 273, 299, **300**, *301*, 303–304; the basis for 271–272; choice modelling (CM) 275; contingent valuation method (CVM) 273–275; cost savings 285–286; direct estimation of OCs 280–282; dose–response 279–280; and environmental macroeconomics 205; game theory methods 277–278; hedonic prices 278–279; of KN in Australia **300**; and market organisation 139; overview of methods of *272*; and PIM 309–313; replacement costs 282–285; and stocks and flows of KN 307–309; threshold values 286–290; travel-cost method (TCM) 276–277; and the use value of KN 296–305, 307; and valuation 276, 290–291
Vanderbilt, T. 333
Vietnam 206–208, *207*, 321
Voegele, E. 79
Volcovici, V. 267

wastes 3, 9, 137, 342–343, 347; and environmental capital 181; and environmental macroeconomics 228; and environmental policies 327, 329; longevity of 21; and the market model 38, 43, 47; and public goods 49; and renewable services 96; and revised economic system 19, 21–23, 27; and valuation 283
water 9, 122–124, 132, 137, 346; and consumer demand 102, 108, 117; and economics of resources 64, 66–67, 69–72, 78; and environmental capital 181, 191; and environmental macroeconomics 206, 210, 220, 224, 242, 248; and environmental policies 323, 329–334; and important concepts in macroeconomics 164; international trade 266; and the market model 42–44; and market organisation 141; and public goods 51–56; and renewable services 86, 88–90, 98; and revised economic system 19, 22, 25; and valuation 273, 275, 282–283, 289–290; water recycling 81–82
Water Resources Engineers 283
Wessehöft, J.-E. 309
West Africa: Guinea 184
Wheeling, K. 62
willingness to accept (WTA) 39, 104, 107, 108, 110, 112, 116, 272, 274, 275, 290
willingness to pay (WTP) 35, 36, 39, 42, 44, 52, 54, 70, 104, 107, 110, 112, 116, 272–275, 290
world price currency units (WCU) 256, 263
World Trade Organisation (WTO) 265
Worrell, A.C. 242, 276, 308
Wright, I. 88
Wyckoff, J.B. 277

Xerox Corporation 334